COLORADO ROCKHOUNDING

A GUIDE TO MINERALS, GEMSTONES, AND FOSSILS

STEPHEN M. VOYNICK

MOUNTAIN PRESS PUBLISHING COMPANY
MISSOULA, MONTANA

Second Printing, February 1995

Cover design by Paulette Livers Lambert
Cartography by Trudi Peek

Library of Congress Cataloging-in-Publication Data

Voynick, Stephen M.
Colorado rockhounding : a guide to minerals, gemstones, and fossils / Stephen M. Voynick.
p. cm.
Includes bibliographical references and index.
ISBN 0-87842-292-7 (pbk.) :
1. Minerals--Colorado--Guidebooks. 2. Precious stones--Colorado--Guidebooks. 3. Fossils--Colorado--Guidebooks. I. Title.
QE375.5.C6V69 1993
549.9788--dc20 93-37542
CIP

MOUNTAIN PRESS PUBLISHING COMPANY
P.O. BOX 2399 • MISSOULA, MT 59806
406/728-1900 • 800/234-5308

EAGLE COUNTY

GILMAN AREA MINES

The inactive company town of Gilman is located on U.S. Highway 24, six miles north of Minturn. Gilman, a former major mining and milling center, has provided many superb specimens of native gold, pyrite, and sphalerite.

Prospectors discovered rich oxidized outcrops of silver-lead ore in the Eagle River Canyon in 1879. Silver miners found gold in the underlying quartzites and in erratic vein and pipe deposits in a three-square-mile area between Gilman and Red Cliff.

In 1887, miners recovered extraordinary native gold specimens from the Ground Hog Mine, where gold and silver occurred in two vertical veins, or "chimneys," ranging in thickness from only a few inches to six feet. One chimney produced massive wires of gold called "ram's horns." The biggest specimen weighed eight ounces, was one inch in diameter, and branched gracefully into five distinct curls. Noting its "curious and rare formation," a mine owner purchased it for just over its bullion value of $150. The specimen passed into several private collections and finally into the collection of the Harvard Mineralogical Museum, where it remains today. Now valued at $750,000, it is the world's largest known specimen of wire gold.

By 1900 production had topped $8 million in gold, silver, and lead. But the real mineral treasure of Gilman was zinc. In 1912 the Empire Zinc Company consolidated many properties and developed the Eagle Mine at Gilman. During the 1930s, the Eagle Mine was Colorado's biggest copper and silver producer, then became the state's top zinc producer in World War II. When the Eagle Mine finally closed in 1978, the Gilman district had produced 310,000 troy ounces of gold, 50 million troy ounces of silver, 180 million pounds of copper, 35,000 tons of lead, and 250,000 tons of zinc worth over $150 million.

Fine specimens of pyrite, sphalerite, siderite, golden barite, and rhodochrosite from the Eagle Mine are still available through mineral dealers. Field collecting opportunities, however, are very limited. Dumps are steep and unstable, and property is posted, as is the company town of Gilman, which clings precariously to the brink of the Eagle River Canyon.

REFERENCES: 18, 19, 23, 66, 67, 73, 81

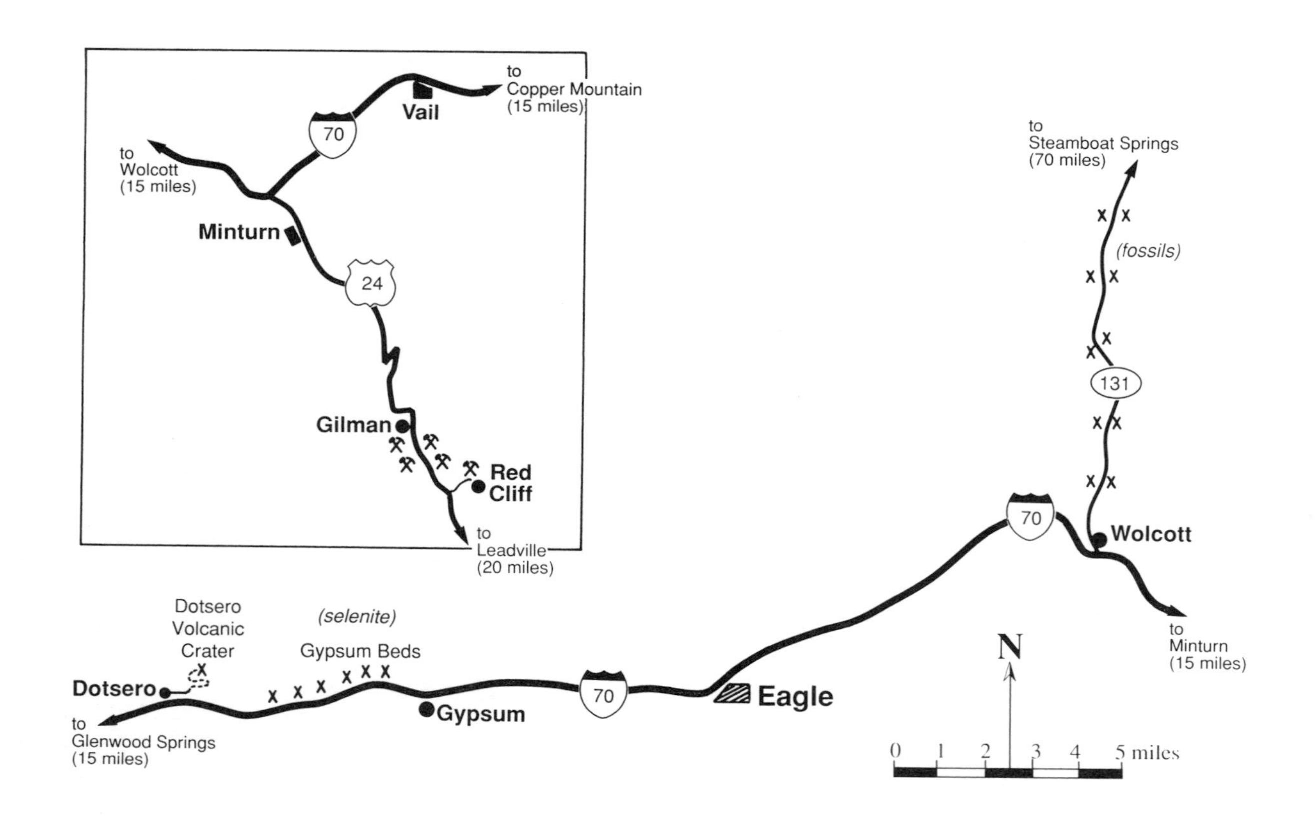
to Copper Mountain (15 miles)
Vail
70
to Wolcott (15 miles)
Minturn
24
Gilman
Red Cliff
to Leadville (20 miles)
to Steamboat Springs (70 miles)
(fossils)
131
70
Wolcott
to Minturn (15 miles)
Dotsero Volcanic Crater
(selenite)
Gypsum Beds
Dotsero
to Glenwood Springs (15 miles)
Gypsum
70
Eagle
N
0 1 2 3 4 5 miles

DOTSERO VOLCANIC CRATER

Three miles east of the town of Dotsero, Interstate 70 passes the Dotsero Crater and an associated black basaltic lava flow. The crater and lava flow, dated at 4,150 years, represent Colorado's most recent known volcanic event.

The crater is one-third of a mile wide and 800 feet deep. The lava flow, which emanated from the base of the cinder cone, is one mile long. Volcanic bombs and lapilli (stony and glassy lava fragments) that were ejected from the crater are scattered about the surrounding hills. Small vesicular cavities in the red scoria are sometimes filled with clear quartz, formed from residual volcanic gases and fluids after solidification of the lava.

The undisturbed western part of the crater is on BLM land, but active mining of lightweight volcanic cinder for local cinderblock manufacture continues to alter the eastern rim. Various groups, including the BLM, have proposed protecting the crater for educational use.

From the Dotsero exit of I-70 (Exit 133), follow the unmarked frontage road on the north side of I-70 east for one-half mile to the end of pavement at the mobile-home residential area. Turn left (north), then right along the cinder road, which ascends 700 vertical feet in one and a half miles to the crater rim.

REFERENCES: 3, 12

The undisturbed north rim of the Dotsero Crater. The crater was formed by Colorado's most recent known volcanic event.

Twelve-inch lapillus, or volcanic bomb, from Dotsero Crater.

EAGLE RIVER GYPSUM BEDS

Along I-70 near Gypsum and Dotsero, the lower Eagle River cuts through Pennsylvanian shales that are folded and distorted by gypsum beds. Gypsum is capable of flowing plastically in a solid state in a manner similar to glacial ice. Many gypsum beds are exposed on the steep slopes along the north side of I-70. They are recognized by light gray color, distorted strata, and lack of covering vegetation. The largest exposures are between mile 140 and 141.

Most of the gypsum occurs in massive form (gypsum rock), but clear and white selenite crystals in deformed blades as long as six inches are present in both the in-situ beds and the weathered material below the slopes.

REFERENCES: 3, 12

FOSSILS

The area north of Wolcott has many exposures of fossiliferous Mancos Shale. Large sections of the papery gray shale are exposed in road cuts along Colorado Route 131 from Wolcott for a distance of eight miles north. Fossils of small pelecypods occur within iron-rich, rust-colored concretions. Most fossils are fragile; a few from certain thin, coaly strata are more durable and retain part of the lustrous shell material. Inch-long calcite crystals occur in seams in the shale.

Fragmented selenite crystal from gypsum-bed exposures on I-70 near Dotsero.

Distorted strata on cliffs along I-70 near Dotsero are gypsum beds with abundant selenite crystals.

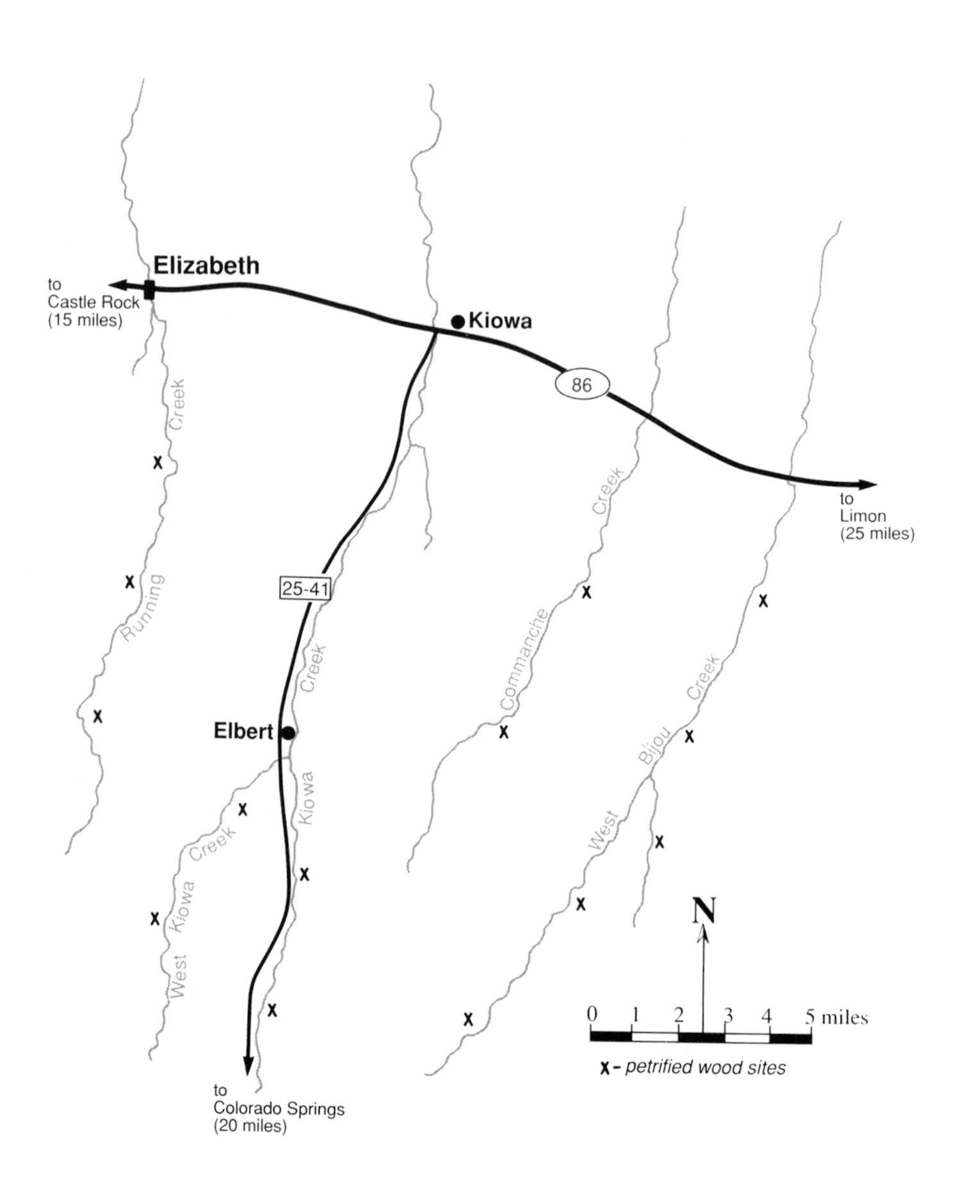
Elizabeth
to
Castle Rock
(15 miles)
Kiowa
86
to
Limon
(25 miles)
25-41
Elbert
Running Creek
Creek
Kiowa
West Kiowa Creek
Commanche Creek
Bijou Creek
West
to
Colorado Springs
(20 miles)
N
0 1 2 3 4 5 miles
x - petrified wood sites

ELBERT COUNTY

PETRIFIED WOOD

Elbert County has provided a great deal of colorful petrified wood and, aside from Florissant (Teller County), some of Colorado's largest petrified logs. Much the petrified wood is found in western Elbert County near Elizabeth and Elbert.

In the 1940s, writers reported petrified logs twenty-five feet long and two feet in diameter washed out of Running Creek gullies a few miles south of Elbert. South of Elbert near the El Paso County line, the gullies of upper Kiowa, Comanche, and Bijou creeks have been good sources of well-agatized and some opalized wood, including specimens with detailed bark structure.

Early collectors told of a petrified-wood locality with "denuded tree stumps up to four feet in diameter" located "thirty miles northeast of Kiowa." Although the location is vague, that site could be near Agate and today's I-70 corridor in eastern Elbert County, where petrified wood is also locally abundant.

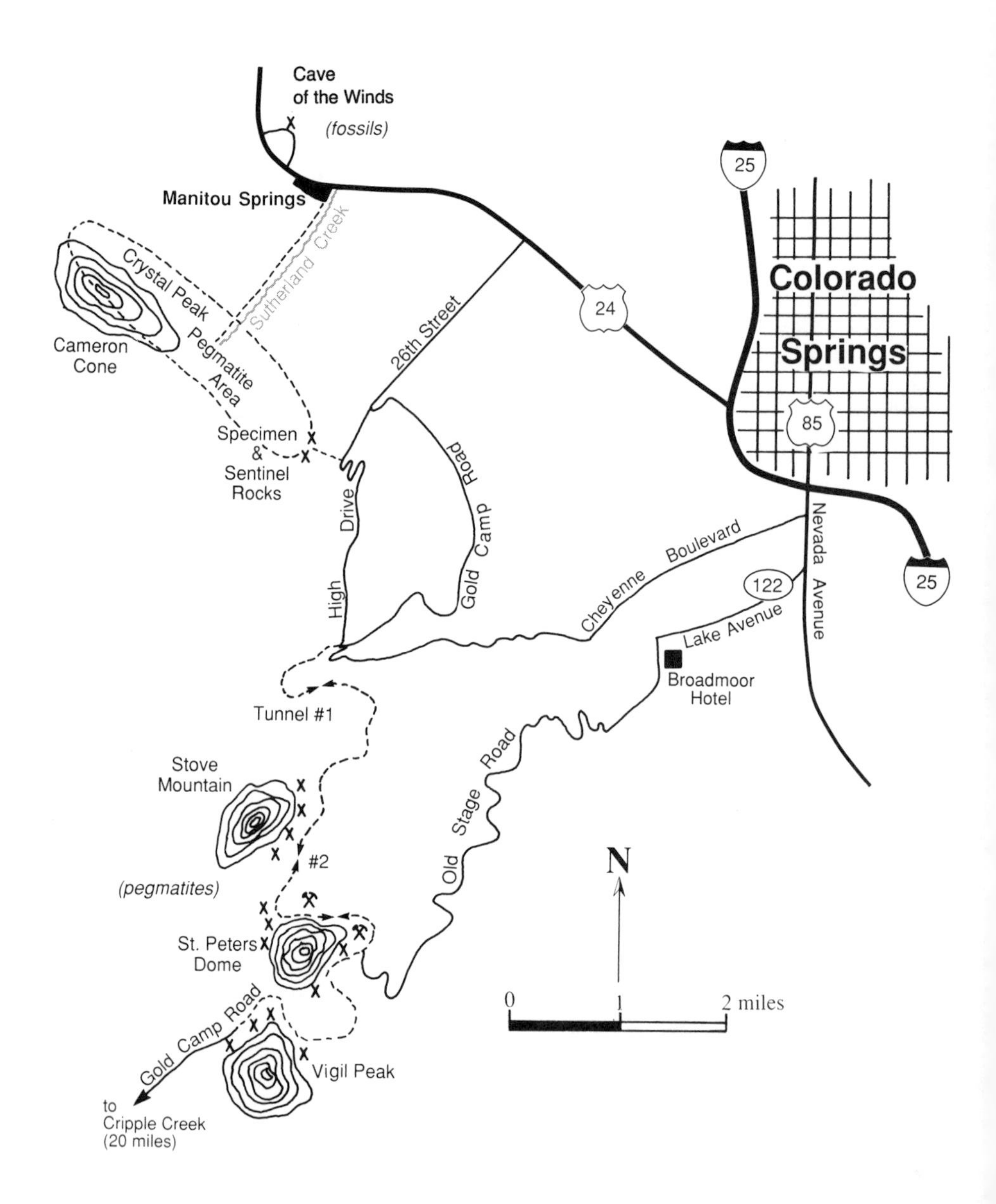

Cave
of the Winds
(fossils)
Manitou Springs
Sutherland Creek
Crystal Peak
Pegmatite
Area
Cameron
Cone
Specimen
&
Sentinel
Rocks
26th Street
24
25
Colorado
Springs
85
High
Drive
Gold Camp Road
Cheyenne Boulevard
Nevada Avenue
122
Lake Avenue
Broadmoor
Hotel
Tunnel #1
Stove
Mountain
#2
(pegmatites)
Old Stage Road
St. Peters
Dome
Gold Camp Road
Vigil Peak
to
Cripple Creek
(20 miles)
N
0
1
2 miles

EL PASO COUNTY

THE PIKES PEAK PEGMATITES

Much of the granite massif of Pikes Peak, including its 14,110-foot summit, lies within El Paso County. Numerous pegmatites, which have yielded fine mineral and gemstone crystals since at least the 1880s, are located west of Colorado Springs.

SENTINEL ROCK AND SPECIMEN ROCK

Smoky quartz and other pegmatite minerals are found at Sentinel Rock and Specimen Rock, two prominent granite outcrops located three miles west of Colorado Springs near the junction of High Drive and Gold Camp Road. In 1885, George Frederick Kunz described Sentinel Rock and Specimen Rock as Colorado's "richest area for smoky quartz" and noted the recovery of thousands of smoky quartz crystals from one inch to four feet in length.

Smoky quartz crystals, two to six inches long and coated with red hematite, are found in pegmatite veins and pockets and loose in talus slopes. Also present are green amazonite, green fluorite,

Smoky quartz crystal from Sentinel Rock.

Sentinel Rock and Specimen Rock, prominent granite outcrops on the eastern slope of Pikes Peak, have yielded thousands of quartz crystals.

bladed hematite crystals up to one-half inch long, and limonite and goethite pseudomorphs after siderite.

The collecting area is on El Paso County Park System watershed and park land. From Colorado Springs, take U.S. 24 west for three miles, then follow 26th Street and Gold Camp Road south for three miles to a hairpin turn and a marked caretaker's house and gate. Beyond the gate, High Drive is one-way (north) only. Eight-tenths of a mile beyond the house, a marked foot trail on the west side of the road leads to Sentinel and Specimen rocks. The strenuous hike ascends nearly a thousand vertical feet in one mile.

El Paso County park rangers close the trail and the Sentinel Rock and Specimen Rock area in June and July to protect nesting peregrine falcons. Temporary restriction notices are posted at the trailhead. For information about restrictions, contact the nearby Bear Creek Regional Park Nature Center.

CRYSTAL PARK

Crystal Park, another pegmatite-collecting locale, is located two miles south of Manitou Springs. It covers five square miles and extends from Sentinel and Specimen rocks three miles northwest to Cameron Cone. The most productive collecting areas are the north and northeast slopes of Cameron Cone and along its southeast-trending ridge, which includes both private and National Forest land.

Mineral crystals occur in pegmatite veins and pockets within a coarse white granite. Crystals include intensely colored blue-green amazonite, smoky quartz, phenakite, and some topaz. Accessory minerals are purple, green, and white fluorite; black tourmaline crystals up to one foot in length in white quartz; and hematite after siderite.

Collecting accounts in the 1920s mention spectacular pockets yielding as many as 500 pounds of amazonite-smoky quartz specimens. Recent Crystal Park recoveries include six- and eight-inch crystals of smoky quartz. There are numerous old pegmatite diggings left from early commercial collecting activity.

From Manitou Springs, follow the road south along Sutherland Creek. Foot trails and rough jeep roads lead to collecting areas.

GOLD CAMP ROAD AREA

Gold Camp Road, four miles due west of and 3,000 feet above Colorado Springs, leads to pegmatite-collecting areas on the slopes of 9,665-foot-high St. Peters Dome and two adjacent peaks, 6,782-foot Stove Mountain and 10,075-foot Mt. Vigil. The old Short Line railroad that connected Colorado Springs with Cripple Creek formerly provided easy access to St. Peters Dome. The Short Line followed a 10,000-foot-high right-of-way through narrow tunnels and along precarious drop-offs. The Short Line was scrapped during World War II and the rail bed converted to Gold Camp Road, an automobile road. The old railroad tunnels were declared dangerous and closed in 1988. Today, the pegmatite areas near the six-mile-long eastern end of Gold Camp Road are accessible only to hikers, mountain bikers, and motorcyclists.

From Colorado Springs, there are two routes to reach Gold Camp Road. From U.S. 85, follow Cheyenne Boulevard west for five and a half miles into Cheyenne Cañon Park to intersect Gold Camp Road. Or follow Colorado Route 122 (Lake Ave.) to the Broadmoor Hotel and Resort, then take Old Stage Road eight miles to Gold Camp Road. Both approaches intersect near the traffic barricades on Gold Camp Road.

From the Old Stage Road intersection, follow Gold Camp Road one-half mile north to the barricade. Tunnel Number Three can be seen ahead. The road and parking area is scattered with purple and white fluorspar from nearby vein outcrops. Seams within the fluorspar contain small crystals of green and purple fluorite.

The old Eureka Mine, once a source of gem-quality transparent zircon in pinks, greens, and yellows, is hidden in the pines 400 feet below the road. Zircon is still present in the dump, usually as opaque brown zircon crystals with characteristic square basal cross sections terminated by four-sided pyramids.

Just beyond Tunnel Number Three (foot and bike trails bypass the old tunnels) a spur road leads to mine dumps with specimens of cryolite and fluorite. Dumps have been well collected and digging is necessary.

One-half mile south of Tunnel Number Two, roughly midway between the two auto intersections, the site of an old railroad station affords a sweeping view of Colorado Springs and the plains to the east. Loose smoky quartz and fluorite crystals occur in the talus slopes above and below the road. Trails lead a quarter-mile up Buffalo Creek to another smoky quartz and amazonite area.

Pegmatite prospectors and collectors have found over a thousand pegmatite veins and pockets on the slopes of St. Peters Dome alone. Off-road prospecting and collecting opportunities are excellent. Although St. Peter's Dome is not far from Colorado Springs, the collecting areas are 3,000 feet higher in elevation. The optimum collecting time extends from late April through October, but sudden, inclement weather is always a possibility.

GLEN COVE

Glen Cove, located in western Teller County, is reached by the Pikes Peak Toll Road from Cascade and U.S. 24 in El Paso County. Glen Cove was a popular collecting locale in the early 1900s, after completion of the automobile road to the summit of Pikes Peak. Collectors have recovered thousands of smoky quartz crystals, many of faceting grade. In 1951, a *National Geographic* article described how climbers rappelled down to crystal-filled cliffside pegmatites. Although exposed in-situ pegmatites on steep cliffs are accessible only to experienced climbers, collectors have found many crystals in the lower talus slopes.

Glen Cove is about two-thirds of the way between Cascade and the Pikes Peak summit on the Pikes Peak Toll Road. The toll road is open from mid-May through mid-October.

REFERENCES: 5, 18, 19, 23, 32, 37, 41, 48, 51, 57, 64, 66, 67, 73, 74, 84

AGATE AND PETRIFIED WOOD

PEYTON-CALHAN-BLACK FOREST AREA

Petrified wood is locally abundant to the north and east of Colorado Springs. Youngsters from the little towns of Peyton and Calhan once set up roadside tables to sell locally collected petrified

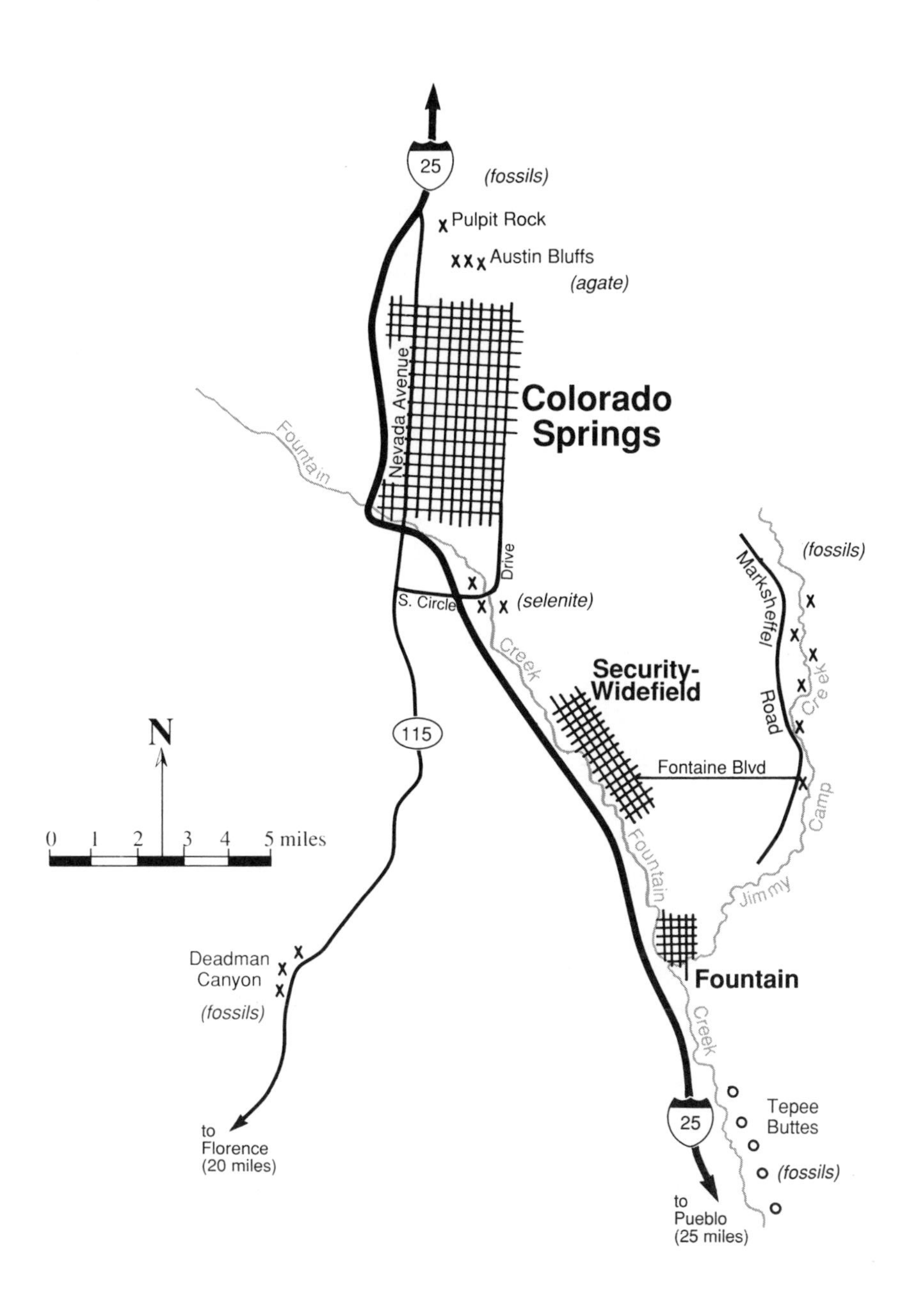
25
(fossils)
Pulpit Rock
Austin Bluffs
(agate)
Nevada Avenue
Colorado
Springs
Fountain
Drive
S. Circle
(selenite)
Creek
Security-
Widefield
Marksheffel
Road
(fossils)
Creek
Fontaine Blvd
Camp
Jimmy
Fountain
N
0 1 2 3 4 5 miles
115
Deadman
Canyon
(fossils)
to
Florence
(20 miles)
Fountain
Creek
Tepee
Buttes
(fossils)
25
to
Pueblo
(25 miles)

wood to tourists traveling on U.S. 24. The Black Forest area, eight miles east of Peyton, is also a good source of petrified wood. Most specimens are found in dry washes and gullies.

AUSTIN BLUFFS

The Austin Bluffs, prominent bluffs of white limestone conglomerate that are part of the city park system, are located in northern Colorado Springs. Banded agate and jasper weather free from the steep limestone exposures. From downtown Colorado Springs, take U.S. 85-87 (Nevada Avenue) three miles north to Austin Bluffs Parkway. Austin Bluffs rises immediately to the northeast and is reached by a number of local roads and trails.

REFERENCES: 32, 33, 64, 65, 66

EL PASO COUNTY FOSSILS

There are many easily accessible fossil occurrences in the foothills and gullies of eastern El Paso County.

TEPEE BUTTES

Dozens of tepee buttes, small, conically shaped hills, are located just east of I-25 from Fountain to the Pueblo County line. The buttes are mounds of resistant limestone in Pierre Shale that originated as reefs on a Cretaceous sea floor. Although only about thirty feet high, they are composed entirely of fossil-rich limestone. Fossils, mainly of the pelecypod *Lucina,* a small clam, are cemented together with gypsum and calcite.

PULPIT ROCK

Pulpit Rock, a prominent rock cliff, is located four miles north of downtown Colorado Springs and just north of the Austin Bluffs agate area. Numerous impressions of Tertiary leaves occur in deteriorating shales near the western base of the rock. Pulpit Rock rises immediately east of U.S. 85-87 (Nevada Avenue), one mile north of Austin Bluffs Parkway.

JIMMY CAMP CREEK

Jimmy Camp Creek cuts through Dawson Arkose, a coarse sandstone, just east of Security-Widefield. Fossil impressions of Tertiary leaves, some measuring eight inches, occur in sandstone fragments and ledges along the creek bed.

From Security-Widefield, six miles southeast of Colorado Springs, take Fontaine Boulevard three miles east to Marksheffel Road, which follows Jimmy Camp Creek north for several miles.

Pulpit Rock, just north of downtown Colorado Springs, is a source of Tertiary fossil-leaf impressions.

DEADMAN CANYON

Deadman Canyon is twelve miles south of Colorado Springs on Colorado Route 115. Little Fountain Creek cuts through sediments of carboniferous Lyons Sandstone rich in fossil impressions of leaves, stems, and branches.

From the north, Colorado Route 115 enters Deadman Canyon just beyond the last Fort Carson gate at mile 36.5. The chalk-colored Cretaceous limestone cliff exposed in the road cut on the west contains numerous small clam fossils. The original calcareous shell material is sometimes replaced by pyrite. A half mile farther south, another road cut exposes sandstone with many stem and branch fossils.

WILLIAMS CANYON (CAVE OF THE WINDS)

Cave of the Winds, Colorado's only commercially developed limestone cave, is two miles west of Manitou Springs near U.S. 24. The cave was created by acidic waters leaching through and dissolving Ordovician limestone in relatively recent Pleistocene times. The alternate exit road returning to Manitou Springs passes through Williams Canyon, where cliff exposures of Ordovician limestone contain many fossils of trilobites, brachiopods, and

Cretaceous limestone exposure at Deadman Canyon is rich in pelecypod fossils, including specimens with pyriticized shell materials.

small early cephalopods. Fossils can be found in limestone fragments in the bed of the creek.

REFERENCES: 5, 12, 13, 21

EL PASO COUNTY SELENITE LOCALITIES

PAINT MINE GULLIES

A deeply eroded gully system, known as the Paint Mines for its brightly colored yellow and green exposed sediments, is located near Calhan and U.S. 24, thirty miles northeast of Colorado Springs. From Calhan, take East Paint Mine Road for three miles to the rim of the gullies. Clear and reddish selenite crystals occur in seams in eroding layers of dark shales within Dawson Formation sediments.

FOUNTAIN CREEK

Clusters of clear selenite crystals occur in clay exposures along the banks of Fountain Creek near the southern limit of Colorado Springs. One site is just north of the South Circle Drive bridge. Most collectors dig the clusters out of seams in the clay, but loose crystals have also been found in the wide creek bed.

REFERENCES: 41, 64

FREMONT COUNTY

THE ROYAL GORGE AREA PEGMATITES

The granite pegmatites of the Pikes Peak Batholith extend into northern Fremont County, providing many opportunities to collect and study pegmatite mineral specimens, especially near the Royal Gorge, eight miles west of Cañon City.

The Royal Gorge is the most topographically spectacular section of the 1,500-mile-long Arkansas River. The ancestral Arkansas River originally flowed south from its source near Leadville into the San Luis Valley. Some 20 million years ago, regional uplift and volcanic action near Poncha Pass blocked its course. Diverted eastward, the river cut a new channel through Cretaceous sediments and into underlying Precambrian granite and gneiss, forming the 1,000-foot-deep, sheer-walled Royal Gorge.

The pegmatites in the Precambrian granite near Royal Gorge range from small lenses to massive dikes over 3,000 feet long and 500 feet wide. Because of high resistance to erosion, the pegmatites form outcrops on hills and ridges. Most are readily identifiable be-

Detailed brachiopod fossils from Pennsylvanian sediments in Arkansas River Canyon near Swissvale.

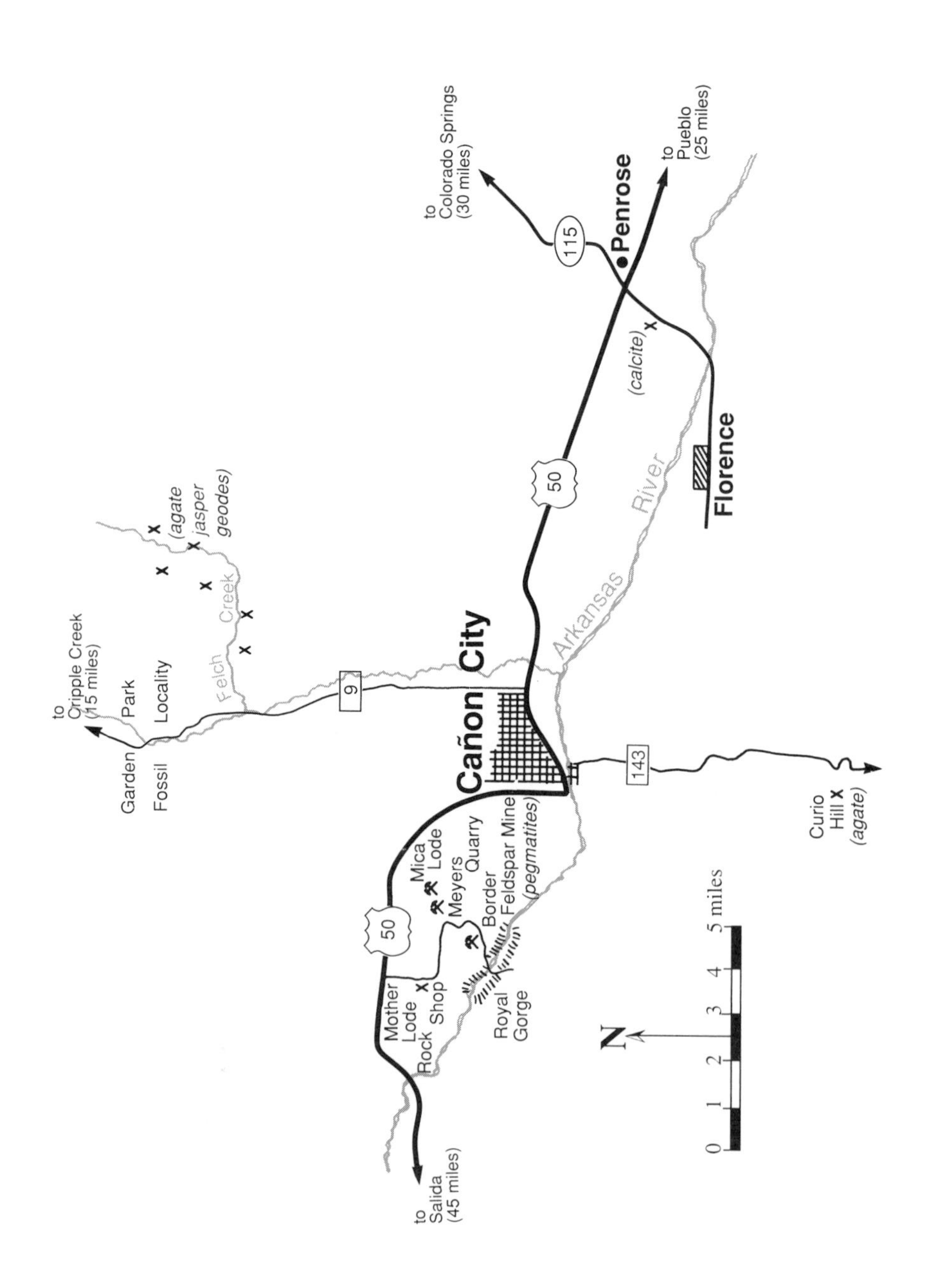
to Colorado Springs (30 miles)
115
Penrose
to Pueblo (25 miles)
(calcite)
Florence
50
Arkansas River
(agate jasper geodes)
Felch Creek
Cañon City
9
to Cripple Creek (15 miles)
Garden Park Fossil Locality
143
Curio Hill (agate)
Mica Lode
Meyers Quarry
Border Feldspar Mine (pegmatites)
Mother Lode Rock Shop
Royal Gorge
to Salida (45 miles)
N
0 1 2 3 4 5 miles

cause of the high reflectivity and light color of their primary minerals—pink feldspar, shiny mica, and white, gray, and rose quartz.

Feldspar and mica mining began about 1900. Miners encountered huge individual crystals of beryl and feldspar, along with eight-foot "books" of muscovite mica. Also present were pockets of gem-quality elbaite tourmaline in pink, lilac, and green. Although pockets were small and erratic, some quarries produced gem tourmaline comparing favorably in beauty and value with tourmaline from Maine and California.

Construction of the famed Royal Gorge Suspension Bridge and associated access roads in 1929 stimulated mining. Miners claimed or leased many pegmatite outcrops, and feldspar and mica mining production peaked. Beryl and rare earth mining began shortly before World War II.

There are about twenty open-cut mines and exploratory digs near the Royal Gorge. Many are privately owned and currently mined by Colorado Quarries, Inc., of Cañon City.

From Cañon City, take U.S. 50 eight miles west to the well-marked Royal Gorge turnoff.

THE MOTHER LODE ROCK SHOP

The Mother Lode Rock Shop, owned by Colorado Quarries, Inc., has an extensive collection of locally quarried pegmatite minerals. From the Royal Gorge entrance on U.S. 50, follow Fremont County Road 3A (Royal Gorge Road) south one mile to the shop.

The parking lot has many large quarried boulders that show remarkable pegmatite crystal development of feldspar, muscovite, and beryl. The shop specimens, mined by Colorado Quarries within a 100-mile radius of the Royal Gorge, include obsidian and green and banded rhyolite from Westcliffe (Custer County); red marble from Cañon City; rose quartz, smoky quartz, and beryl from the big Devil's Hole Quarry near Texas Creek; and white marble from the Monarch Quarry (Chaffee County). The Mother Lode Rock Shop is open from mid-May through September. Even during the off-season, the parking-lot specimens alone are worth a visit.

Colorado Quarries, Inc., is headquartered in Cañon City on 15th Street, one block south of U.S. 50. Specimens of pegmatite minerals are sold for as little as thirty cents per pound, and arrangements can sometimes be made to visit nearby operating quarries.

THE MICA LODE

One mile beyond the Mother Lode Rock Shop and opposite the Buckskin Joe "Wild West" attraction, an unmarked gravel road heads east a short distance toward the Mica Lode. The mines are the open cuts on the hillside ahead, with light, almost brightly colored dumps characteristic of pegmatite mining. These quarries are posted and gates are locked. Permission to visit the quarries must be obtained from the Colorado Quarries office in Cañon City.

The Mica Lode and the adjacent Meyers Quarry, two of Colorado's biggest pegmatite quarries, are located on the same 1,400-foot-long pegmatite dike. Sections of the dike "blew out," as the miners say, to widths of over 400 feet. These two quarries have produced 500,000 tons of microcline feldspar, 60,000 tons of muscovite mica, and large amounts of greenish beryl, much as well-formed crystals several feet in length. Miners also recovered columbite-tantalite, valuable for its niobium-tantulum content. Sections of the pegmatite contain lithium, apparent in the lepidolite and gem elbaite tourmaline found in the early years.

The nearby gravel roads sparkle with mica, and gullies are littered with tons of pegmatite dump material, including boulders with foot-long feldspar and mica crystals and masses of white, gray, and rose quartz. Several small, in-situ pegmatites crop out on the road shoulders.

THE BORDER FELDSPAR MINE

The Border Feldspar Mine produced mica and feldspar, but most collectors seek black tourmaline. Prisms six inches long and one-half inch in diameter are common. The black tourmaline, ranging in color from jet black to dark greenish black, occurs in a snow white quartz matrix and makes attractive display specimens. Specimens are found loose in the dumps or in situ in the quarry walls.

From the Mother Lode Rock shop, follow the Royal Gorge Road south 1.3 miles. Immediately beyond the wooden sign marking the entrance to Royal Gorge Park, turn right into the unmarked parking area. This was formerly a loading area, and fragments of pegmatite minerals are scattered about. The quarry is located 200 yards away on the opposite (north) side of the hill. Walk along the old, overgrown access road, which is covered with shiny mica flakes. This property is now posted.

REFERENCES: 1, 64, 99

THE GARDEN PARK RESEARCH NATURAL AREA

Garden Park, eight miles north of Cañon City on Fremont County Road 9, is one of North America's most important late-Jurassic vertebrate localities. Charles Felch discovered local fossilized dinosaur bone in 1876, but the huge bones were thought to be petrified tree trunks. Kansas State Geologist Benjamin Mudge correctly identified them, attracting the attention of the prominent paleontologists Othniel Marsh and Edward Drinker Cope. The fossils excavated at Garden Park in the 1880s, along with those at Morrison (Jefferson County) and Como Bluff, Wyoming, helped revolutionize dinosaur paleontology.

Excavation of the Morrison Sandstone fossil beds has since yielded twenty genera and nineteen described species of fossil fish, turtles, crocodiles, dinosaurs, and mammals. Garden Park is the type locality for eight species of vertebrates, including the dinosaur genera *Allosaurus, Camarasaurus, Ceratosaurus*, and *Diplodocus*.

Garden Park dinosaur bone is typically well silicified as bright red, yellow, and gray jasper. Early commercial collectors cut and polished the bone for sale to tourists as "watch charms."

In the 1930s, a Cañon City high school teacher and his students discovered the superb *Stegosaurus* skeleton now displayed at the Denver Museum of Natural History. Garden Park dinosaur fossils are exhibited at many major natural history museums, and paleontological excavation continues today. Paleontologists from the Denver Museum of Natural History made several important discoveries in 1992 and 1993.

The Garden Park fossil area is unusual because dinosaur remains occur throughout the entire vertical extent of the Morrison Formation, unlike most other dinosaur localities where fossils are stratigraphically limited. Garden Park also produces many freshwater mollusk and arthropod fossils, and it's the type locality for thirteen species of freshwater pelecypods and gastropods.

The Bureau of Land Management has designated Garden Park a research natural area and an area of critical environmental concern because of current excavations and future paleontological potential. Several historic dinosaur quarries and active digs are located within the protected area. Fossil collecting or removal of any materials is prohibited.

Garden Park is located seven miles north of Cañon City on Fremont County Road 9 and is part of the BLM's "Gold Belt Scenic Byway." Interpretive signs opposite the old Cleveland Museum of Natural History dinosaur quarry on Fourmile Creek explain the local paleontological history.

The 600-member Cañon City Paleontology Society has proposed establishment of the Garden Park Fossil Area, which will include a $25 million visitors center. The area, to be jointly managed by the BLM and the society, would preserve the historic quarries and offer tours and exhibits.

The BLM now offers tours of the active dinosaur quarries by special arrangement during the summer months. For information, contact the BLM District Office in Cañon City. Other significant vertebrate and invertebrate fossil localities in Fremont County include Spring Gulch, Twin Mountain, and Indian Springs, all of which have been proposed for special management and protection.

***REFERENCES:* 3, 63, 64, 68**

FREMONT COUNTY AGATE, GEODES, AND CALCITE

FELCH CREEK AREA

Felch Creek joins Fourmile Creek just south of the Garden Park fossil area. The eroded badlands east of Fourmile Creek drained by Felch Creek are a source of agate, red jasper (most of it dinosaur bone), and geodes containing small, well-formed crystals of clear quartz or white calcite.

The eroded bluffs and gullies where the agate, jasper, and geodes occur is about a one-mile hike east of Fourmile Creek, but private property along Fremont County Road 9 must be crossed to reach the collecting area.

Agate from Curio Hill.

CURIO HILL

Seven miles south of Cañon City on Fremont County Road 143, Curio Hill is a source of delicately banded blue and multicolored agate. Curio Hill, actually an isolated, steeply tilted limestone section of hogback ridge, rises from juniper-scrub oak woodlands just west of the county road. It is a half mile south of a small ranch and 200 yards before a cattle guard at the top of a low hill.

Curio Hill once provided commercial quantities of agate for the Cañon City tourist trade, and an open-ended short tunnel can be seen near the top. The agate occurs in thin veins within the limestone near the top of the ridge. Although any visible veins appear to have been worked out, the base and slopes of the ridge have many scattered agate fragments. Curio Hill has been heavily collected since the turn of the century.

PENROSE CALCITE

Thick calcite veins occur in a gully system that cuts through weathered limestone near Penrose. From the Penrose bridge over U.S. 50, take Colorado Route 115 south for 1.4 miles. Park on the shoulder and walk west into the gullies, keeping to the south side. Collectors' diggings and scattered calcite crystals can be seen in the first small side gully. Some calcite veins are one foot thick. Well-formed crystals of clear, white, and light yellow calcite occur in hollow seams within the veins. Digging is fairly easy in the weathered limestone.

REFERENCES: 19, 64, 73

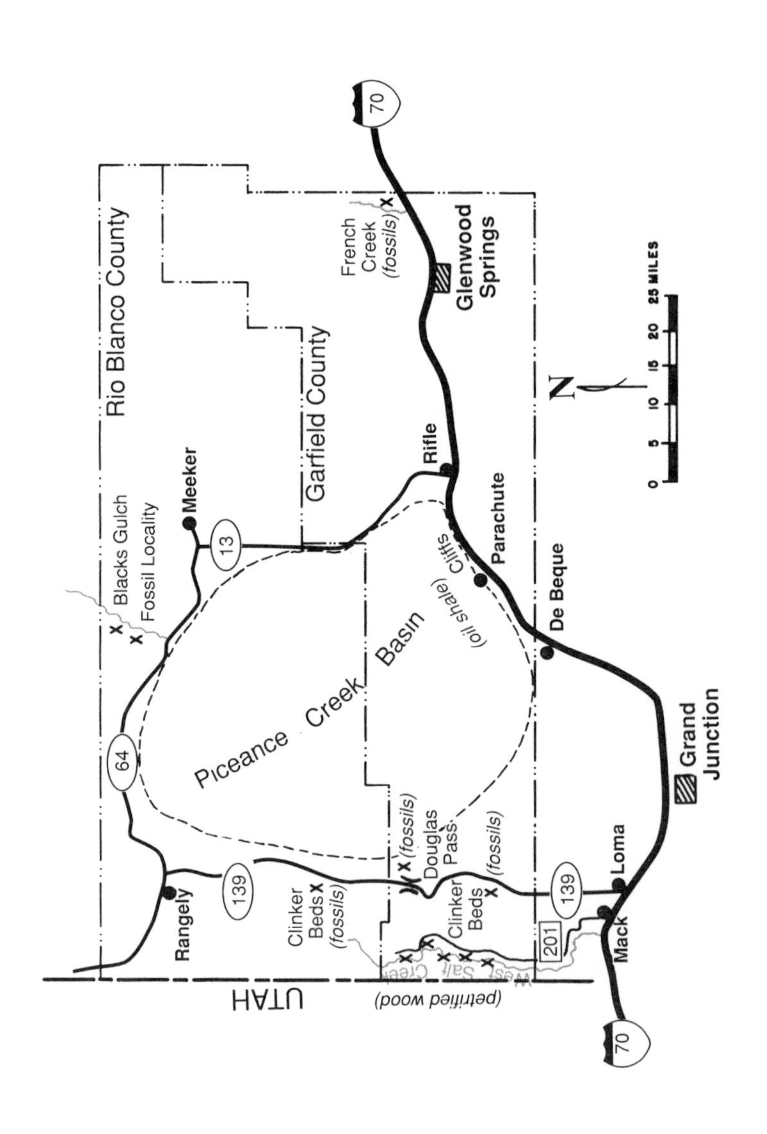
Rio Blanco County
Garfield County
Blacks Gulch
Fossil Locality
Meeker
13
Rifle
French
Creek
(fossils)
Glenwood
Springs
70
Parachute
Cliffs
(oil shale)
De Beque
Piceance Creek Basin
Grand
Junction
64
139
Rangely
Clinker
Beds
(fossils)
Douglas
Pass
(fossils)
Clinker
Beds
(fossils)
Loma
Mack
201
West Salt Creek
UTAH
(petrified wood)
70
N
0 5 10 15 20 25 MILES

GARFIELD COUNTY
RIO BLANCO COUNTY

OIL SHALE

Garfield and Rio Blanco counties cover 4,000 square miles, more than the state of Connecticut. Half of that area is the Piceance Creek Basin, the heart of Colorado's oil shale country. Interstate 70 passes the most impressive exposure of Green River oil shales, the Roan Cliffs, near the "oil shale towns" of Rifle, Parachute, and DeBeque.

The Roan Cliffs extend from Rifle thirty miles west to DeBeque. The cliffs, as high as 2,000 feet, have colors ranging from buff and tan to mahogany. The darker colors contain higher grades of kerogen. The old government Anvil Points oil shale mines and retort plant appear near the cliffs just west of Rifle.

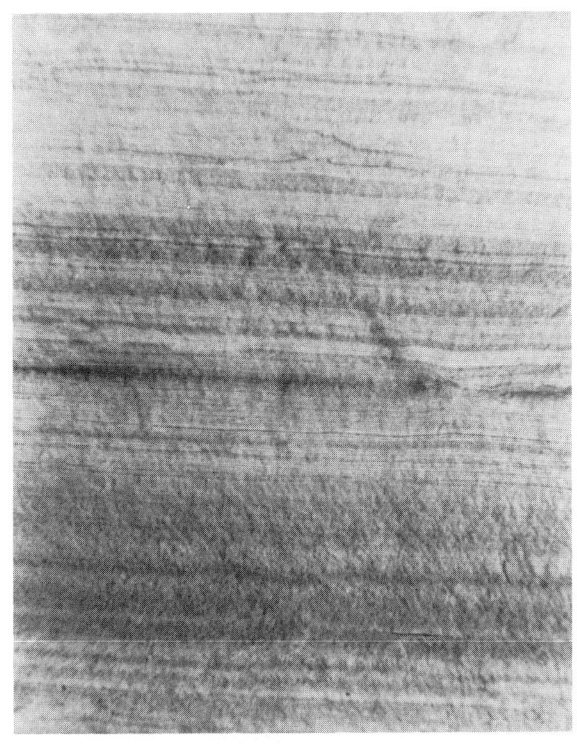

Delicately banded oil shale is among the most attractive of all sedimentary rocks.

Parachute (formerly Grand Valley) was the center of the past oil shale booms. Oil shale boulders are displayed at the Parachute Visitor Center. North of Parachute, Parachute Creek cuts deeply through the Roan Cliffs. Garfield County Road 215 leads north for six miles toward some of the inactive oil shale mines. In places the road bed is built on oil shale earth fill, and specimens in many patterns and colors may be collected at culverts and gullies.

REFERENCES: 63, 81, 92

GARFIELD-RIO BLANCO COUNTY FOSSILS

FRENCH CREEK

French Creek is located just west of Dotsero along I-70 near the mouth of Glenwood Canyon, one mile west of the Garfield-Eagle county line. The brown cliffs on the north side of I-70 at the mouth of French Creek are Ordovician dolomite containing numerous marine fossils, including those of trilobites, brachiopods, and gastropods. The fossils are found in the talus at the base of the cliffs and in the bed of French Creek.

French Creek was a popular fossil-collecting locale for many years when U.S. 6 afforded a more leisurely transit of Glenwood Canyon. Recent construction of I-70 through Glenwood Canyon, however, now restricts access to French Creek. There is no highway shoulder, and parking or stopping near French Creek is prohibited.

REFERENCES: 12, 13

BLACKS GULCH

Blacks Gulch, in northern Rio Blanco County thirty miles west of Meeker near Colorado Route 64, is Colorado's best early Eocene vertebrate-fossil locality. University paleontologists have studied the site for many years, recovering hundreds of fossilized vertebrate bones, including those of early primates.

The Bureau of Land Management will designate Blacks Gulch a research natural area with unauthorized fossil collecting prohibited. Specific information regarding Blacks Gulch paleontological sites can be obtained from the BLM office in Craig.

REFERENCES: 3

DOUGLAS PASS LEAF- AND INSECT-FOSSIL LOCALITY

Douglas Pass, elevation 8,268 feet, is located in northwestern Garfield County on Colorado Route 139 midway along the 75-mile-long stretch (no services) between the towns of Loma and Rangely. A nearby Green River Formation exposure is a classic collecting locality for Eocene plant and insect fossils.

A collector at work at the Douglas Pass fossil locality.

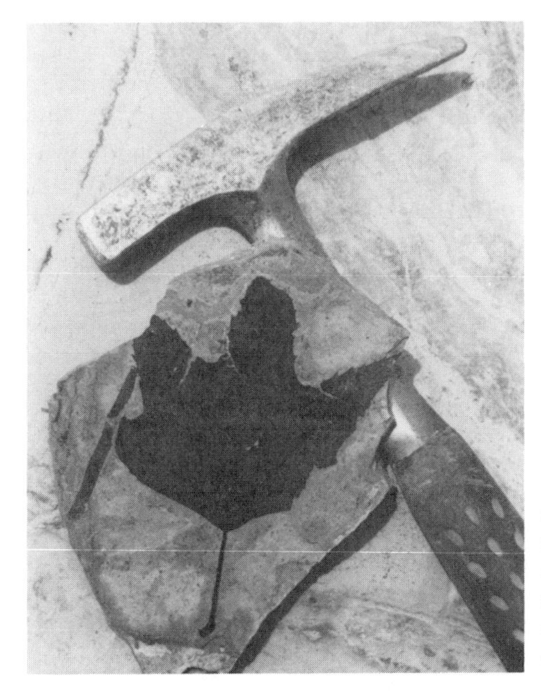

Eocene maple-leaf fossil impression from Douglas Pass fossil locality. Leaf fossils from this area are characterized by superb detail and high contrast.

High contrast in plant fossils at the Douglas Pass fossil locality is caused by carbonization of the original organic material.

The Douglas Pass sediments, deposited by Lake Uinta, a "migrating" Eocene lake, consist of calcareous to dolomitic siltstones, sandstones, and organic marlstones (oil shales). Fossil leaves and insects are abundant within a thin strata of "mahogany bed" oil shale. The fine-grained sediments produce fossils with extraordinary detail.

Four general groups of plant fossils are present, each representing a different paleoenvironment. Lakeshore sediments contain willow, sycamore, poplar, fern, and cattail; sediments from better-drained low-lying areas have fossilized remains of hackberry, oak, and sumac; sediments from higher slopes contain fossils of maple, hickory, and certain conifers; and sediments from well-drained hill and mountain slopes have leaf and stem fossils from hardwood deciduous and coniferous forests.

Paleontologists have collected over 300 species of fossilized insects from the Green River Formation, many from Douglas Pass. Douglas Pass fossils are displayed at the Museum of Western Colorado in Grand Junction.

At the summit of Douglas Pass, turn east on the graded gravel road that begins at the Highway Department maintenance shed. In 5.5 miles the road climbs nearly 800 feet, ending at an operating

Plant fossils are abundant in Mesaverde Formation clinker-bed quarries.

FAA radar dome. The primary site is the high road cut immediately below the radar station. Collectors have left heaps of broken shale along the road shoulder.

The original carbonized organic matter produces superbly detailed leaf and stem fossils that contrast sharply against the light gray shale, making striking display pieces. Fossil fragments can be found by searching through the broken shale, but most collectors dig fresh sections of in-situ shale, then split the bedding planes to expose fossil specimens. Hammers and sharp-edged chisels or knives are useful in splitting the shale.

The road from the Douglas Pass summit to the collecting site is passable for highway cars in dry weather. Douglas Pass is thirty-five miles from the nearest services.

REFERENCES: 3,16,17

CLINKER-BED PLANT FOSSILS

Plant fossils are abundant at two clinker-bed quarries in the lower Mesaverde Formation along Colorado Route 139 near Douglas Pass. One quarry is twelve miles south of Douglas Pass (twenty miles north of Loma); the other is thirteen miles north of Douglas Pass (twenty-four miles south of Rangely). Both quarries, with prominent brightly colored red and yellow rock walls, are easily accessible on the west side of Colorado Route 139.

Clinker beds are sedimentary rock altered by local heat from natural combustion of adjacent coal beds. Clinker beds provide a durable, reddish rock for road surfacing and decorative use. Plant fossils, mostly stems and branches up to three inches wide and two feet long, are colored red and make attractive display specimens against the buff-colored matrix rock.

No digging is necessary, as most of the broken quarry rock contains plant fossils. Collectors sort through broken rock to find suitable specimens. The fossils are a record of obviously profuse vegetation that thrived about 70 million years ago.

REFERENCES: 16

WEST SALT CREEK PETRIFIED WOOD

West Salt Creek, in western Garfield County just east of the Utah line, contains many large fragments of petrified wood. From Mack on I-70 (Mesa County), take County Road 201 (West Salt Creek Road) ten miles northwest. The road follows West Salt Creek for the next fifteen miles. Petrified wood is locally common throughout the area. West Salt Creek Road (no services) is passable for highway cars in dry weather.

Naturally reliefed Cretaceous palm fossil from Mesaverde Formation clinker-bed quarry.

A collector searches a Mesaverde Formation clinker-bed quarry for plant fossils.

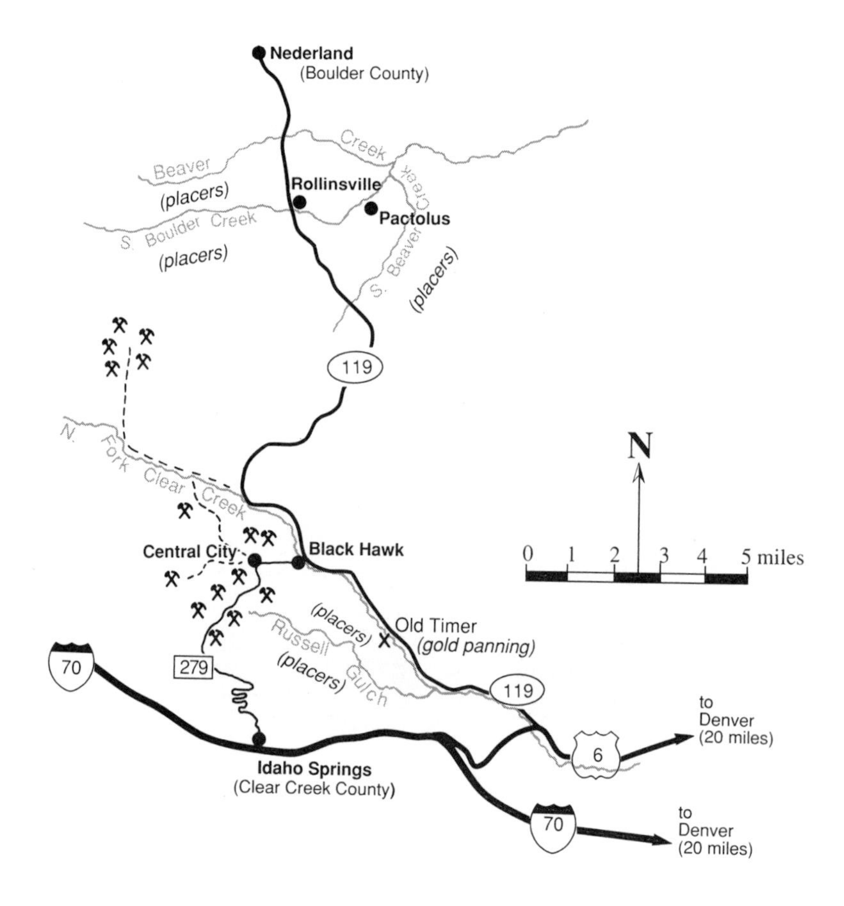
Nederland
(Boulder County)
Beaver Creek
(placers)
Rollinsville
Pactolus
S. Boulder Creek
(placers)
S. Beaver Creek
(placers)
119
N. Fork Clear Creek
N
0 1 2 3 4 5 miles
Central City
Black Hawk
(placers)
Old Timer
(gold panning)
Russell Gulch
(placers)
70
279
119
to
Denver
(20 miles)
6
Idaho Springs
(Clear Creek County)
70
to
Denver
(20 miles)

GILPIN COUNTY

PLACER GOLD

Gilpin County ranks sixth among Colorado counties in placer gold production. Miners have recovered about 50,000 troy ounces from the upper drainages of the North Fork of Clear Creek and South Boulder Creek.

In May 1859, John Gregory followed a trail of "color" up the North Fork of Clear Creek to the gulch that now bears his name. In Gregory Gulch, just above the town of Black Hawk, he discovered outcrops of gold-bearing quartz veins. Other prospectors rushed in and found placer gold along the entire length of the North Fork of Clear Creek and in many western tributaries, such as Russell and Nevada gulches. Gregory's strike helped "save" the Pikes Peak rush. Among a dozen early camps, Central City and Black Hawk emerged as mining, milling, and commercial centers.

The discovery placers were extraordinarily rich in coarse gold. Within weeks of Gregory's strike, miners staked every foot of three-mile-long Gregory Gulch. Many panned two troy ounces of gold

Visitors learn the basics of gold panning at the Old Timer, Colorado's most popular fee panning area.

every day. By August 1859, 900 men mining two-mile-long Russell Gulch washed out $35,000 in gold each week. In the first three months of 1860, one miner with a rocker washed $2,400 in gold from Nevada Gulch.

Other placers along South Boulder Creek (Boulder County) extended into northeast Gilpin County near the boom camps of Pactolus and Rollinsville. Commercial placer mining ended in the 1960s, but recreational gold panning remains very popular along the North Fork of Clear Creek.

OLD TIMER

The Old Timer, Colorado's busiest fee panning site, is located on Colorado Route 119 five miles below Black Hawk in the historic gravels of the North Fork of Clear Creek. Established in 1961, the Old Timer now hosts about 5,000 visitors each year from mid-May through September. Because of proximity to the original lode sources, the gold is quite coarse. The gravels are dug directly from a terrace placer deposit bypassed by the fifty-niners.

The basic rate includes use of a pan, expert personal instruction, guaranteed results, and the right to pan all day. For information contact: The Old Timer, P.O. Box 387, Black Hawk, Colorado 80422.
REFERENCES: 23, 53, 105, 108

CENTRAL CITY AREA MINES

Colorado's lode-mining industry started with John Gregory's 1859 discovery of gold-bearing quartz outcrops. Miners first crushed the rich oxidized quartz ores with hammers, then recovered the gold by sluicing and amalgamation. At the Gregory, Bobtail, Bates, Mammoth, and Hunter lodes, miners recovered $100 in gold per day, phenomenal earnings for the era. To crush ores, miners turned next to *arrastres* (circular depressions in the bedrock around which mules pulled heavy boulders), then to steam-powered stamp mills, but gold recovery from the crudely crushed ores remained inefficient.

Miners quickly exhausted the shallow, oxidized portion of the outcrops and encountered complex sulfide mineralization. The sulfides interfered with amalgamation, dropping gold recovery efficiency so low that mining nearly stopped.

James E. Lyon built Colorado's first smelter at Black Hawk in 1864, profitably recovering gold, lead, and silver. In 1867, Nathaniel P. Hill's improved Black Hawk smelters separated gold, silver, and copper. Gilpin County milling innovations were adopted through-

Clear quartz from Central City.

out the Colorado Mineral Belt and the mining West.

Mineralogists detected uranium in ores of Central City's Wood Mine in 1871. For years, the Wood's small quantities of hand-cobbed uraninite (pitchblende) supplied most of the world's radium. In 1900, pitchblende production peaked at thirty-six tons, but attention had already shifted to the carnotite deposits of the Uravan Mineral Belt. By the 1920s, Gilpin County mine production surpassed $100 million in gold, silver, copper, and lead. Most lode mining ceased in the 1950s.

Gravel county roads lead to hundreds of mines, mine dumps, and prospect holes around Central City. The Central City ore bodies are part of a mineralized zone extending south to Idaho Springs (Clear Creek County). Central City ores usually contain pyrite, chalcopyrite, and galena, and often gold and silver. Mineralogists classify the gold-silver ores into four types: pyritic, galena-sphalerite, a transitional pyritic-galena-sphalerite, and tellurides.

Quartz is the most abundant gangue mineral, occurring in attractive composites with metal sulfide minerals. Well-developed clear quartz prisms and scepters as long as three inches are found at several mines, most notably the Patch Mine, locally known as the "Glory Hole." Selective mining at the Patch in the early 1980s provided many fine specimens of crystalline quartz associated with metal sulfides. Pale amethyst is also found on dumps near Nevadaville.

Central City's Lost Gold Mine offers a walk-in underground mine tour. The Gilpin County Historical Museum displays early mining equipment and local ores and minerals.

The John Gregory Memorial Plaque, just west of Black Hawk on Colorado Route 279, marks one of the most historic sites in Colorado mining. The hollowed vertical crevice directly across Gregory Gulch was among the first gold-quartz outcrops ever mined in Colorado.

REFERENCES: 10, 23, 24, 32, 42, 64, 66, 67, 81, 98, 108

GRAND COUNTY

PLACER GOLD

Grand County has produced about 100 ounces of placer gold from the Willow Creek tributaries of Gold Run, Denver, Elk, and Kauffman creeks and from Stillwater Creek. The creeks all originate on Gravel Mountain. Prospectors discovered gold on Willow Creek in 1871.

Gravel Mountain, elevation 11,769 feet, is twelve miles northwest of Granby. Colorado Route 125 leads to Willow Creek, and its gold-bearing tributaries are reached by Colorado Route 125.

REFERENCES: 54

THE GREEN RIDGE PEGMATITE

The Green Ridge pegmatite area is located south of Lake Granby in northeastern Grand County. Prospectors discovered the pegmatite, actually a geologically related series of hundreds of individual pegmatites, in the 1930s. They staked claims and dug numer-

Green Ridge (in middle distance) has hundreds of pegmatites and has yielded many fine specimens of beryl and black tourmaline.

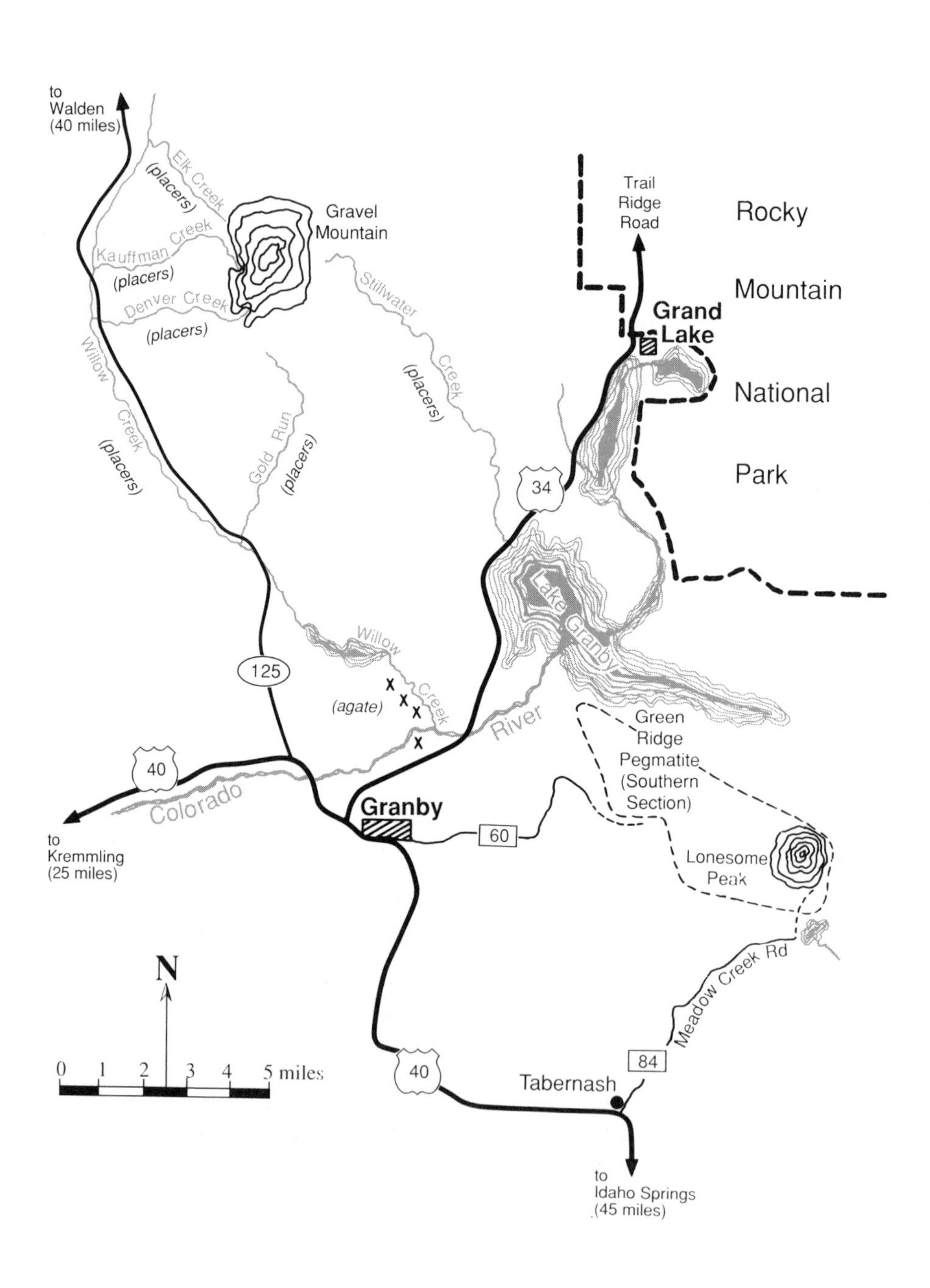
to
Walden
(40 miles)
Elk Creek
(placers)
Gravel
Mountain
Kauffman Creek
(placers)
Denver Creek
(placers)
Stillwater
Creek
(placers)
Willow
Creek
(placers)
Gold Run
(placers)
Trail
Ridge
Road
Grand
Lake
Rocky
Mountain
National
Park
34
Lake Granby
Willow
Creek
(agate)
125
River
Green
Ridge
Pegmatite
(Southern
Section)
40
Colorado
Granby
60
to
Kremmling
(25 miles)
Lonesome
Peak
Meadow Creek Rd
N
0 1 2 3 4 5 miles
40
84
Tabernash
to
Idaho Springs
(45 miles)

ous prospect holes, but never mined commercially. Collectors have recovered many interesting specimens.

The Green Ridge pegmatite is exposed in two sections, each covering about twelve square miles. Most of the northern section, located just south of Grand Lake, is within the Rocky Mountain National Park boundary and closed to collecting. The six-mile-long southern section extends from Walden Hollow southeast to Lonesome Peak (elevation 10,588 feet). The pegmatite is not continuous, but varies repeatedly from pegmatite to granite, granite-gneiss and schist, then back to pegmatite. The coarsest pegmatite crystallization occurs near Lonesome Peak.

Pegmatite minerals include quartz, orthoclase and microcline feldspar, biotite mica, black tourmaline, and poorly crystallized garnet. Portions of the pegmatite contain sufficient magnetite to locally alter compass readings. Beryl is present in greenish white crystals, some as large as six inches in diameter and two feet long.

Several county roads pass near Green Ridge, which is mostly within Arapaho National Forest. To reach the Lonesome Peak area, take U.S. 40 from Granby eight miles south to Tabernash. Follow Grand County Road 84 (Meadow Creek Road) east for six miles to Meadow Creek Reservoir. Lonesome Peak, two miles to the north, is reached by rough jeep roads and hiking trails. Collectors will find heavily wooded Green Ridge a challenging prospecting opportunity.

REFERENCES: 27

MIDDLE PARK AGATE

In his 1867 classic, *The Mines of Colorado*, Ovando Hollister wrote that Middle Park provided the "finest moss agate in Colorado." The official annual report of the Colorado state geologist in 1881 went into greater detail:

> Beautiful moss agates are found in the Middle Park, and many fine gems have been and are to be found there. Quite an important trade has been carried on during past years in obtaining these agates and shipping them to the Eastern States, where, as well as here, they are highly esteemed and much worn as settings for brooches, earrings, sleeve buttons, etc. After being cut they sell for prices varying from 50 cts. to $5 each, and occasionally one of extraordinary beauty, and showing some rare peculiarities, will sell for $10 to $20. The dendritic delineations are usually of a brownish black color, but green, red, yellow, and white are not uncommon colors. They are found in many locations in Colorado, but so far as our knowledge extends, no fine gems have been found outside of the Middle Park localities.

The Colorado RIver near Parshall has been a source of agate for over a century.

Among the collecting areas mentioned in early reports is the confluence of Willow Creek and the Colorado River two miles north of Granby. Another is along the Colorado River from Hot Sulphur Springs six miles downstream (west). The first two miles of that section pass through a canyon where the original channel has been altered for highway and railroad construction.

Another site is Williams Fork, a tributary that joins the Colorado River at Parshall. Early reports indicate that fine moss agate was collected from the mouth of Williams Fork for a distance of two miles upstream, or to the present site of the Williams Fork Reservoir dam.

Fortification and moss agate, as well as bloodstone, chrysoprase, and jasper have been found in these areas along river and creek channels, adjacent terraces, and the slopes and tops of the low surrounding ridges.

REFERENCES: 2, 32, 65, 73, 74

KREMMLING AREA FOSSILS AND PETRIFIED WOOD

Fossiliferous exposures of Pierre Shale and overlying Tertiary sediments are common near Kremmling. Many low hills and ter-

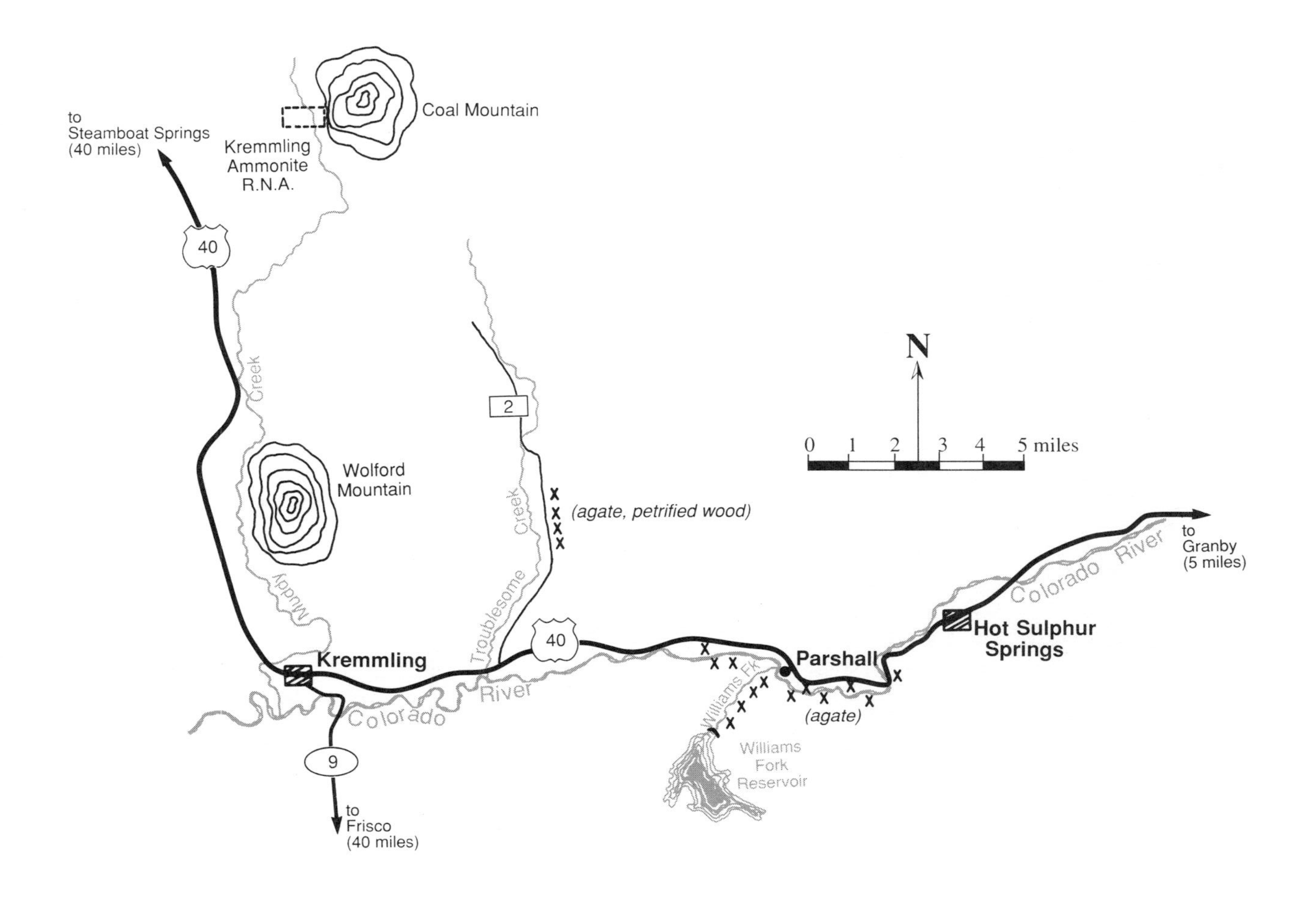
to
Steamboat Springs
(40 miles)
Kremmling
Ammonite
R.N.A.
Coal Mountain
40
Creek
2
Wolford
Mountain
Creek
x
x
x
x
(agate, petrified wood)
Muddy
Troublesome
N
0 1 2 3 4 5 miles
to
Granby
(5 miles)
Colorado River
Hot Sulphur
Springs
40
Kremmling
Parshall
River
Colorado
Williams Fk
(agate)
9
Williams
Fork
Reservoir
to
Frisco
(40 miles)

Petrified wood from Troublesome Creek. Specimen on right has been cut and polished.

races east of Kremmling are fossil-rich Tertiary lake-bed sediments.

Sediments along upper Troublesome Creek, which joins the Colorado River five miles east of Kremmling, have yielded Tertiary vertebrate fossils, including those of Oligocene, Miocene, and lower Pliocene mammals. Unauthorized collecting of vertebrate fossils on public land is prohibited.

There are excellent collecting sites for banded agate and petrified wood along Troublesome Creek. From Kremmling, take U.S. 40 east four miles, then turn north on Grand County Road 2 along Troublesome Creek. Just east of mile markers three and four are heavily eroded badlands. Search the slopes and tops of bluffs for banded gray and light blue agate, red and brown jasper, and petrified wood. The fragments of petrified wood, up to a foot in length, retain much of the original grain pattern and bark structure. Many pieces have a jet black interior that takes a beautiful polish.

The Pierre Shale exposures north of Kremmling along Muddy Creek on the lower slopes of Wolford Mountain and Coal Mountain have yielded large numbers of Cretaceous marine fossils, such as clams, gastropods, and, most notably, ammonites, including the giant ammonite, *Placenticeras meeki*.

THE KREMMLING CRETACEOUS AMMONITE RESEARCH NATURAL AREA

The Kremmling Cretaceous Ammonite RNA may be Colorado's most outstanding marine-fossil locality. For decades, commercial collectors have illegally removed large quantities of ammonite fossils for sale on markets as far away as Europe. Because of uncontrolled commercial collecting, which has damaged some surface features, the Bureau of Land Management has designated the site a research natural area reserved for research and educational purposes. The Kremmling Cretaceous Ammonite RNA is now protected and closed to mineral and land entry. The maximum criminal penalty for unauthorized removal of fossils is a fine of $1,000 and twelve months imprisonment.

Some fossil specimens of the giant ammonite, *Placenticeras meeki*, are two feet in diameter. Many retain the lustrous mother-of-pearl shell material. The fossils occur in a muddy, crumbly sandstone, typically within large sandy calcareous concretions. The muddy sandstone originated as marine sandbars along the western shore of a 75-million-year-old Cretaceous sea. Paleontologists believe the giant ammonites were washed onto sandbars after dying en masse following storms or sudden climatic changes. The presence of a broad variety of other fauna, as well as terrestrial plant debris, indicates a shallow near-shore marine depositional environment.

Also present are more than a hundred species of mollusks, including other ammonoids, nautiloids, clams, gastropods, bryozoans (moss animals), brachiopods, crabs and lobsters, vertebrates such as fish and marine reptiles, and shore and estuary types of terrestrial plants. The Kremmling Cretaceous Ammonite RNA covers 200 acres near Muddy Creek and Coal Mountain about ten miles due north of Kremmling. Those wishing to visit the site, or to collect invertebrate fossils in the Kremmling area, should check first with the BLM's Kremmling Resource Area office in Kremmling.

The Colorado State Land Board is currently considering a proposal to protect an additional 1,040 acres of state land adjacent to the Kremmling Cretaceous Ammonite RNA.

REFERENCES: 3, 67

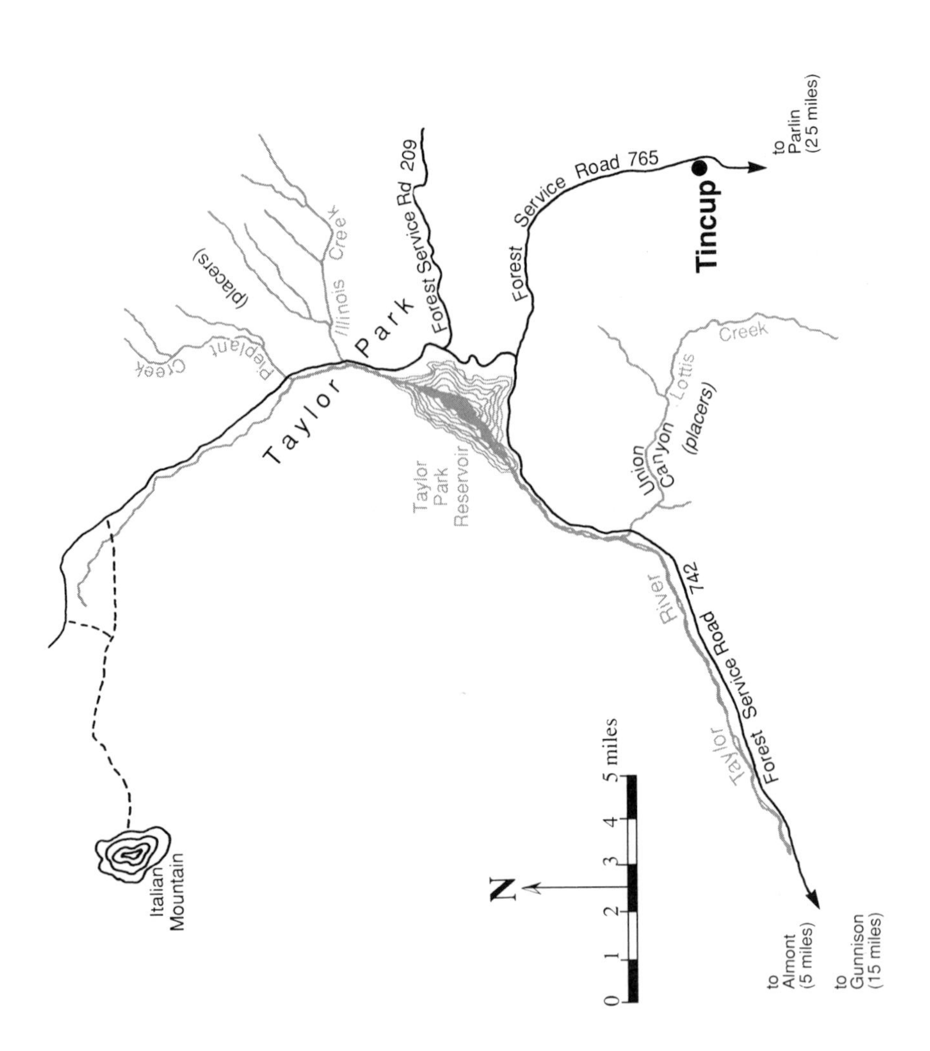
Tincup
to
Parlin
(25 miles)
Forest Service Road 765
Forest Service Rd 209
Illinois Creek
(placers)
Pieplant Creek
Taylor Park
Taylor
Park
Reservoir
Union
Canyon
Lottis
Creek
(placers)
Forest Service Road 742
Taylor River
to
Almont
(5 miles)
to
Gunnison
(15 miles)
Italian
Mountain
N
0
1
2
3
4
5 miles

GUNNISON COUNTY

PLACER GOLD

Gunnison County has produced 10,000 troy ounces of placer gold, mostly from Washington Gulch near Crested Butte and the Taylor River headwaters near Tincup in Taylor Park. Gold also occurs in Gold Creek near the town of Ohio.

Prospector James Taylor discovered the upper Taylor River placers in 1860. The camp of Tincup was supposedly named after the tin cup Taylor used to pan the first gold. Taylor Park miners recovered gold from Illinois and Pieplant creeks and from all the gulches between them. A small floating-bucketline dredge operated from 1913 to 1915. Taylor Park miners also worked the Union Canyon section of Lottis Creek.

Prospectors discovered phenomenally rich placers in Washington Gulch in 1861 that yielded $25 to $50 in gold per pan. Although the bonanza gravels were quickly exhausted, small-scale commercial placer mining survived until the 1950s. Recreational gold panning is a popular summer activity in the Taylor Park area.

REFERENCES: 54

NORTH ITALIAN MOUNTAIN

THE BLUE WRINKLE LAPIS LAZULI MINE

North Italian Mountain is located in rugged country twelve miles northeast of Crested Butte and just west of Taylor Park. Despite unusual geology and mineralization, there are few local commercial metal mines. North Italian Mountain is best known as North America's premier source of lapis lazuli.

Carl Anderson, a miner and part-time prospector, discovered lapis lazuli high on 13,378-foot North Italian Mountain in 1939. Anderson, after checking his lead and grossular garnet claims, was descending the mountain in a cold drizzle when he noticed rain-intensified blue coloration on a piece of talus rock. Suspecting the presence of copper, he collected samples, but mineralogists at the Colorado School of Mines at Golden identified the blue mineral as high-quality lapis lazuli.

Anderson used an old prospector's trick to discover the in-situ source of his sample. From different locations, he rolled similar

Carved lapis lazuli ornaments from the Blue Wrinkle Lapis Lazuli Mine on North Italian Mountain.

sized rocks down the steep slopes. When the rocks came to rest where he had found his sample, he knew he was close. Digging through the talus, he uncovered a lapis lazuli vein. Although the vein was erratic and varied from mere inches to several feet in thickness, Anderson traced it on the surface for several hundred feet. The vein was metamorphic, separating a quartz intrusion from a body of unaltered gray limestone.

Anderson staked the vein, named it the Blue Wrinkle, then spent years casually investigating it. Although he mined just enough lapis to sell in Gunnison and Denver, both the Denver Museum of Natural History and the National Museum of Natural History (Smithsonian) acquired specimens.

Carl Anderson died in 1969 without leaving a will. His son Ande, who helped him work the claims, persuaded a county clerk to quietly transfer them to his name. Like his father, Ande was quiet, reclusive, and worked alone. He maintained the claims and built a small cabin below the mine site, living there during the short alpine summers.

Collector interest in the Blue Wrinkle lapis lazuli grew steadily, but Ande Anderson had few problems with trespassing and unauthorized collecting. He wore a heavy revolver and posted his claims with signs reading: *This Property Belongs To A Madman—He's A Dead Shot—No Digging!*

Although he appeared a rough, eccentric prospector, the real Ande Anderson was different. During his infrequent visits to Gunnison, he spent much time at the tiny Gunnison Public Library reading, enjoying music, or listening to the children's storytelling programs. Meanwhile, disruption of the world supply of lapis lazuli was making Anderson's claims valuable. Afghanistan had traditionally been the leading source of quality lapis lazuli. But Soviet occupation in 1979 curtailed supply and drove prices upward. Oklahoma oilman Paul Schulz, interested in opening a domestic lapis lazuli source, struck a deal with Ande Anderson that year and purchased the Blue Wrinkle claims for more than $60,000 in cash.

Schulz began mechanized mining in the mid-1980s and contracted with The House of Art, a Gunnison shop specializing in mineral specimens and fine jewelry, to handle grading, slabbing, and distribution. Lapis lazuli is graded on color, the amount of matrix (usually calcite), and the distribution pattern of any pyrite present. The top grade, No. 1, is free of calcite matrix and has evenly distributed, tiny bits of glittering, brassy pyrite. Most Blue Wrinkle lapis lazuli is carved in Idar-Oberstein, Germany, or fashioned into cabochons in Taiwan. The House of Art also designs and creates its own 14k gold lapis lazuli jewelry.

After Ande Anderson sold his claims, he lived unpretentiously in a small cabin at Almont, north of Gunnison. Immediately after his death in 1981, relatives searched his cabin for a suspected horde of cash. But Ande Anderson had fooled them. A bank announced it held $70,000 in certificates of deposit in Anderson's name, along with a specific payment order. Upon Anderson's death, the $70,000 was paid to the Gunnison Public Library for construction of music/reading and children's storytelling rooms. The Gunnison Public Library used the gift to construct a 1,500-square-foot music/reading room and a small alcove dedicated to the memory of Ande Anderson. Displays include a scrapbook of pictures of Anderson and the Blue Wrinkle Mine, a bronze plaque in his honor, a pen-and-ink portrait of Ande, and a collection of rough and worked Blue Wrinkle lapis lazuli.

The House of Art, 134 North Main Street in Gunnison, displays worked and mounted lapis lazuli from the Blue Wrinkle Mine. The Blue Wrinkle Mine on North Italian Peak is under claim and operates during the brief alpine summers. Collecting and trespassing are prohibited.

Over eighty minerals occur on Italian Mountain, many as veinlets, fracture fillings, and coatings. They include epidote, actinolite, calcite, white and golden barite, grossular and andradite garnet, pyrite, sphalerite, and stibnite.

REFERENCES: 74, 79, 85

THE YULE MARBLE QUARRY

The historic Yule Marble Quarry is located in northern Gunnison County three miles south of Marble. Magmatic doming within 11,495-foot Whitehouse Mountain provided the necessary heat and pressure to metamorphose thick beds of Mississippian-aged Leadville limestone into a fine-grained, snow white marble. The 240-foot-thick marble beds outcrop for 4,000 feet along Yule Creek on the northeast slope of Whitehouse Mountain. Although chert bands occur within the bed, large sections are flawless.

Prospectors discovered the marble in 1873, and the first quarry opened a decade later. Production at the Yule Marble Quarry peaked in 1912. Marble was then a booming town of 2,000 people, and the Yule mill was the world's largest marble cutting, grinding,

The Yule Marble Quarry, now reopened, has been a source of fine marble since 1900.

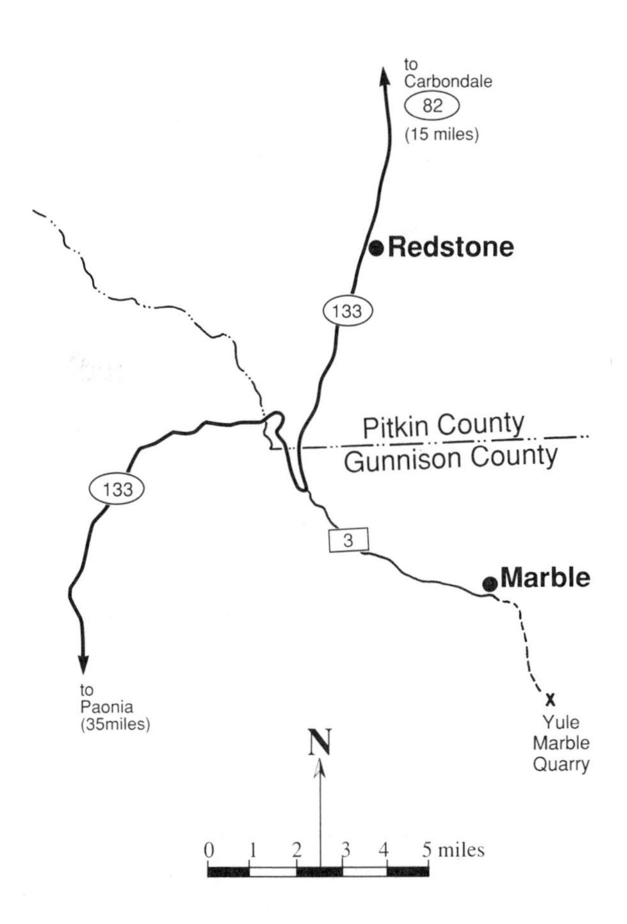
to
Carbondale
82
(15 miles)
Redstone
133
Pitkin County
Gunnison County
133
3
Marble
to
Paonia
(35miles)
Yule
Marble
Quarry
N
0 1 2 3 4 5 miles

Interior of the reopened Yule Marble Quarry.

and polishing plant. Yule marble ranked with the finest Italian marble and was in great demand for construction of public buildings across the United States. After a 1925 fire destroyed the mill, the competing Vermont Marble Company purchased and operated the quarry until the disastrous 1941 Yule Creek flood. Rebuilding during World War II was impossible and the quarries closed. By then, Yule marble was used in more than sixty public buildings, including Denver's City and County Building, the Federal Reserve Bank, and the State Capitol Annex. Yule marble was also selected for the Tomb of the Unknown Soldier in Arlington National Cemetery and the 46-foot-high fluted columns of the Lincoln Memorial.

Aside from an unsuccessful mining attempt in the 1950s, the quarries were inactive for nearly a half century. Tourists and mineral collectors carried away specimens, and sculptors gathered large blanks, pieces of flawless white marble weighing hundreds of pounds. During the decades of inactivity, the huge quarries, with ropes still dangling from wooden derricks, had a ghostly aura. With half-finished marble blocks and columns scattered about, visitors compared the quarries to Grecian ruins and Easter Island.

Backed by a British loan, two Denver businessmen reopened the quarries in 1990 and now ship 1,000 5x5x8-foot, 17-ton marble blocks each year. The marble is cut and polished in foreign mills.

The Art Student's League of Denver conducts annual summer outdoor sculpting workshops at the historic quarry. Sculptors employ power tools in the outdoor "studio" to quickly shape the marble.

From Carbondale and Colorado Route 82, take Colorado Route 133 seventeen miles south to Redstone. Five miles beyond Redstone, turn east on Gunnison County Road 3 and proceed five miles to Marble. The Yule Marble Quarry is three miles farther on a well-marked road. For safety, follow posted traffic instructions carefully. The road is used by heavy trucks, has no guard rails, and is sometimes a single lane.

Public traffic is restricted to a parking area at the entrance to the quarries. A half-mile-long foot trail leads along the Yule Creek chasm and through quarry dumps strewn with huge blocks of marble, then climbs to the quarry itself. From a fenced observation point, visitors may look into the working quarry, which is enclosed entirely within the mountain. Bring a camera for the scenery, the ruins, and the view into the quarry.

Collectors and stoneworkers can find snow white pieces and blocks of glistening marble in unlimited quantities in the dumps, as well as specimens representing the entire metamorphic transition from gray Leadville limestone to fine Yule marble. Clear calcite occurs as one-inch rhombohedrons. Bright green epidote is present as fracture fillings, occasionally as attractive quarter-inch crystals.

REFERENCES: 18, 41, 64, 81

THE GUNNISON "GOLD BELT" MINES

The Gunnison "Gold Belt," a twenty-by-six-mile mineralized area south of Gunnison, is one of Colorado's minor and now largely forgotten mining districts. Although never a bonanza, mine dumps still provide interesting mineral specimens.

Prospectors discovered gold near Cochetopa Creek (Saguache County) at the eastern end of the gold belt in 1880, then in 1893 in Wildcat Gulch in Gunnison County, triggering a small rush to the camps of Vulcan, Spencer, and Midway. Mining shifted from gold to copper after 1900, then ceased after World War I.

To reach the western section of the Gunnison Gold Belt from Gunnison and Blue Mesa Reservoir, take Colorado Route 149 south

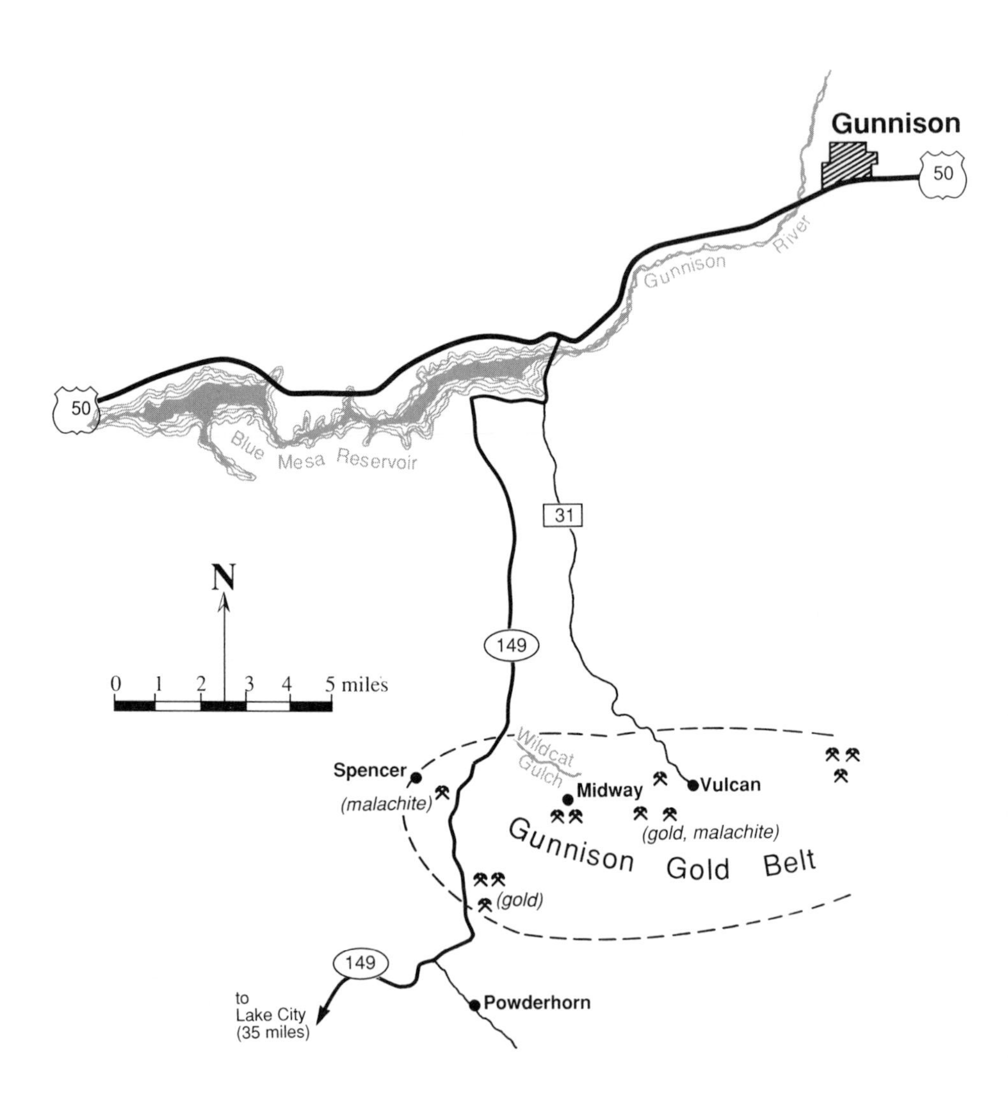
Gunnison
50
50
Gunnison River
Blue Mesa Reservoir
31
N
0 1 2 3 4 5 miles
149
Wildcat Gulch
Spencer
(malachite)
Midway
Vulcan
(gold, malachite)
Gunnison Gold Belt
(gold)
149
to
Lake City
(35 miles)
Powderhorn

An abandoned mill in the Gunnison "Gold Belt."

for 13.3 miles. The old headframe on the west side of the highway marks a mine dump containing specimens of white quartz and schist-laced bright green veinlets of malachite. The nearby sites of Spencer and Midway can be reached on unmarked ranch roads.

Two miles south of the malachite dump on Colorado Route 149 (about fifteen miles from U.S. 50), several gold mines, dumps, and prospects identify an area of shallow, oxidized gold-quartz veins on the east side of the highway.

The Vulcan area has other interesting copper-gold mine dumps. From Blue Mesa Reservoir and Colorado Route 149, follow County Road 31 south for thirteen miles.

The Gunnison Gold Belt's two types of ore are sulfide-bearing auriferous chert and massive sulfides, which include pyrite, arsenopyrite, chalcopyrite, and small amounts of galena. In the early 1980s, exploration geologists found dump ore specimens grading 2 percent copper, twenty troy ounces of silver, and one-half ounce of gold per ton.

REFERENCES: 50

THE QUARTZ CREEK PEGMATITES

The Quartz Creek pegmatites, twenty miles east of Gunnison, cover ten square miles and extend from lower Quartz Creek north-

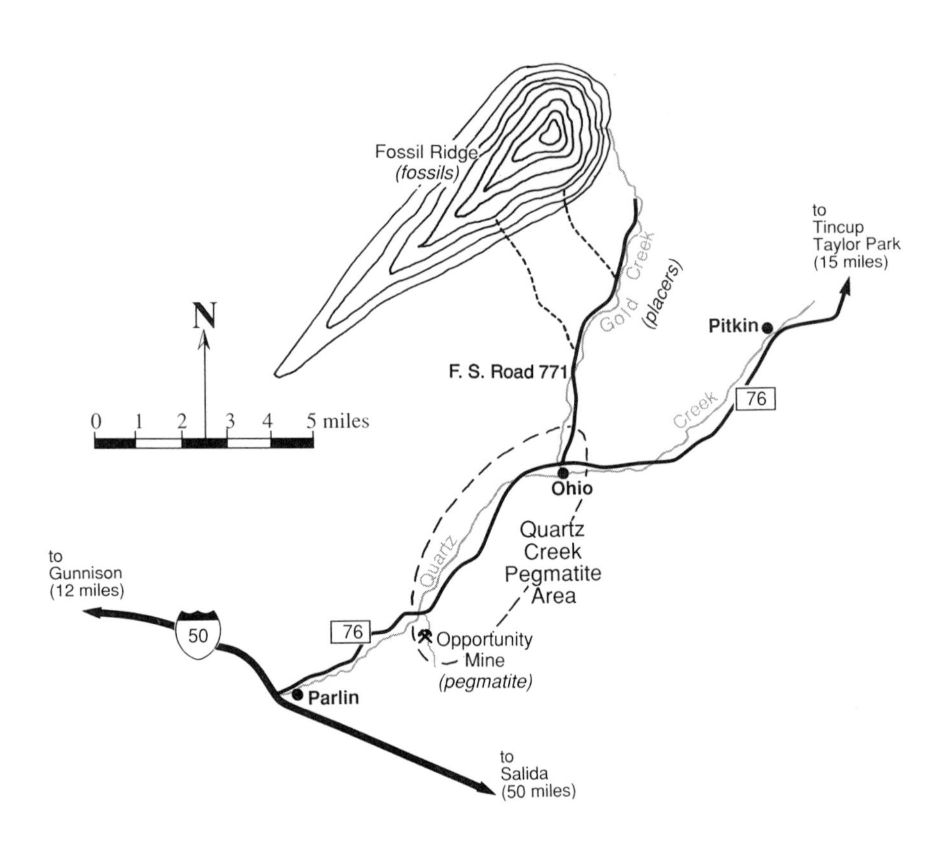
Fossil Ridge
(fossils)
N
0 1 2 3 4 5 miles
F. S. Road 771
Gold Creek
(placers)
to
Tincup
Taylor Park
(15 miles)
Pitkin
Creek
76
Ohio
Quartz
Creek
Pegmatite
Area
Quartz
to
Gunnison
(12 miles)
50
76
Opportunity
Mine
(pegmatite)
Parlin
to
Salida
(50 miles)

Mine dumps near this old headframe in the Gunnison Gold Belt are filled with malachite.

east to old Ohio City. The area is reached by following Gunnison County Road 76 northeast from Parlin and U.S. 50.

Numerous pink-colored granite dikes and pegmatites, ranging from short, narrow stringers to pegmatite bodies one mile long, intrude the granite country rock. Most of the 1,800 Quartz Creek pegmatites have extremely coarse crystallization of mica, quartz, and feldspars. Over 230 contain beryl, 17 contain lepidolite, and 20 contain columbite-tantalite. Several were mined for beryl, feldspar, and columbite-tantalite in the 1940s and 1950s. Others contain blue beryl, topaz, and black tourmaline.

To reach the old Opportunity mine site from Parlin, follow County Road 76 north for just over four miles, then turn east on the unmarked gravel road. Proceed a half mile to a rough track leading to several small open cuts on the hillside 200 yards to the right. Although inactive and well-collected since the 1950s, the dumps and in-situ trench walls still yield inch-long crystals of green beryl in cleavelandite, lilac-colored lepidolite, and long, thin prisms of black tourmaline in white quartz.

Continuing north toward Ohio City on County Road 76, pegmatite diggings appear on hillsides on both sides of the road. Several small pegmatites are exposed in road cuts. The Quartz Creek area is a good pegmatite-prospecting opportunity.

The Brown Derby Mine, although not open to collecting, is interesting as one of Colorado's richest sources of lithium minerals, specifically lepidolite and tourmaline. The mine was first worked during World War II for beryl and has been inactive since the 1950s.

Global Subsurface Products, Inc., of Golden, has recently acquired the property and reopened the underground workings for selective gemstone and ornamental stone mining. Lepidolite associated with clear quartz is abundant in massive lilac-colored blocks suitable for ornamental use. Also present is opaque green and pink elbaite tourmaline in long, delicate prisms.

REFERENCES: 64, 81

GUNNISON COUNTY FOSSILS

FOSSIL RIDGE

The long southwestern ridge of 12,749-foot Fossil Mountain, Fossil Ridge is located west of Quartz Creek. The higher sections of the well-faulted ridge contain many broken and exposed sediments representing 100 million years of geologic time, from Pennsylvanian to Cretaceous. Fossil brachiopods and early cephalopods are locally common in the higher sediments.

From the town of Ohio, take Forest Service Road 771 (Gold Creek Road) north along Gold Creek for about four miles. Fossil Ridge rises to the northwest and is reached by several foot trails.

REFERENCES: 68

HINSDALE COUNTY

LAKE CITY AREA MINES

Members of Col. John C. Fremont's U.S. Army exploration party reported gold near the site of Lake City in 1848, a decade before the Pikes Peak rush. The region was still Ute territory when prospectors discovered the Ute and Ulay veins in 1871. Development followed the 1873 Brunot Treaty, which legalized mineral entry in the San Juan region.

By 1876, Lake City was the commercial center for dozens of silver-lead mines to the west along Henson Creek and southwest along the Lake Fork of the Gunnison River. Many mines closed after the 1893 silver-market crash, but Lake City survived on gold production, mostly from the Golden Fleece.

Although discovered in 1874, the Golden Fleece Vein (originally

Sulfide minerals can often be found near ore-loading facilities, such as this structure near Lake City.

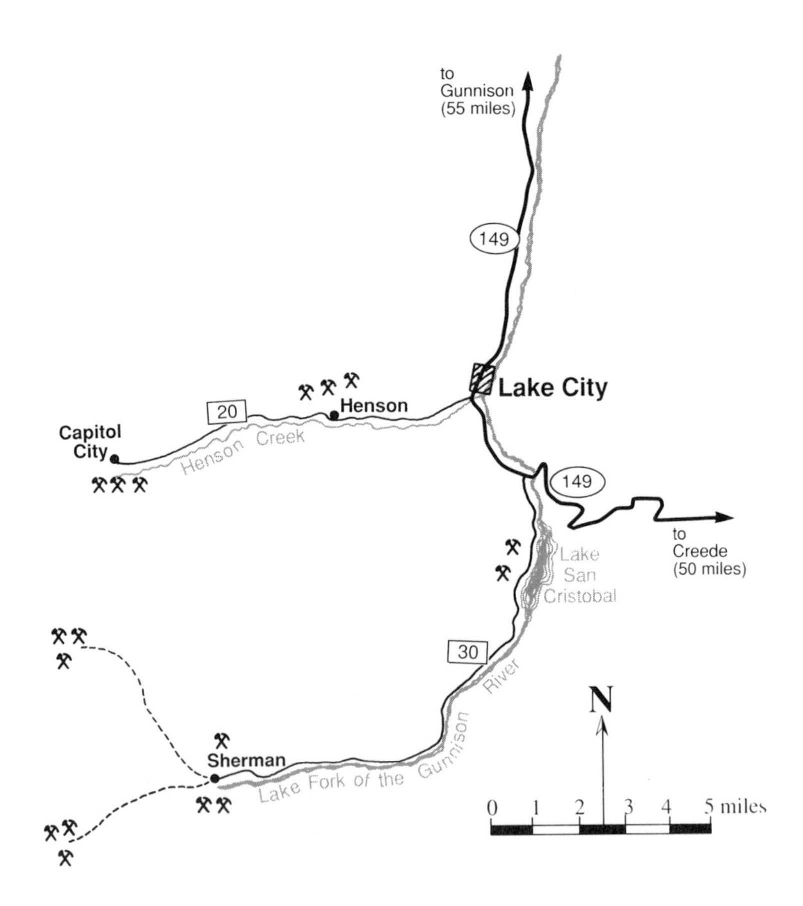
to
Gunnison
(55 miles)
149
Lake City
Henson
20
Capitol
City
Henson Creek
149
to
Creede
(50 miles)
Lake
San
Cristobal
30
River
N
Sherman
Lake Fork of the Gunnison
0 1 2 3 4 5 miles

the Hotchkiss Vein) had only break-even production until 1889, when miners encountered rich tellurides, including petzite ores grading over 100 ounces of gold and 1,000 ounces of silver per ton. By 1904, the Golden Fleece had produced 75,000 troy ounces of gold worth $1.5 million.

Lake City's best years were over, but sporadic zinc and copper mining added substantially to district production. By 1924, Lake City mines had produced $10 million in gold, silver, lead, zinc, and copper.

The three types of Lake City ores included tetrahedrite-rhodochrosite, quartz-galena-sphalerite, and tellurides. Mineralization, as in other San Juan Mountain districts, followed formation of a caldera, or collapsed volcanic system. Ores occurred as veins, fracture fillings, and cementing material in breccias.

In mines like the Hidden Treasure, Monte Queen, and Champion, vein fillings were largely crystalline and massive rhodochrosite. The Champion, inactive since World War I, provided commercial collectors with superb rhodochrosite specimens in the 1970s.

Lake City has interesting mine dump collecting. Ore minerals include pyrite, galena, sphalerite, chalcopyrite, stibnite, several tellurides, argentite, proustite, tetrahedrite, cerussite, bornite, malachite, and azurite. Gangue minerals include rhodochrosite, quartz, calcite, fluorite, and barite. Collectors have found clear and white quartz crystals to two inches in length on mine dumps near Lake San Cristobal.

Lake City is located fifty-five miles south of Gunnison on Colorado Route 149. From Lake City, take County Road 20 to mines along Henson Creek and near the sites of Henson and Capitol City; take County Road 30 to mines along Lake San Cristobal, the Lake Fork of the Gunnison River, and the site of Sherman.

REFERENCES: 18, 23, 25, 66, 67, 81

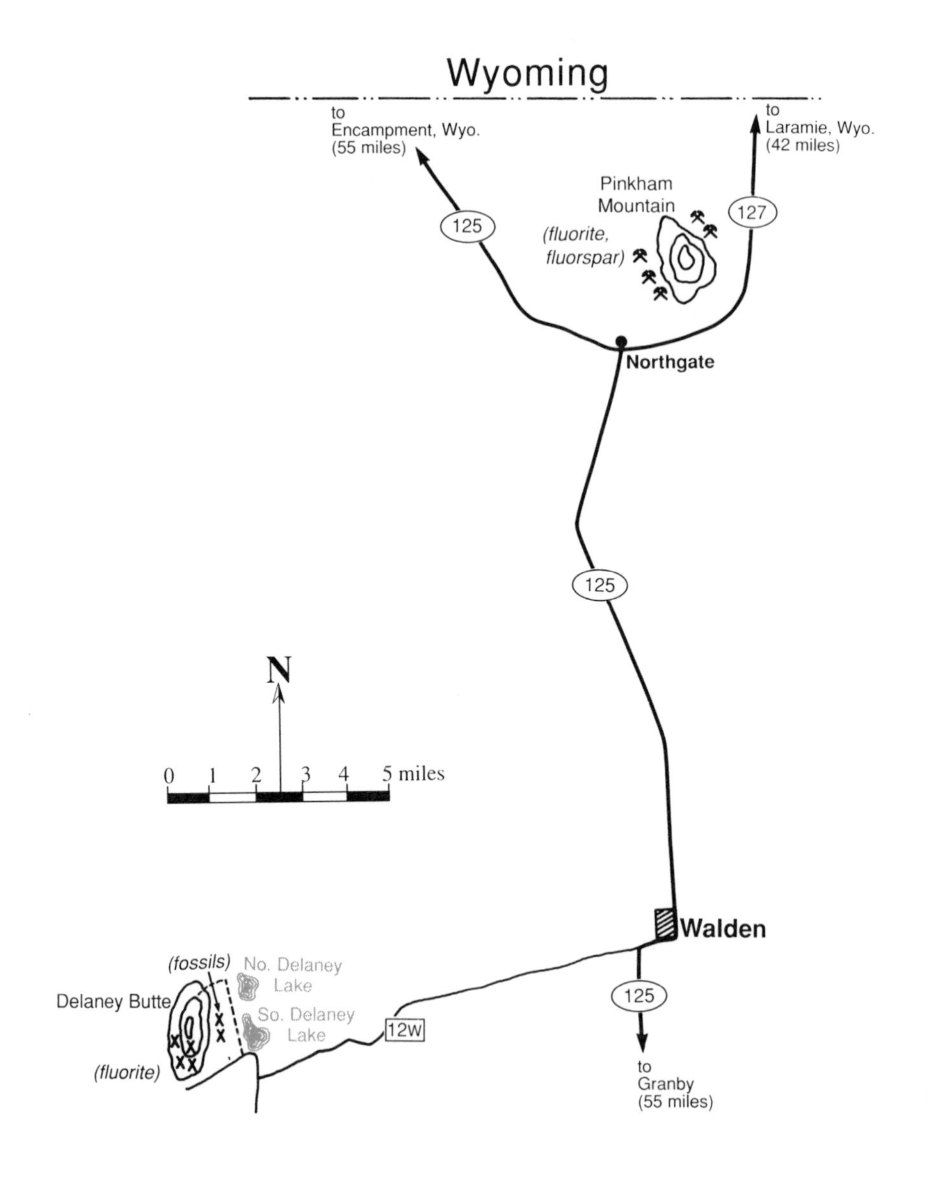

Wyoming
to
Encampment, Wyo.
(55 miles)
to
Laramie, Wyo.
(42 miles)
Pinkham
Mountain
(fluorite,
fluorspar)
125
127
Northgate
125
N
0 1 2 3 4 5 miles
Walden
(fossils)
No. Delaney
Lake
Delaney Butte
So. Delaney
Lake
12W
(fluorite)
125
to
Granby
(55 miles)

JACKSON COUNTY

PINKHAM MOUNTAIN FLUORITE

The Northgate district in northern Jackson County is a former major fluorspar source. Northgate is twelve miles north of Walden on Colorado Route 125 just south of the Wyoming line.

Prospectors discovered the Northgate vein deposits in 1900. Open-pit and underground mining lasted until the early 1970s.

Extensive surface workings on the southwest and northeast sides of Pinkham Mountain can be seen from Colorado Route 127. Deep chasmlike cuts expose vertical vein walls and indicate the impressive size of the original veins.

Some fluorspar veins are two miles long and fifty feet thick. The high-silica, coarse-grained, massive fluorspar was primarily yellowish and greenish and contained chalcedony and white and clear comb quartz.

White, green, and purple fluorite occurred in mammillary, stalactic, and botryoidal forms, and less frequently as crystals up to one inch in size. In the 1950s lapidaries created ornamental objects from attractive specimens of massive purple and green fluorite banded with pyrite and marcasite.

The mine properties are owned by the Ozark-Mahoning Company; fences are posted and gates are locked. Permission to visit the property is required.

REFERENCES: 25, 80, 81

DELANEY BUTTE FLUORITE AND FOSSILS

Delaney Butte, ten miles west of Walden, is a prominent North Park landmark. The top of the granite butte, especially the southern rim and slope immediately west of South Delaney Lake, has thin fluorite veins marked by numerous diggings and prospect holes. Although the fluorite veins are not commercial, small vugs and seams contain clear and pale green octahedral fluorite crystals up to an inch in size.

Several rough jeep roads and tracks ascend the northeast side of the butte. From South Delaney Lake, on the Delaney Butte Lakes State Wildlife Area, the hike to the top takes one hour.

The lower eastern slope of Delaney Butte, near South Delaney Lake, has gully exposures of fossiliferous Tertiary sandstone and

Pelecypod fossils in Tertiary sandstone from Delaney Butte.

shales. Fossils of small pelecypods are locally abundant. The sandstone, while durable, splits easily along planes; some surfaces are covered with well-preserved, detailed shell fossils that make nice display specimens.

From Walden, take Jackson County Road 12W west and follow the signs to Delaney Butte Lakes State Wildlife Area.

REFERENCES: 25

JEFFERSON COUNTY

PLACER GOLD

Jefferson County's Clear Creek placer mines have produced 14,000 troy ounces of gold. Eight years before the Pikes Peak rush, California-bound Cherokees discovered gold on Ralston Creek, a tributary of Clear Creek, near the present Wadsworth Avenue bridge in Arvada. A diary entry for June 20, 1850, noted: "Gold found," and "We called this Ralston's Creek because a man of that name found gold here."

Most of Jefferson County's placer gold has been recovered since the 1950s as a by-product of sand- and gravel-mining operations near Golden.

Clear Creek Canyon, just west of Golden along U.S. 6, is very popular among panners and recreational gold miners. Amateur

Clear Creek Canyon is one of Colorado's most popular panning and recreational sites.

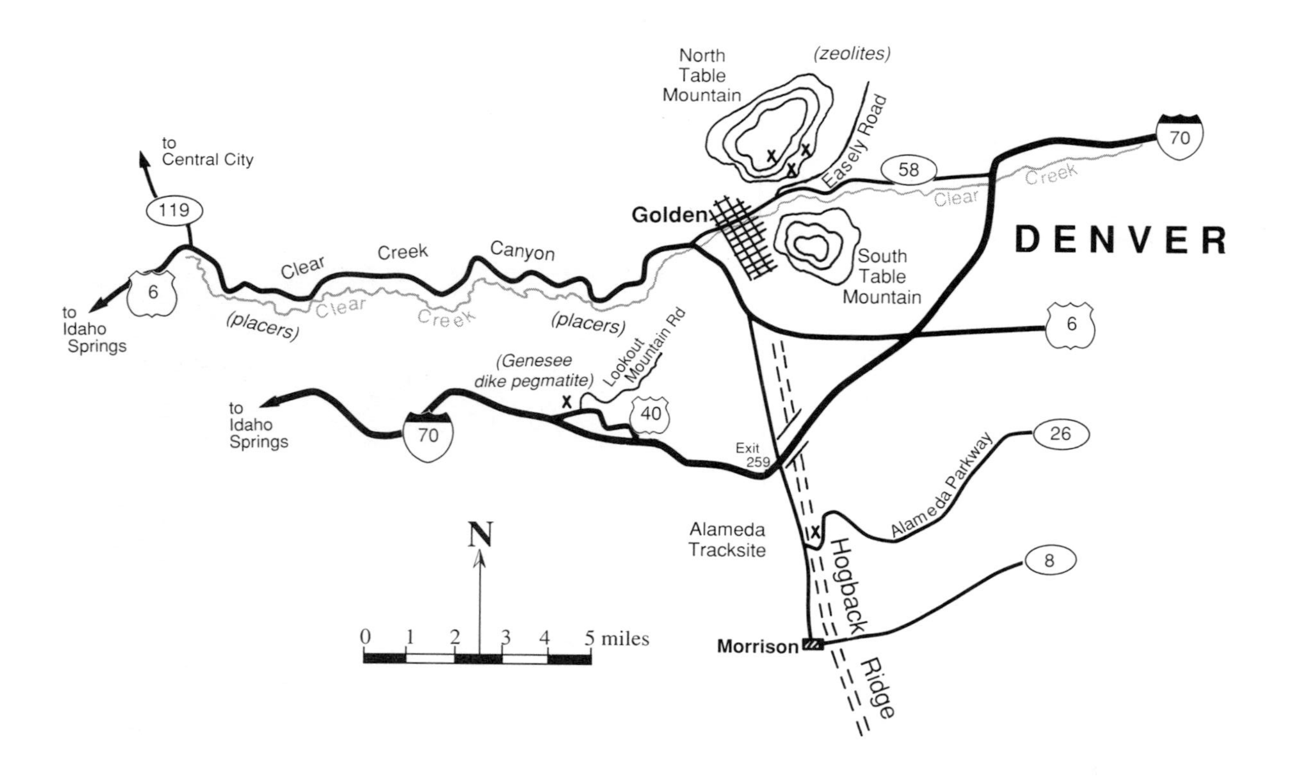
North Table Mountain
(zeolites)
Easely Road
to Central City
119
58
70
Golden
Clear Creek
DENVER
Clear Creek Canyon
South Table Mountain
6
to Idaho Springs
(placers)
Clear Creek
(placers)
Lookout Mountain Rd
(Genesee dike pegmatite)
40
to Idaho Springs
70
Exit 259
26
Alameda Parkway
Alameda Tracksite
N
8
Hogback Ridge
0 1 2 3 4 5 miles
Morrison

prospectors recently found concentrations of coarse placer gold on bedrock terraces well above the present channel north of the highway. One site is on the shoulder of U.S. 6 two miles east of the Colorado Route 119 junction. The miners remove overburden with picks and shovels, scrape the pay dirt from bedrock crevices, then pan or sluice it in Clear Creek. Recoveries have included nuggets as large as one-quarter troy ounce.

The diggings along the highway are on State Highway Department property. Some panners, however, have dug on the claimed nearby hillsides, drawing complaints from claim holders and the attention of county sheriff's officers.

REFERENCES: 23, 53

DINOSAUR RIDGE FOSSILS AND TRACKWAYS

THE I-70 HOGBACK ROAD CUT

The Dakota hogback ridge near Morrison, eight miles west of Denver, is one of Colorado's most interesting paleontological and geological areas. The road cut, at the Morrison exit (Exit 259) of Interstate 70, is 400 feet high and a half mile long. Removal of 25 million tons of rock has exposed 20 million years of geologic sedi-

The I-70 road cut near Morrison exposes 20 million years of Jurassic and Cretaceous sediments.

ments in colorful, clearly delineated strata ranging from coaly blacks and dull browns to bright yellows and reds. Paved walkways along the road cut have interpretive signs identifying and explaining the various sandstones, mudstones, siltstones, and shales. The layers are the lithified sedimentary accumulations of different paleoenvironments, including coastal plains, sea bottoms, beaches, bars, freshwater marshes, and tidal flats.

The exposed sediments provide a fascinating perspective in geologic time. At one point, a human hand can span the sharp contact between the red Morrison Formation sandstone and the overlying yellow Dakota Formation sandstone. The contact sediments, deposited 135 million years ago, represent the transition between the Jurassic and Cretaceous periods of Mesozoic time.

DINOSAUR FOSSILS AND TRACKWAYS

Arthur Lakes, a professor for whom the Lakes Library at the Colorado School of Mines in Golden is named, made one of North America's greatest paleontological discoveries on the Dakota hogback. While searching for fossil leaves in Dakota Formation exposures two miles north of Morrison in March 1877, Lakes found a 14-inch-thick fossilized femur and a large vertebrae in the underlying sandstone. He shipped fossils to two prominent paleontolo-

The Dakota hogback is known for its dinosaur trackways and fossil bones.

gists, Othniel Marsh and Edward Drinker Cope. Both Marsh and Cope, bitter professional rivals, soon arrived in Colorado to make spectacular dinosaur-fossil discoveries.

For several years, Lakes and others excavated hogback sites and "trotted out the menagerie," which included fossils of several dinosaur species and genera. Today, this section of the hogback, appropriately named Dinosaur Ridge, is best known for dinosaur trackways. Dakota Sandstone on the hogback's eastern slope is exposed as slickrock, with visible dinosaur trackways.

Although discovered in the 1880s, paleontologists did not document the fossil footprints as dinosaur tracks until the 1930s, and have only recently studied them in detail. Traditionally, dinosaur tracks were considered paleobiologic indicators of taxonomic identification, paleogeographical species range, type and approximate speed of locomotion, and certain behavioral patterns. Tracks also provided insight into herding tendencies and paleopathology, the study of wounded, malformed, or otherwise abnormal dinosaurs.

Dinosaur ichnology, the modern study of dinosaur tracks, is now revealing much about the physical nature of the paleoenvironment. In mid-Cretaceous times, the Morrison area was the western edge of the Interior Seaway, a warm, salty sea with a broad coastal plain supporting a thriving dinosaur population. Most

Dinosaur tracks in Dakota Sandstone at the Alameda tracksite at Dinosaur Ridge.

trackways were created in a narrow zone of moist, vegetation-free sediments typical of mud flats or marine shorelines. Thirty Colorado dinosaur tracksites are known within the same stratigraphic horizon of the Dakota Sandstone. When correlated, these individual trackways form a major trackway system extending from the Boulder area south into New Mexico.

Dinosaur ichnology also provides information on water depth, current direction, gradient, and water content of the sediments when the tracks were made. Dinosaurs covered and compacted small plants and animals, an effect paleontologists call dinoturbation. The study of dinoturbation offers much paleoenvironmental information, for small fossils within dinosaur tracks are a definitive record of life-forms that existed at the precise moment the track was made.

The easily accessible Alameda tracksite, located on Alameda Parkway (Colorado Route 26) on the east side of the hogback ridge, has been fenced and marked with interpretive signs. Locations of other nearby tracksites are not publicized, to protect them.

For further information or to schedule field trips to trackways, contact the Dinosaur Ridge Natural Resource Center at the Morrison Museum in Morrison.

REFERENCES: 13, 68, 101

TABLE MOUNTAIN ZEOLITES

North Table Mountain, a prominent lava-capped mesa just east of Golden, has been one of North America's best zeolite-collecting locales for over a century. North and South Table mountains, rising about 700 feet above Golden, are remnants of three early Tertiary lava flows that covered Denver Formation sediments. The lava formed a durable, basaltic rock rich in potassium feldspar, and the two uppermost flows are distinguishable today as cliff outcrops. The lava solidified with many oval-shaped gas cavities that later filled with a variety of zeolite mineral crystals. The cavities vary in size from one inch to several feet.

Mineralogists have identified fourteen zeolite minerals at North Table Mountain. The most abundant species are thomsonite, analcime, chabazite, natrolite, and mesolite. Calcite and small amounts of opal are also present.

Since 1900, North and South Table Mountain quarries have provided basalt for railroad ballast, concrete aggregate, building blocks, and monument stones. Quarrying exposed huge sections of basalt containing zeolite-filled cavities, yielding many museum-

Golden's North Table Mountain, a basaltic lava flow, is a noted source of zeolite minerals.

grade zeolite specimens. Although commercial quarrying ceased in the 1960s, North Table Mountain is still a fine source of zeolite mineral specimens.

Many different crystal habits represent some fifteen separate periods of deposition of mineral-bearing solutions. Thomsonite usually occurs as delicate, colorless radiating crystals; chabazite as single and twinned white-pink rhombohedrons; and mesolite as fibrous "tufts." Clear, dogtooth calcite crystals are also present.

Collectors find zeolite minerals in lava exposures on both Table Mountains. The primary collecting area is on the southeast side of North Table Mountain in the upper section of the middle basalt outcrop near several old quarries. This site is a mile and a half east of the Washington Avenue-Colorado Route 58 overpass, and due north of and about 400 feet above Easley Road.

To find unweathered, intact crystals, collectors must break the durable basalt to expose fresh cavities. Heavy sledges, chisels, and eye protection are necessary. Both the climb to the quarries and the physical breaking of rock are strenuous.

The lava from the Table Mountain flows has protected sections of the underlying Denver Formation sediments. The mudstones and shales on the lower slopes of both North and South Table

Mountain contain locally abundant petrified wood and fossils of leaves and stems.

Land on the Table Mountains is privately owned. Although some zeolite areas are generally considered "open," collectors should always check first with landowners.

REFERENCES: 5, 19, 31, 41

GEOLOGY MUSEUM OF THE COLORADO SCHOOL OF MINES

The Colorado School of Mines, one of the world's foremost colleges of mining engineering, is located in Golden. The Colorado School of Mines originated in 1868 as a short-lived Episcopal school. In 1874, the Colorado territorial legislature reestablished the school's Department of Mines as the Colorado School of Mines (CSM) to teach basic mining skills to support the territory's growing mineral economy. Professors used the original mineral collection, assembled from many Colorado mines, as a tool for teaching mineral identification and ore assaying. The first curator of the CSM collection was Arthur Lakes, who discovered fossilized dinosaur skeletons at nearby Morrison in 1877.

The Geology Museum at the Colorado School of Mines has excellent displays of Colorado gold, minerals, and gemstones.

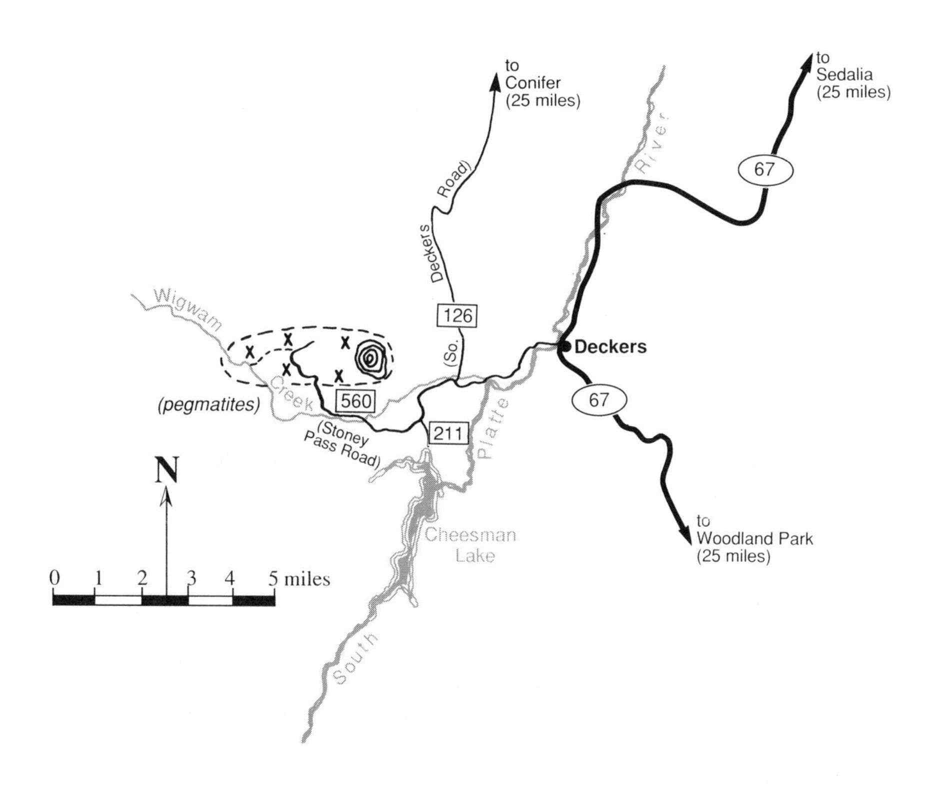
to
Conifer
(25 miles)
to
Sedalia
(25 miles)
River
67
Road)
Deckers
(So.
126
Wigwam
Creek
(pegmatites)
560
(Stoney
Pass Road)
211
Deckers
67
Platte
N
Cheesman
Lake
0 1 2 3 4 5 miles
South
to
Woodland Park
(25 miles)

Over the decades, CSM acquired many fine private mineral collections. By 1916, CSM curators had catalogued 66,000 specimens. In 1967, the Colorado state legislature transferred to CSM control of the Colorado state collection, a collection assembled since 1895 by the Colorado Scientific Society and the Colorado Bureau of Mines. Today, that collection has an unusual significance: It contains the only known specimens of ore and gangue minerals from hundreds of Colorado mines that are now abandoned, flooded, or collapsed.

The CSM Geology Museum, located in Berthoud Hall on the CSM campus, has superb exhibits of Colorado, North American, and world minerals and gemstones, as well as mining, paleontology, and geology displays. The Frank C. Allison collection, with its outstanding Colorado native gold specimens, is displayed in the lobby of the nearby Arthur Lakes Library.

REFERENCES: 83

WIGWAM CREEK PEGMATITES

The Wigwam Creek-Sugarloaf Peak area in southern Jefferson County just west of Cheeseman Lake has granite pegmatites containing smoky quartz, muscovite, orthoclase, and some pale blue amazonite and purple fluorite. Collectors have recently recovered very large, but clouded and flawed, quartz crystals and smaller crystals of gem quality.

The collecting area covers three square miles near the Wigwam Creek trailhead along Forest Service Road 560 (Stoney Pass Road), and extends east to Sugarloaf Peak. Several pegmatites are under claim.

From Deckers on Colorado Route 67, follow Jefferson County Road 126 (South Deckers Road) three miles west to Forest Service Road 211 (Wigwam Creek Road.) Continue on Road 211 for three miles, then turn west on Forest Service Road 560 (Stoney Pass Road) and proceed three miles to the collecting area near the marked Wigwam Creek trailhead.

REFERENCES: 32

LOOKOUT MOUNTAIN PEGMATITE

An easily accessible road-cut pegmatite is located on the shoulder of U.S. 40 near Lookout Mountain. The site is six-tenths of a mile west of the intersection of Lookout Mountain Road on Highway Department property.

The U.S. 40 road cut exposes about fifty feet of granite pegmatite

The Genesee Mountain Dike, an exposed pegmatite on the shoulder of U.S. 40, is a source of garnet, epidote, scheelite, and titanite.

and an associated quartz-garnet dike. Grossular garnet is common in small bright brown-red crystals, but exceptional specimens may be one foot in diameter. Other minerals present include titanite, scheelite, and epidote. Well-formed small crystals are not difficult to find. Nice micromount and miniature specimens may be found by screening the broken rock left behind by other collectors.
REFERENCES: 5, 64

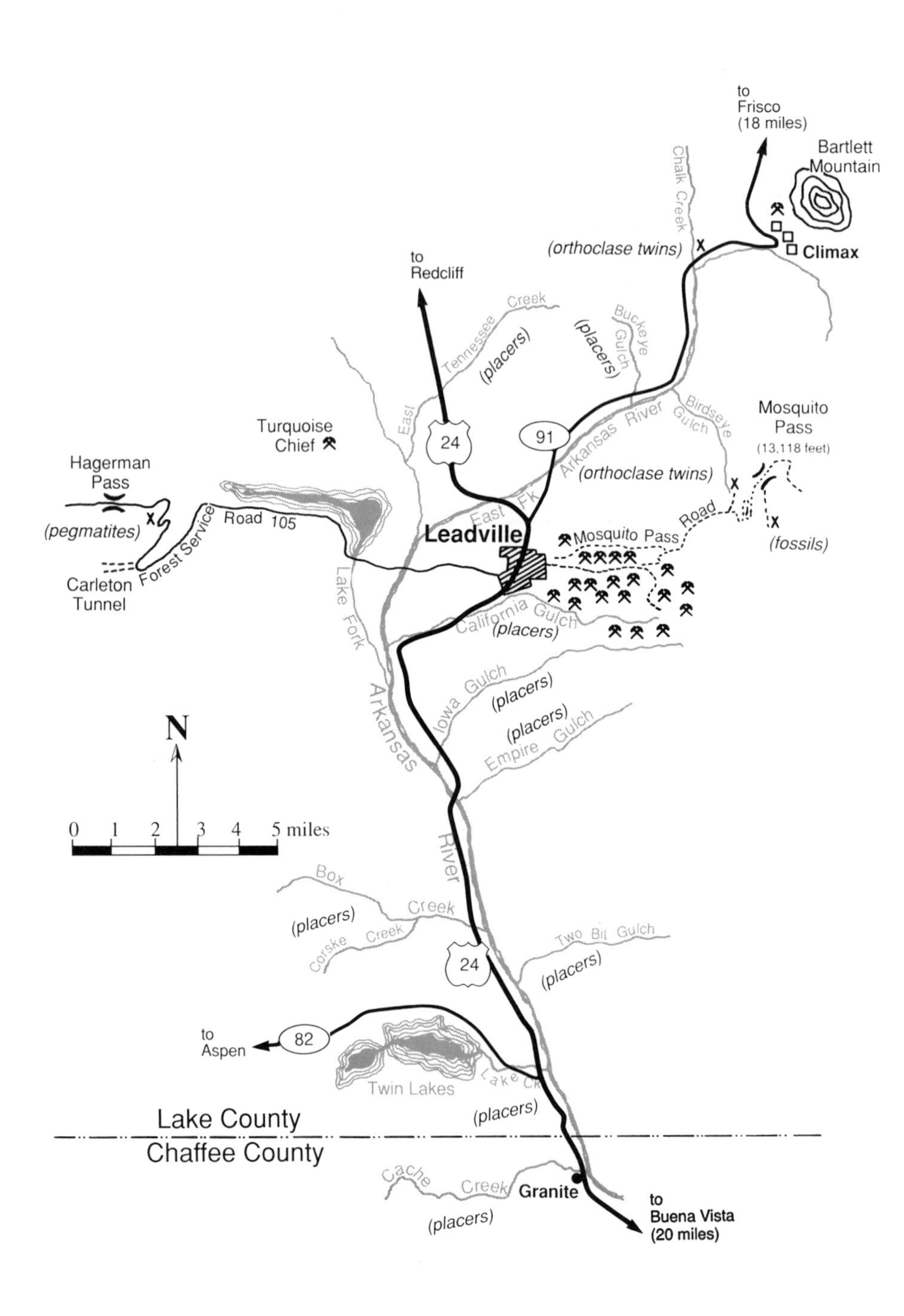
to
Frisco
(18 miles)
Bartlett
Mountain
Chalk Creek
(orthoclase twins)
Climax
to
Redcliff
Tennessee Creek
(placers)
(placers)
Buckeye Gulch
East
Birdseye Gulch
Mosquito
Pass
(13,118 feet)
Turquoise
Chief
24
91
Arkansas River
Hagerman
Pass
(orthoclase twins)
(pegmatites)
Road 105
East Fk
Mosquito Pass Road
Forest Service
Leadville
Mosquito Pass
(fossils)
Carleton
Tunnel
Lake Fork
California Gulch
(placers)
Iowa Gulch
(placers)
Arkansas
(placers)
Empire Gulch
N
0 1 2 3 4 5 miles
River
Box
Creek
(placers)
Corske Creek
Two Bit Gulch
24
(placers)
to
Aspen
82
Twin Lakes
Lake Ck
(placers)
Lake County
Chaffee County
Cache Creek
Granite
to
Buena Vista
(20 miles)
(placers)

LAKE COUNTY

PLACER GOLD

Lake County has produced over 360,000 troy ounces (about twelve tons) of placer gold, second in Colorado only to Summit County. Prospectors discovered placer gold along the upper Arkansas River in the fall of 1859. The following April, they struck it rich in California Gulch near the present site of Leadville. By May 1860, 10,000 people milled about the boom camp of Oro City in upper California Gulch.

Miners staked the entire seven-mile-length of California Gulch into 339 100-foot-long claims, the best yielding 3,000 troy ounces of gold in the first season. By 1865, miners recovered over 200,000 troy ounces of gold worth $4 million.

Prospectors found gold in the main channel of the Arkansas River and in such tributaries as Lake, Corske, and Box creeks and Iowa, Colorado, Twobit, Birdseye, Empire, Buckeye, East Tennessee, Thayer, and Little Fryingpan gulches.

Cache Creek, near Granite on the Lake-Chaffee county line, is among Colorado's least-known major placer districts. By the 1940s, miners had employed ground sluicing, hydraulicking, and draglining to recover 80,000 troy ounces of coarse placer gold worth $1.5 million. The Denver Museum of Natural History displays a collection of Cache Creek nuggets mined about the turn of the century.

Recreational gold mining and panning is popular, especially near Lake Creek and Granite. Local placers owned by Twin Lakes Associates, Inc., are open to panners. President Dennis O'Neill asks that panners check in and pay a token fee for panning, sluicing, or dredging. Suction dredges are limited to intake diameters of two inches or less. About 200 panners and recreational miners come to Lake and Cache creeks each year. O'Neill resides near the Balltown settlement at the junction of U.S. 24 and Colorado Route 82, where panners may inquire for directions.

REFERENCES: 23, 54, 97

THE LEADVILLE MINING DISTRICT

Lake County, one of the nation's greatest mining centers, has produced billions of dollars in molybdenum, gold, silver, lead, zinc, and lesser amounts of tin, tungsten, bismuth, copper, and iron.

Collecting in the historic Leadville mining district.

In the 1860s, California Gulch placer miners cursed the heavy black sands that fouled their sluice boxes. In 1875, mineralogists identified the sand as lead-silver carbonate. Miners built a small smelter and named their new camp Leadville, after the metal that seemed to offer economic salvation. Prospectors traced the source of the heavy sands to hills near upper California Gulch, where they discovered a world-class, shallow lead-silver sulfide ore body.

By 1880, Leadville was the biggest city between St. Louis and San Francisco. The population neared 30,000 and 400 mines produced 10 million ounces of silver annually.

Leadville's leading citizen was Horace Austin Warner Tabor. Tabor and his wife, Augusta, had arrived at booming California Gulch in 1860. In April 1878, Tabor, who still operated a small general store, grubstaked two prospectors with $64 worth of groceries and supplies. Weeks later, they sank a shallow shaft into the Little Pittsburg ore body, which contained 50 percent lead and 300 troy ounces of silver per ton. Tabor parlayed his Little Pittsburg profits into other mines like the Chrysolite and Matchless, becoming a millionaire in just three years.

In 1879, the Robert E. Lee Mine produced $118,000 in silver in just eighteen hours. The richest ore was almost half silver, grading 11,000 troy ounces per ton.

By 1893, Leadville had produced 150 million troy ounces—over 5,000 tons—of silver. Although hundreds of silver mines closed after the 1893 silver-market collapse, the district survived due to the diversity and richness of its mineralization.

In 1896 miners struck rich gold ore at the Little Jonny Mine. After 1900, Leadville helped satisfy growing industrial demand for zinc, lead, and copper. By the 1920s, Leadville had produced 1.6 million ounces of gold, 230 million ounces of silver, 1 million pounds of copper, 1.2 million tons of lead, and 1.5 million tons of zinc with a cumulative value of a half-billion dollars.

Leadville mines boomed again during World War II and the Korean War. But by 1960, the smelters and most mines had closed. Today, the Black Cloud Mine, an important producer of zinc-lead-silver-gold since 1974, is the only continuously active mine.

The twenty-square-mile Leadville mining district is a fascinating graveyard of frontier mining. A U.S. Bureau of Mines survey documented 1,330 shafts, 155 tunnels, 1,600 prospect holes, and approximately 200 miles of old underground workings.

Leadville's four types of ores included silicate-oxide deposits, mixed sulfide veins, sulfide replacement bodies in dolomite limestone, and gold-bearing quartz-pyrite veins. Over a hundred different minerals occur at Leadville, including chlorargyrite, chrysocolla, azurite, malachite, goethite, native gold, argentite, chalcocite, pyrite, sphalerite, galena, smithsonite, native silver, proustite, tetrahedrite, calcite, quartz, barite, fluorite, siderite, and rhodochrosite. Collectors find many nice specimens of pyrite, galena, sphalerite, and other metal sulfide minerals, especially in miniature and micromount sizes.

The Black Cloud Mine is a source of golden barite and sphalerite-galena-pyrite on white dolomite. Trespassing on Black Cloud Mine property is prohibited, however. Specimens of local ore and gangue minerals are available in local shops.

From Leadville, follow Fifth Street, Seventh Street, or County Road 2 east into the sprawling mining district. Many important mine sites are marked, and the local chamber of commerce sells auto-tour maps of the district.

Other sites of interest are the National Mining Hall of Fame and Museum, which has a fine mineral collection and a superb underground mine re-creation; the Heritage Museum, with an original stamp mill and local minerals; The Mining Gallery, a shop specializing in mining artifacts, art, books, and cut-and-polished ore specimens; and The Rock Hut, which has the largest collection of Colorado minerals in the central Rockies.

REFERENCES: 18, 19, 23, 25, 66, 67, 73, 81, 89

THE CLIMAX MOLYBDENUM MINE

The Climax Mine, located on Colorado Route 91 atop 11,318-foot Fremont Pass fourteen miles north of Leadville, was the world's largest underground mine. Bartlett Mountain, overlooking Fremont Pass, contains the world's largest known deposit of molybdenite.

Prospector Charles Senter discovered unfamiliar mineralization on Bartlett Mountain in August 1879 and staked three claims hoping it contained gold. In 1900, mineralogists finally identified Senter's mineral as molybdenite, the only known ore of molybdenum. Molybdenum had no value or use until French metallurgists discov-

Section of five-inch pyrite cube from Climax Mine.

ered its ability to toughen steel. By World War I, German scientists had developed superior weapons-grade molybdenum steels.

In 1917, the Climax Molybdenum Company won control of Bartlett Mountain and adapted the new flotation separation process to treat the very low-grade ore. Mining halted after Armistice Day. The mine reopened in 1924 and engineers developed an extraordinarily efficient block-caving mining system that slashed costs and won control of the world molybdenum market. During World War II, the Climax Mine, one of the nation's highest-priority defense installations, worked around the clock providing virtually all the molybdenum vital to the Allied war effort.

In the 1970s, employment reached 3,000, work proceeded on four major underground levels and a huge open pit, and daily production topped 50,000 tons of ore. In sixty years of full operation, the Climax Mine produced over $3 billion in molybdenum and by-product tin and tungsten, one of the world's highest single-mine production figures. Full production ceased in 1982. Environmental reclamation, downsizing, and limited mining continues with a greatly reduced work force.

Climax has provided superb specimens of pyrite cubes as large as ten inches on a side; green and purple fluorite; clear quartz crystals; and deep red rhodochrosite, some of gem quality.

A Climax historical exhibit is located on the west side of the Fremont Pass summit opposite the main mine gate. Collecting is not permitted on Climax property, but the large boulders at the roadside exhibit are molybdenite ore from the open pit. The ore is a gray, sugary-textured silicified porphyry laced with thin veins of blue-gray molybdenite and speckled with tiny pyrite crystals. Molybdenite has a characteristic greasy feel. Some specimens are dusted with bright yellow molybdite, or molybdenum oxide, the oxidation product of molybdenite.

REFERENCES: 18, 19, 23, 81, 89

THE TURQUOISE CHIEF MINE

The Turquoise Chief Mine, one of Colorado's three commercial turquoise deposits, is located in the St. Kevin mining district northwest of Leadville and one mile north of Turquoise Lake.

Indians may have worked the deposit in prehistoric times, and the turquoise certainly attracted the attention of district silver miners in the 1880s. The deposit was first mined commercially in the 1930s to supply Navajo jewelry makers. During the summer of 1935 and 1936, two Navajo miners recovered 1,000 pounds of gem-quality turquoise from an open cut 100 feet wide and 25 feet deep. Subsequent commercial mining was sporadic; in the 1970s, the Turquoise Chief Mine was a popular fee collecting site. Owners closed the property in 1982 and collecting is no longer permitted.

The deposit was formed by circulating groundwater leaching and concentrating turquoise mineral constituents, then precipitating them in fractures and shear zones of granite. The attractive blue-green turquoise occurs as thin veins and occasional nodules within the weathered, buff-colored granite. Specimens are sometimes available in Leadville shops.

REFERENCES: 19, 56, 59, 64

ORTHOCLASE CARLSBAD TWINNED CRYSTALS

CHALK CREEK

Well-formed, twinned orthoclase crystals up to two inches in length occur in abundance north of Leadville near Colorado Route 91. The crystals are cream-colored, glassy, and sharp.

From the junction of U.S. 24 and Colorado Route 91 near Leadville, follow Colorado Route 91 north for 8.5 miles (or 3 miles south from the summit of Fremont Pass). Park on the wide shoulder just above the Chalk Creek culvert. Drive (four-wheel-drive only) or walk 200 yards north up the rough road to the first hairpin turn at the base of the small hill. Search up the hill toward the weathered gray outcrops of quartz monzonite porphyry. Twinned and occasional single orthoclase crystals are abundant at and immediately below the monzonite outcrops; many may be extracted directly from the weathering monzonite with a rock pick.

MOSQUITO PASS

Two adjacent outcrops of quartz monzonite porphyry with twinned orthoclase crystals are located six miles east of Leadville high in the Mosquito Range. The site is just north of Mosquito Pass Road near the headwaters of Birdseye Gulch, not far below the 13,118-foot-high summit of Mosquito Pass.

From Leadville, take Mosquito Pass Road 4.5 miles east to the Diamond Mine turnoff. Follow the signs to Mosquito Pass (four-wheel-drive only) and proceed 1.5 miles, then follow Birdseye Gulch Road north for one-quarter mile to the light gray outcrops on both sides of the road. The cream, two-inch-long, twinned orthoclase crystals weather free from the monzonite porphyry exposures and in places cover the ground.

Twinned orthoclase crystals from Chalk Creek.

Twinned orthoclase crystals are abundant at Chalk Creek.

Crinoid fossils from 13,000-foot-high Mosquito Ridge.

MOSQUITO RIDGE CRINOID FOSSILS

Crinoid fossils are present in exposures of Pennsylvanian sandstone one mile south of Mosquito Pass. The site, at an elevation of about 13,000 feet, demonstrates the dramatic uplifting of marine sediments during the formation of the modern Rockies.

From Leadville, follow four-wheel-drive Mosquito Pass Road eight miles east to the 13,118-foot summit of Mosquito Pass. Drive or hike one mile south along the crest of the ridge to a small automated radio-transmission facility. Fossils are found about 100 yards east of the crest of the ridge in talus and in-situ sandstone. Mosquito Pass is accessible in summer only.

REFERENCES: 69

HAGERMAN PASS AREA PEGMATITES

Black tourmaline in a white quartz matrix occurs in thin pegmatite veins one mile north of the east portal of the old Carlton tunnel. The site is about eight miles west of Leadville and just below the Continental Divide.

From the Turquoise Lake Dam four miles west of Leadville, follow Turquoise Lake Road for three and a half miles along the west side of the lake. Turn west on Forest Service Road 105 (Hagerman Pass Road) and proceed four miles to the Carlton Tunnel portal. Four-wheel-drive is recommended beyond this point.

Proceed two miles beyond the Carlton Tunnel to the second of two hairpin turns. Above and slightly north of the second hairpin turn and the powerline, search the granite exposures for light-colored pegmatite veins. The first vein is about fifty yards above the road.

The thin and erratic pegmatite veins are primarily feldspar that "blossom" into sections of white quartz with two-inch "books" of muscovite and numerous pencil-like prisms of black tourmaline. Some searching is necessary to locate veins, but local collectors have recovered black tourmaline crystals nearly one foot long and two inches thick. The black tourmaline is brittle, and hammers and chisels are needed to extract the crystals intact within their quartz matrix. Some veins have eroded away, and vein sections can also be found in talus boulders.

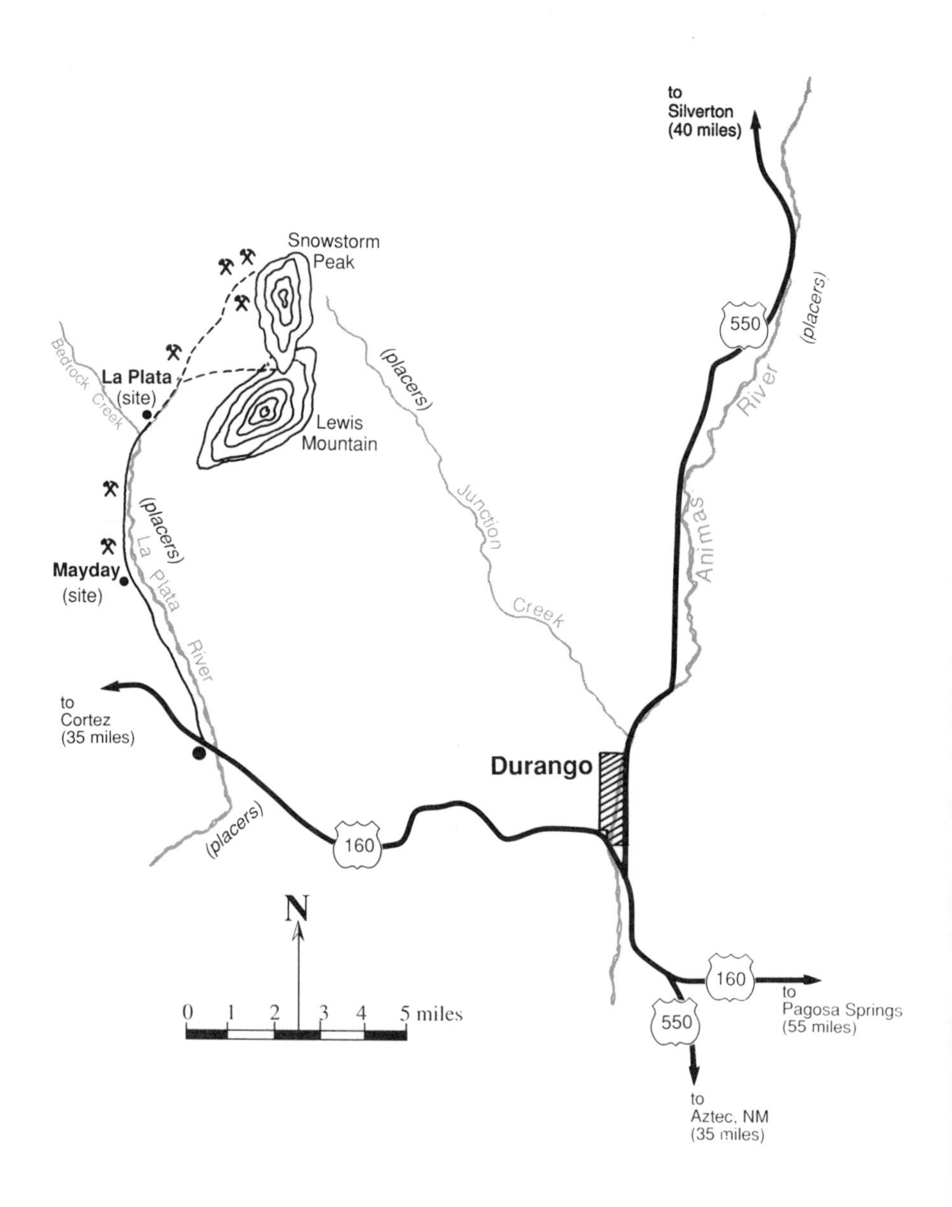
to
Silverton
(40 miles)
Snowstorm
Peak
Bedrock Creek
La Plata
(site)
Lewis
Mountain
(placers)
550
(placers)
River
Junction
Creek
Animas
(placers)
La Plata
River
Mayday
(site)
to
Cortez
(35 miles)
Durango
(placers)
160
N
0 1 2 3 4 5 miles
160
to
Pagosa Springs
(55 miles)
550
to
Aztec, NM
(35 miles)

LA PLATA COUNTY

PLACER GOLD

La Plata County placer miners have produced several hundred troy ounces of gold from the gravels and terraces of the Animas River near the site of Baker's Bridge, sixteen miles north of Durango on U.S. 550.

Gold also occurs in the La Plata River from Bedrock Creek and the site of La Plata City eight miles downstream to Hesperus. The La Plata River, ten miles west of Durango, is reached on County Road 124.

REFERENCES: 54

THE LA PLATA MOUNTAIN MINES

The La Plata Mountain mineral deposits, northwest of Durango, are within the southwestern limit of the Colorado Mineral Belt. Prospectors arrived in the La Plata Mountains in 1861, but were discouraged by the Utes. The government legalized mineral entry in 1873; by the 1880s, miners made strikes from Mayday to the La Plata River headwaters and Snowstorm Peak. Extreme elevation, avalanches, isolation, and unusually complex mineralization all hindered development.

After 1900, development of the Mayday and Idaho mines boosted district production. By the 1920s, cumulative production, primarily of silver and gold, neared $5 million.

The five general types of local mineralization include chalcocite veins, proustite (ruby silver) veins, telluride-bearing replacement bodies and veins, low-grade chalcopyrite veins with small amounts of native palladium and platinum, and pyrite-rich sulfide veins and replacement bodies containing native gold and silver.

Ore minerals are chalcopyrite, pyrite, galena, sphalerite, tetrahedrite, chalcocite, and the tellurides sylvanite, calaverite, and petzite. The primary gangue minerals are barite, quartz, and calcite. In a few mines, small occurrences of mercury were found.

Bedrock Creek, just south of the La Plata town site, has eroded away a noncommercial but interesting mineralized vein system. The vein material is primarily gray and light blue quartz and green chalcedony containing bright, tiny crystals of pyrite and chalcopyrite. Rounded fragments of vein material can be found in the bed of the rushing creek. Bits of native copper, precipitated from copper-bearing solutions and occurring as irregular grains as long as a quarter-inch, can be panned from the dark swampy bog sediments at the southern end of the La Plata City town site.

The Copper King Glory Hole is located just above La Plata City. Beyond that the increasingly rough and steep road (four-wheel-drive only) climbs to timberline at Tomahawk Basin at the base of Snowstorm Peak. Nice ore and gangue mineral specimens can be found on the dumps of higher mines. Some mines high on Lewis Mountain and Snowstorm Peak are on exploration and standby status and are posted.

REFERENCES: 19, 23, 25, 67, 81

LARIMER COUNTY

OWL CANYON ALABASTER

Owl Canyon, seventeen miles north of Fort Collins on U.S. 287, is a good collecting area for the selenite, satin spar, and alabaster varieties of gypsum, and calcite after aragonite. South of Owl Canyon, U.S. 287 passes between two hogback ridges. The west ridge is a Pennsylvanian hogback, named for its capping of 300-million-year-old red Pennsylvanian sandstone; the east ridge is a Dakota hogback, capped by 100-million-year-old Dakota Sandstone. Gypsum-rich Lykins Formation sediments are exposed in the Dakota hogback.

Large, shallow, slowly evaporating basins of retreating Permian seas deposited the Lykins sediments 250 million years ago. Calcium sulfate, the first sea salt to precipitate, accumulated in unconsolidated beds that eventually became strata of gypsum rock and alabaster.

From the highway junction of Owl Canyon on U.S. 287, follow Larimer County Road 72 (Owl Canyon Road) east for one-third of

In-situ seams of satin spar in gypsum beds at Owl Canyon.

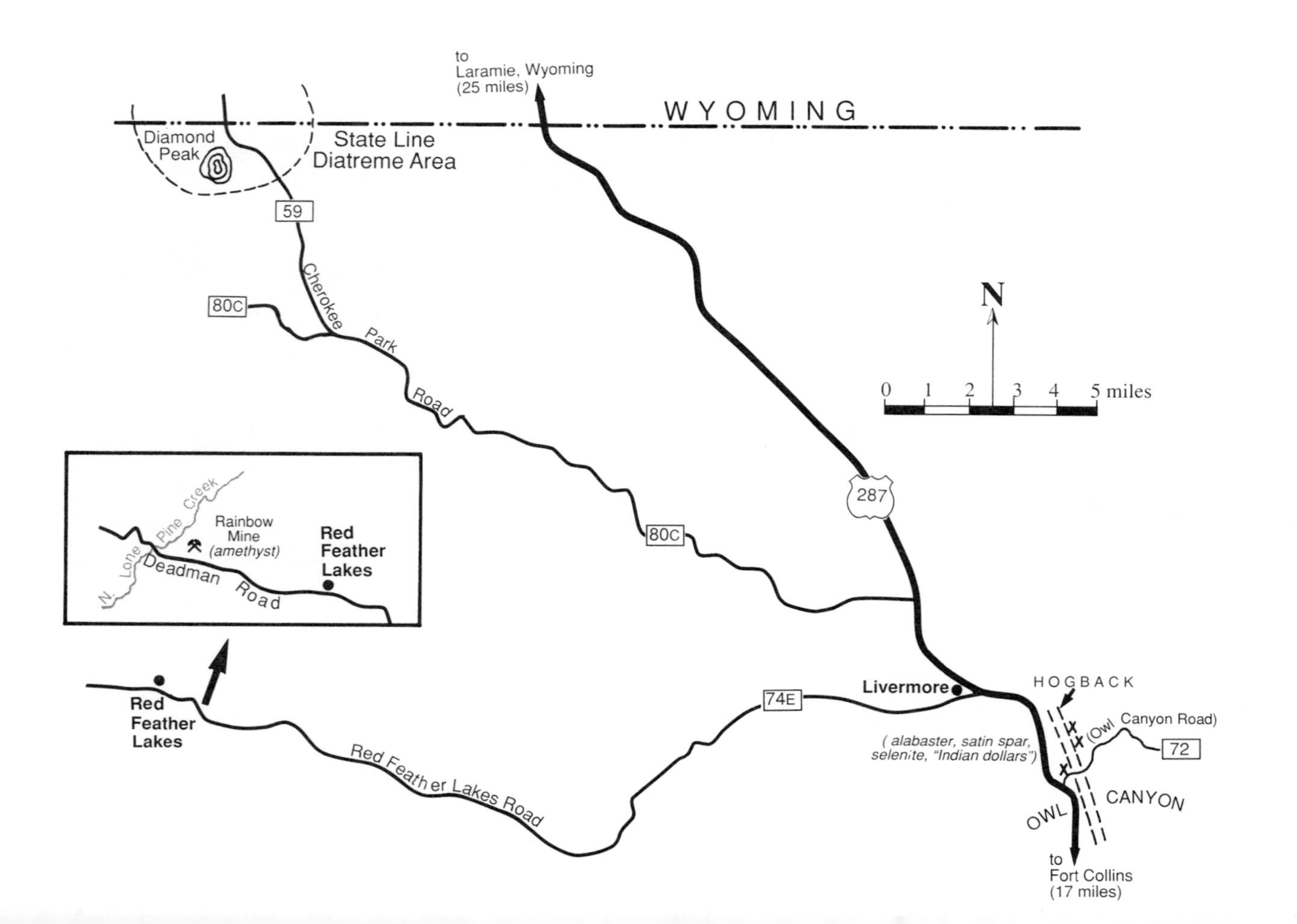

to
Laramie, Wyoming
(25 miles)
WYOMING
Diamond Peak
State Line
Diatreme Area
59
Cherokee Park Road
80C
80C
287
N
0 1 2 3 4 5 miles
Rainbow Mine (amethyst)
Red Feather Lakes
N. Lone Pine Creek
Deadman Road
Livermore
HOGBACK
74E
(Owl Canyon Road)
72
(alabaster, satin spar, selenite, "Indian dollars")
OWL CANYON
Red Feather Lakes
Red Feather Lakes Road
to
Fort Collins
(17 miles)

Satin spar from Owl Canyon exposures of Lykins Formation gypsum beds.

a mile to several small, alabaster exposures and quarry cuts north of the road.

Commercial alabaster quarrying at Owl Canyon began in the 1880s and peaked during the late 1940s and 1950s, when Fort Collins craftsmen "turned" (lathed) the alabaster into lawn ornaments, ashtrays, lamp bases, high-relief decorative panels, and bookends.

Alabaster occurs in irregular horizontal beds from one to six feet thick. Most is snow white and nearly translucent; some is a mottled light pink and greenish color. Its dense, fine-grained texture is similar to that of quality marble.

The alabaster exposures are bordered with or contain seams of white and pink satin spar up to five inches thick. Selenite seams have transparent crystals up to three inches long. The satin spar and selenite seams formed when part of the original gypsum rock and alabaster dissolved, then recrystallized. The calcium sulfate solutions have also leached below the gypsum beds to form isolated seams and pockets of satin spar and selenite in the lower sandstones and shales.

Collectors find numerous "Indian dollars," flat, hexagonal-shaped crystals of calcite derived from aragonite. Calcite and aragonite are forms of calcium carbonate with different molecular structures. They formed when weak carbonic acid in groundwater dissolved calcium sulfate from the gypsum beds, precipitating it as

This "Indian dollar" from Owl Canyon is calcite after aragonite. Original twinned aragonite crystal accounts for the pseudohexagonal shape.

the aragonite form of calcium carbonate, which was later replaced by calcite.

Calcite crystallizes in the rhombohedral system. Aragonite crystallizes in the orthorhombic system, often as twinned crystals with pseudohexagonal cross sections. The weathered gypsum beds free many calcite after aragonite "Indian dollars," all with hexagonal cross sections and some as large as silver dollars.

Collecting sights are also located on the eastern slope of the Dakota hogback. Continue east another mile (1.3 miles from U.S. 287) on Owl Canyon Road and park. A half mile to the north, several deep, narrow gullies cut the hogback slope. The gully alluvia contain tons of alabaster float, satin spar, and selenite. Farther up the gullies, the alabaster beds are exposed as naturally polished bedrock.

Alabaster's great workability accounts for its popularity as an ornamental stone. Like steatite (soapstone), alabaster is easily "carved" with a knife. Amateur sculptors and stoneworkers use most of the alabaster collected today. Home power tools, such as grinders, wire wheels, drills, and routers, quickly work alabaster into almost any conceivable shape.

REFERENCES: 18, 64, 81, 103

RED FEATHER LAKES AMETHYST

A Colorado State University geology professor from Fort Collins discovered veins of clear, milky, and amethyst quartz west of Red Feather Lakes in the 1920s. In 1939, *The Mineralogist* reported the site had yielded hundreds of pounds of crystals. By then the diggings had progressed twelve feet into a mass of disintegrated rock and clay. Miners recovered the crystals by washing and screening, and later recovered small amounts of eluvial gold by adding a sluice to the washing and screening line. The site, originally known as the Pennoyer Amethyst Mine, was later a popular fee collecting area.

Excavation has since progressed into solid rock, where foot-thick quartz veins cut the Precambrian granite. Typical veins have milky quartz and amethyst crystals growing inward to a clay-filled center. Amethyst crystals are medium to dark purple, well developed, and sometimes doubly terminated. Average length is about one inch, but some individual crystals are as long as three inches. Many crystals have a thin outer layer of milky quartz masking the interior purple, as well as a coating of iron oxides. Many crystals are fractured, limiting the size of cut gems to three or four carats.

The deposit, now called the Rainbow Lode, was again a popular fee collecting site in the 1980s. In 1987, mechanical excavation yielded hundreds of pounds of amethyst crystal clusters.

From Fort Collins, follow U.S. Route 287 twenty-four miles north to Livermore, then take Larimer County Road 74E (Red Feather Lakes Road) twenty-four miles west to Red Feather Lakes. Continue west on Larimer County Road 162 (Deadman Road) five miles to the crossing of North Lone Pine Creek. A nearby, unmarked road leads a short distance north to the site.

In recent years, fee collecting has been conducted on an informal basis by prior arrangement only. For information call (303) 669-2589.

REFERENCES: 19, 32, 40, 41, 42, 67, 73

THE CRYSTAL MOUNTAIN PEGMATITES

Crystal Mountain, Colorado's northernmost pegmatite area, is located in the Front Range foothills twenty driving miles west from Fort Collins along Larimer County Roads 38E and 27. Crystal Mountain is an eastern ridge of 10,626-foot Lookout Mountain.

Three types of Crystal Mountain pegmatites occur within schists metamorphosed from Precambrian granite: barren, that is, containing only quartz, feldspars, and mica; lithium-bearing; and

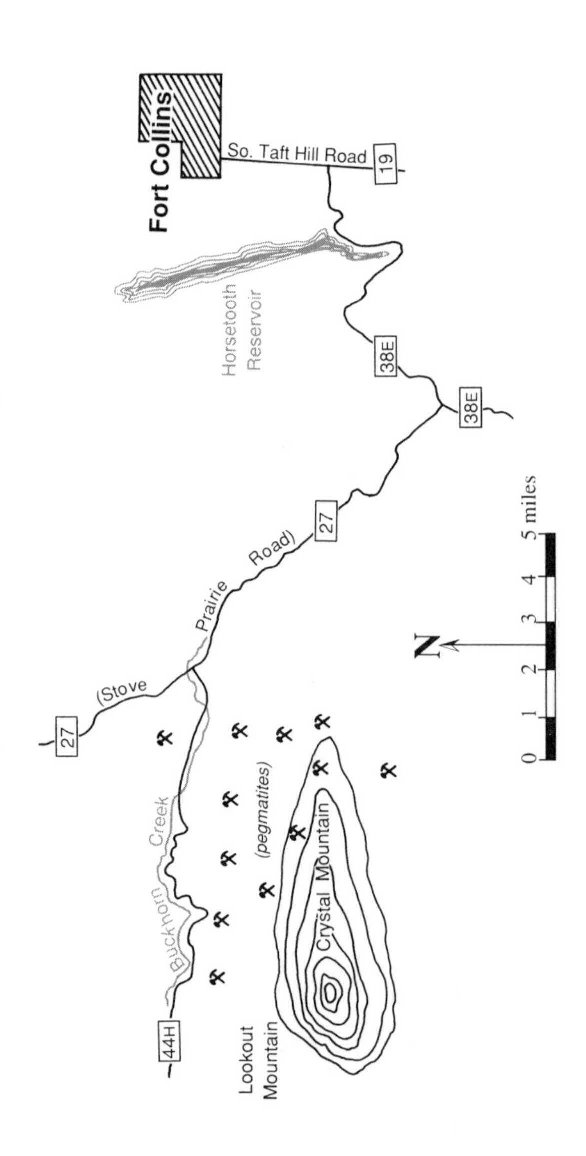
Fort Collins
So. Taft Hill Road
19
Horsetooth Reservoir
38E
38E
27
(Stove Prairie Road)
27
Buckhorn Creek
44H
(pegmatites)
Crystal Mountain
Lookout Mountain
N
0 1 2 3 4 5 miles

beryl-bearing. The pegmatites have few mariolytic cavities and thus are not known for large, well-formed individual crystals. Nevertheless, many minerals are present and Crystal Mountain is an excellent pegmatite prospecting and collecting opportunity.

Mica mining at Crystal Mountain began in the 1880s. Prospectors, searching for increasingly valuable rare earth minerals, discovered beryl and tantalum in pegmatites in the 1930s. Miners produced beryl during World War II, when the USGS explored for rare earth and other strategic minerals. Mining was halted in the 1960s.

Most of Crystal Mountain's twenty major pegmatites have been mined or at least partially excavated. Several hundred smaller pegmatites occur within a fifteen-square-mile area. Most are located south of Buckhorn Creek and Larimer County Road 27 on the north and east slopes of Crystal Mountain.

Over forty minerals occur at Crystal Mountain, and pegmatite mineralization varies considerably. Crystal Mountain is Colorado's best source of apple green chrysoberyl, which occurs in inch-wide contact twins usually associated with small "books" of muscovite. Columbite-tantalite crystals are as long as two inches. Beryl, the most common accessory mineral, occurs in light greens, blues, and whites. Beryl miners reported crystals one foot in diameter and six feet long. Collectors have also found colorless topaz in terminated crystals up to four inches in length. Other interesting minerals include albite, bismuthinite, fluorite, spessartine garnet, scheelite, spodumene, lepidolite, zircon, and uraninite.

Most of the larger pegmatites are on private land or under claim. Many smaller pegmatites and prospects, however, are within Roosevelt National Forest and open to prospecting and collecting.

REFERENCES: 18, 28, 29, 41, 67

STATE LINE DIAMONDS

Since the 1970s, when mineralogists identified a series of kimberlite pipes, or diatremes, northern Larimer County has been one of few interesting diamond prospects in the United States. Kimberlite is the only commercial diamond-bearing rock known in the earth's upper crust.

Placer miners have found diamonds in several North American gold-mining districts. But until the discovery of the South African diamond fields in 1870, prospectors believed diamonds occurred only in alluvial gravels. Excitement over South Africa's diamonds played a role in Colorado's first experience with "diamonds" (see Moffat County) in 1872.

One local quarry mined a serpentine-like decorative stone until

1960, when a large, expensive carborundum cutting wheel was inexplicably damaged. Years later, the "serpentine" was correctly identified as an altered kimberlite.

In the early 1970s, USGS researchers discovered diamonds after damaging a grinding plate attempting to take thin rock sections of the kimberlite. Analysis revealed that diamonds had scored the hard plate. "Diamond fever" raged briefly when prospectors discovered and staked over a hundred small kimberlite occurrences.

The so-called "State Line" kimberlite deposits have since yielded numerous diamonds. Most are colorless octahedrons and dodecahedrons less than two millimeters in diameter. A few specimens have weighed over three carats, and colors have also included yellow, light orange, blue-white, and nearly black. Pyrope garnet crystals up to six millimeters in size are also present in the kimberlite.

Most larger known kimberlite pipes in the State Line diatreme area are under claim, and permission should be obtained before collecting. Adjacent likely kimberlite areas are open for prospecting.

The State Line diatreme area covers twenty square miles and is located about fifty miles northwest of Fort Collins, or thirty miles south of Laramie, Wyoming. From Fort Collins, take U.S. 287 twenty-four miles north to Livermore. Then follow Larimer County Road 80C (Cherokee Park Road) northwest for thirty miles to the Diamond Peak area near the Wyoming line.

REFERENCES: 15

LAS ANIMAS COUNTY

STONEWALL JASPER AND SATIN SPAR

A distinctive bright green, jadelike jasper is found near The Stonewall, a steeply tilted section of the Dakota hogback ridge, in western Las Animas County. The collecting areas are creek beds near Stonewall Gap, where the Middle Fork of the Purgatoire River cuts through the Dakota hogback.

Lykins Formation gypsum beds are exposed at several points in the hogback between Stonewall Gap and Monument Lake. Seams of white and pink satin spar are as thick as six inches.

The Stonewall area is twenty-five miles south of La Veta and twenty-five miles west of Trinidad on Colorado Route 12.

REFERENCES: 65, 67

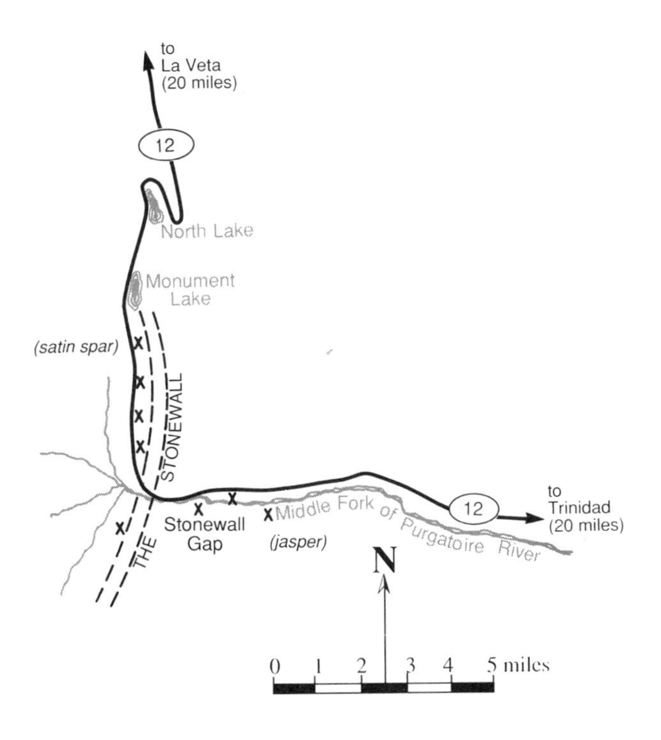

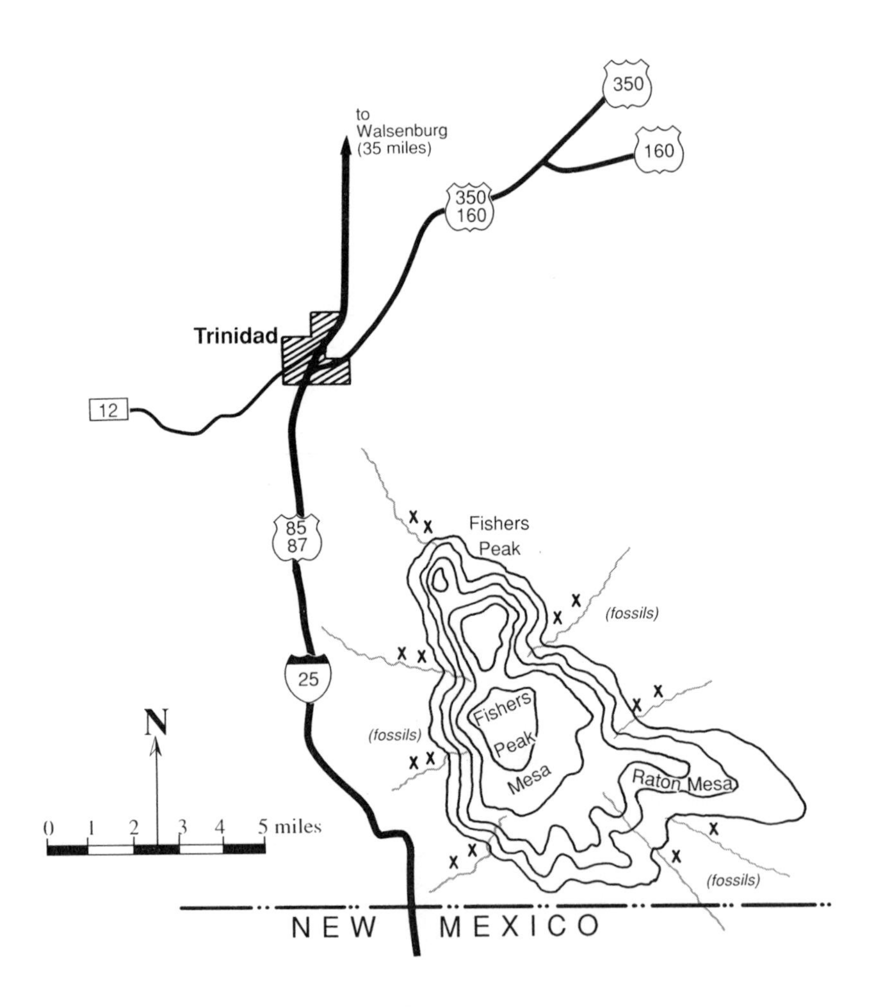

to
Walsenburg
(35 miles)
350
160
350
160
Trinidad
12
85
87
25
Fishers
Peak
(fossils)
Fishers
Peak
Mesa
Raton Mesa
(fossils)
(fossils)
N
0 1 2 3 4 5 miles
NEW MEXICO

TRINIDAD AREA FOSSILS

Fishers Peak, Fishers Peak Mesa, and Raton Mesa rise 3,000 feet above Trinidad to the south and southeast. Numerous creeks draining the mesa slopes expose fossil-rich late Cretaceous and early Tertiary sediments interspersed with thin coal seams. A U.S. Army exploration party gathered some of the first Colorado fossils ever collected for scientific purposes from Raton Mesa in January 1846. Reporting on his 1873 government survey, Ferdinand V. Hayden wrote: "On the hills surrounding Trinidad are great quantities of deciduous leaves in the rocks. The most conspicuous, as well as abundant fossil, is a species of fan palm. . . . "

Fossils are locally abundant in exposed sediments and in creek beds and gullies draining the mesas.

REFERENCES: 34

APISHAPA CANYON FOSSILS

Fossils of Dakota Formation leaves are found in Apishapa Canyon, northwest of Thatcher, where the Apishapa River exposes Cretaceous Carlisle Shale and Dakota Sandstone. Much of the Apishapa Canyon is within the Apishapa State Wildlife Area.

Agate, jasper, and fossils can be locally abundant near gully exposures of the Morrison Formation.

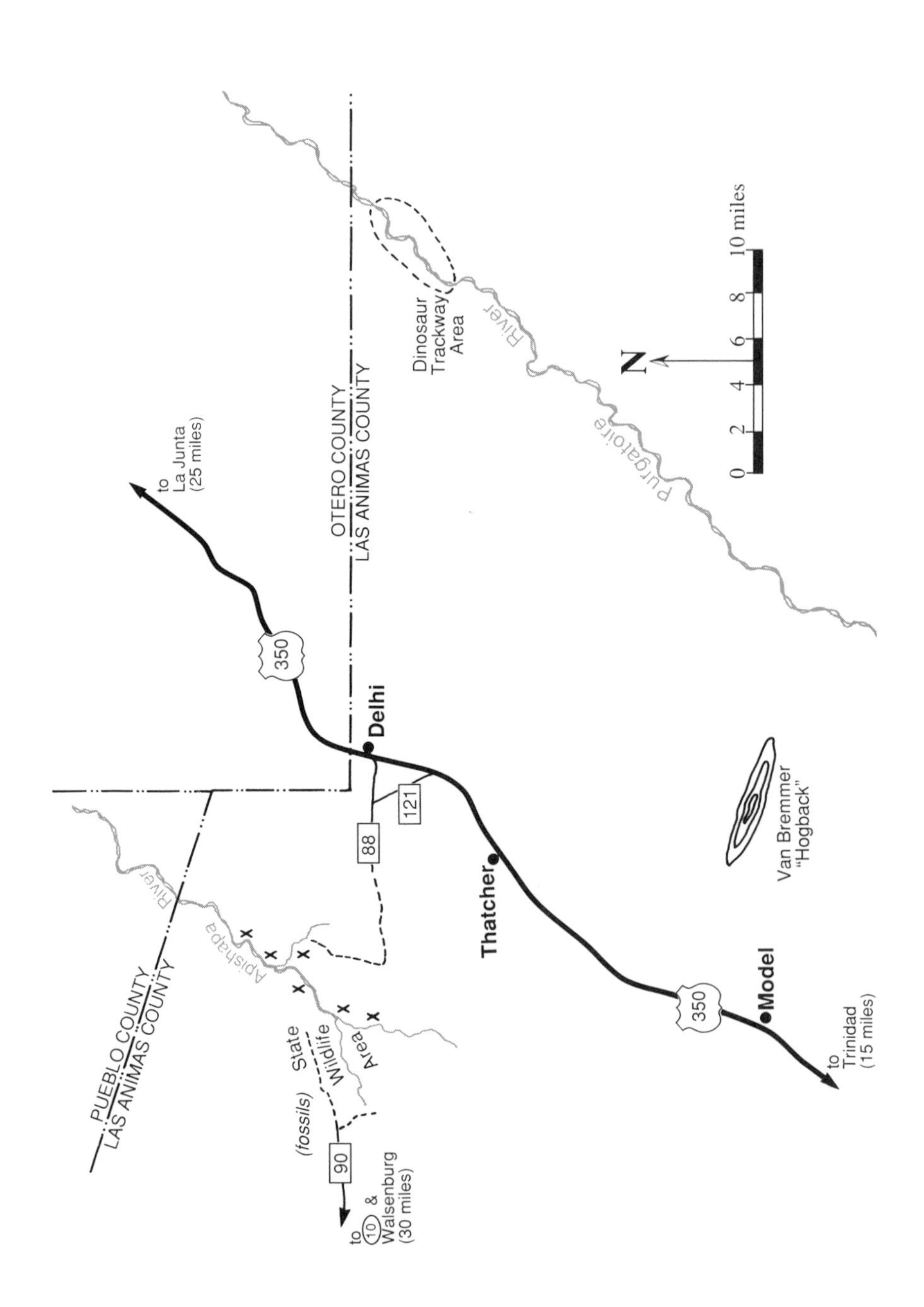
to
La Junta
(25 miles)
OTERO COUNTY
LAS ANIMAS COUNTY
Dinosaur
Trackway
Area
Purgatoire
River
N
0
2
4
6
8
10 miles
350
Delhi
121
88
Thatcher
Van Bremmer
"Hogback"
Model
350
to
Trinidad
(15 miles)
PUEBLO COUNTY
LAS ANIMAS COUNTY
Apishapa
River
State
Wildlife
Area
(fossils)
90
to 10 &
Walsenburg
(30 miles)

From Thatcher, take U.S. 350 nine miles north to Delhi junction on the Otero County line, then follow 88.0 Road west for ten miles into the state wildlife area. Take any of the unmarked dirt roads a short distance farther to Apishapa Canyon's rim and gullies. Fossils are locally abundant in canyon-wall sedimentary exposures.
REFERENCES: 78

THE VAN BREMMER HOGBACK

Located twenty-five miles northeast of Trinidad and six miles east of the town of Model on U.S. 350, the Van Bremmer Hogback was formerly a popular collecting locale for calcite and fossils. This site is now permanently closed.

The Van Bremmer "hogback" is actually a basalt dike intruded into Cretaceous sediments. The resistant basalt forms a low but prominent southeast-northwest trending ridge. Surrounding Cretaceous sediments, including Carlisle Shale and Greenhorn Limestone, are exposed on the sides of the dike.

Calcite occurs as vein fillings, concretions, and septaria in the Carlisle Shale. In seams and cavities, the calcite forms small, well-developed, translucent crystals ranging in color from gray and brown to greenish blue. The lower Greenhorn Limestone contains many marine fossils, including small ammonites and clams that occasionally retain lustrous shell material.

In the late 1980s, the Van Bremmer Hogback and large tracts of land to the north were acquired by the U.S. Army for expansion of the Fort Carson field-maneuver area. The Van Bremmer Hogback is now fenced and posted against entry.
REFERENCES: 64

PURGATOIRE RIVER DINOSAUR TRACKWAYS

Paleontologists have identified North America's longest known dinosaur trackway in the remote Purgatoire River Canyon country of northern Las Animas County. Paleontologists have documented more than 1,200 individual dinosaur tracks in a mile-long exposure of Cretaceous Dakota Sandstone.

Through recent land exchanges between the U.S. Army and the United States Forest Service, the dinosaur trackways are now part of Comanche National Grassland. To protect the site for paleontological study, location of the trackway is not generally publicized. However, Forest Service-guided public tours to the site began in 1992. The tours leave from La Junta and advance reservations and personal four-wheel-drive vehicles are required. For information, contact the U.S. Forest Service ranger station in La Junta.

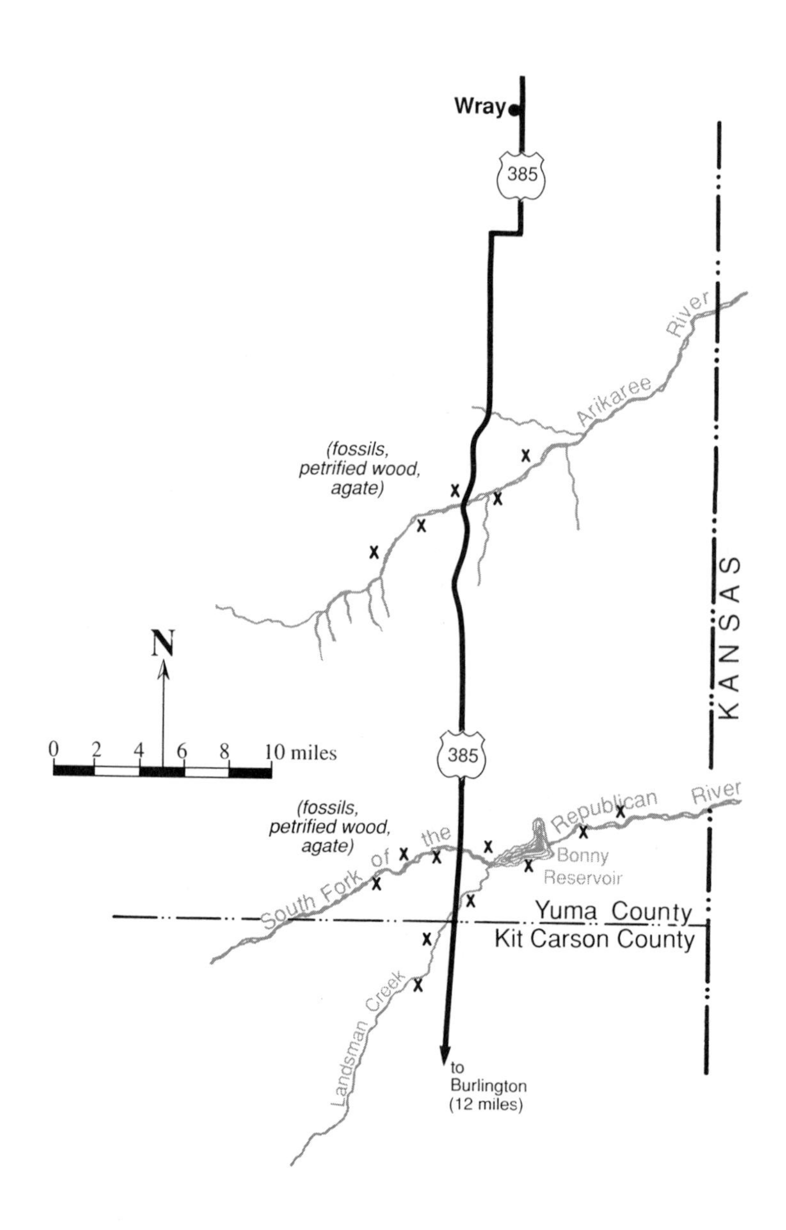
Wray
385
Arikaree River
(fossils, petrified wood, agate)
N
0 2 4 6 8 10 miles
385
KANSAS
(fossils, petrified wood, agate)
South Fork of the Republican River
Bonny Reservoir
Yuma County
Kit Carson County
Landsman Creek
to Burlington (12 miles)

LINCOLN, CHEYENNE, KIT CARSON & YUMA COUNTIES

Lincoln, Cheyenne, Kit Carson, and Yuma counties cover 7,000 square miles of grasslands and farmlands on Colorado's eastern plains.

FOSSILS, PETRIFIED WOOD, AND AGATE

SOUTH FORK OF THE REPUBLICAN AND ARIKAREE RIVERS

In eastern Yuma County, the upper drainage systems of the South Fork of the Republican River and the Arikaree River occasionally cut through overlying Tertiary gravels to expose fossiliferous Pierre Shale. Cretaceous marine fossils that weather free from the shale include pelecypods and cephalopods.

The condition of marine fossils from Colorado's Pierre Shale exposures varies greatly. Many crumble upon exposure, while others are durable casts retaining parts of lustrous shell material. Even limited alluvial movement, however, will abrade the lustrous shell surface or destroy it entirely. The finest fossils are collected directly from weathering shale outcrops.

Agate, jasper, and fragments of petrified and some opalized wood are also found in the gully systems of the South Fork of the Republican and Arikaree rivers, along the wide shoreline of Bonny Reservoir, and along the South Fork and Landsman Creek upstream from Bonny Reservoir into Kit Carson County. U.S. 385 crosses these drainages between Burlington and Wray.

REFERENCES: 67

FLAGLER AREA FOSSILS

In western Kit Carson County, I-70 and U.S. 24 (the parallel frontage road) cross the South Fork of the Republican River. East of Flagler, many gullies have small exposures of Cretaceous Pierre Shale containing a variety of marine fossils, including those of clams, oysters, ammonites, and shark's teeth.

The Pierre Shale exposures are not prominent, but appear as dark gray, crumbling sections of the gully walls. The frontage road, U.S. 24, provides easy access to the gullies near I-70. Fossiliferous Pierre Shale exposures are found in many gullies north of I-70 and west of Colorado Route 59.

REFERENCES: 12

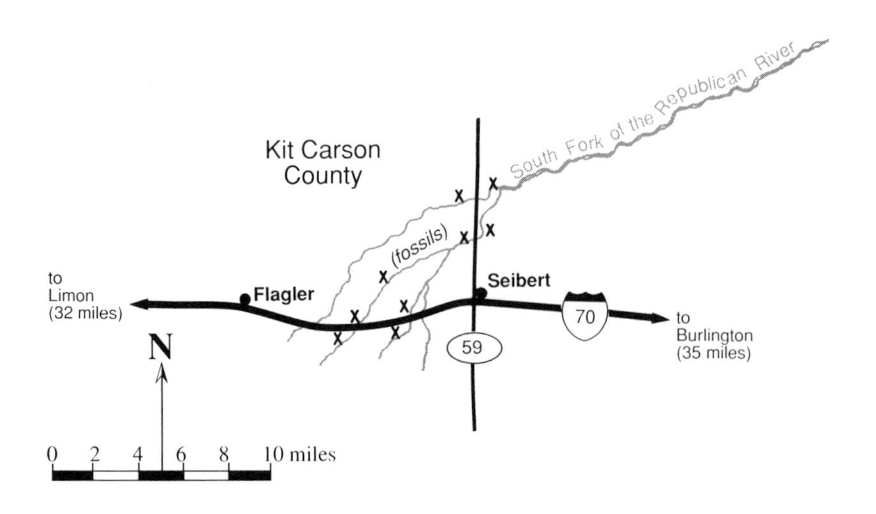
South Fork of the Republican River
Kit Carson
County
(fossils)
to
Limon
(32 miles)
Flagler
Seibert
70
59
to
Burlington
(35 miles)
N
0 2 4 6 8 10 miles

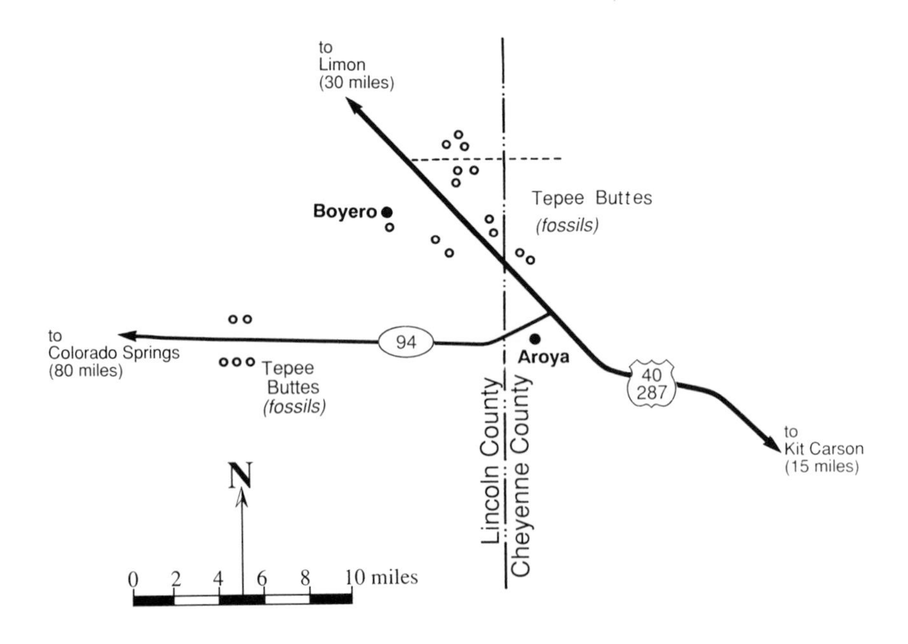
to
Limon
(30 miles)
Boyero
Tepee Buttes
(fossils)
to
Colorado Springs
(80 miles)
94
Aroya
Tepee
Buttes
(fossils)
40
287
to
Kit Carson
(15 miles)
Lincoln County
Cheyenne County
N
0 2 4 6 8 10 miles

Fossil clams from Pierre Shale exposures near Flagler and I-70 often retain lustrous shell material.

Tepee buttes near Boyero are rich in fossil clams and calcite.

LINCOLN–CHEYENNE COUNTY TEPEE BUTTES

Numerous tepee buttes—small, conical hills that were once Cretaceous sea-bottom reefs—dot the prairies east of U.S. 40-287 near the little town of Boyero. The buttes are not prominent and can be easily missed from the highway. From Boyero, take Road 2J east for three miles to a group of buttes on both sides of the road. Composed primarily of a gray limestone, the buttes contain many small fossil clams and thin seams of calcite and gypsum. Another group of similar but more prominent tepee buttes is located fourteen miles east of Aroya on Colorado Route 94.

REFERENCES: 12

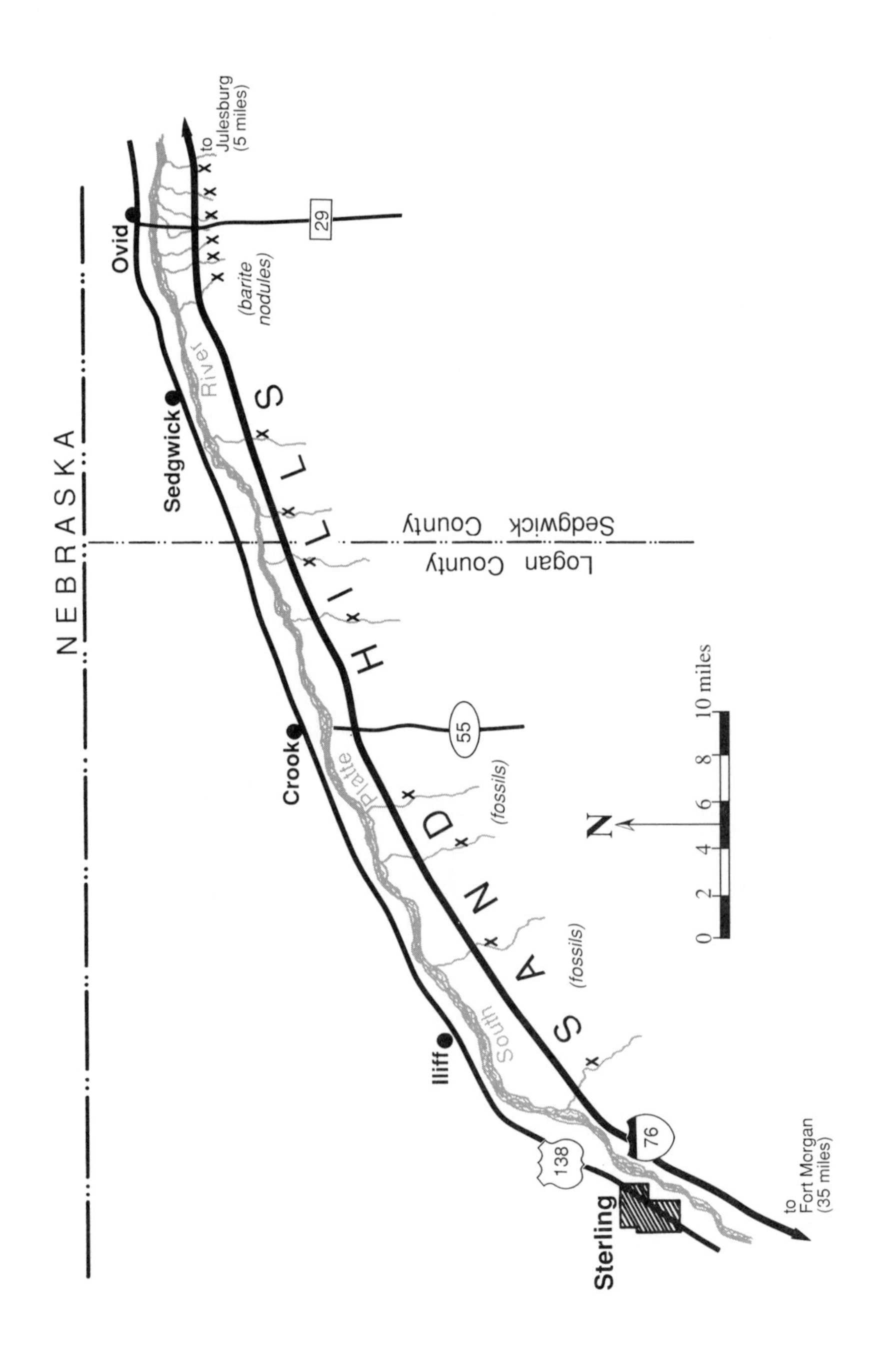
NEBRASKA
Ovid
Sedgwick
Crook
Iliff
Sterling
to Julesburg (5 miles)
to Fort Morgan (35 miles)
29
55
76
138
(barite nodules)
(fossils)
(fossils)
SAND HILLS
South Platte River
Sedgwick County
Logan County
N
0 2 4 6 8 10 miles

LOGAN COUNTY SEDGWICK COUNTY

OVID BARITE NODULES

In northeast Colorado, the South Platte River cuts a broad channel through the Tertiary rock and loosely consolidated sediments that form the surface of the plains. Numerous intermittent streams cut gullies through the higher terraces south of the South Platte River bottom, exposing horizontal interbedded layers of sand, silt, and clay. Nodules containing small, bright barite crystals often occur in the clay layers.

The nodules, gray to tan with a dull luster, are noticeably heavy. Most are about one to two inches in diameter. They have a subradial structure with stellate cavities lined with small, well-formed barite crystals ranging in color from white to pale yellow and blue. Under ultraviolet light, the barite fluoresces with a bright light yellow color.

One collecting site is near Ovid, six miles south of Julesberg in Sedgwick County. From I-76 exit 172, take County Road 29 south about one-half mile. Search the nearby gullies on the south side of I-76 for dark brown, nodule-bearing clay layers. Exposures of nodule-bearing clay layers are also found in gullies several miles east and west of the Ovid site. Collectors gathered many barite nodules during construction of I-76 in the early 1970s.

REFERENCES: 45

LOGAN COUNTY FOSSILS

Through Logan County, the South Platte River channel cuts through Tertiary surface sediments and into the underlying fossil-rich Cretaceous Pierre Shale. Shale exposures, however, are not numerous because of soil cover on the river bottom and sand cover on the higher lateral terraces.

Most exposures are found along the southeast side of I-76, where gullies cut through the barren sand hills. The shale exposures, typically a dark gray, have a weathered, papery appearance. Some form unstable near-vertical walls at the bottom of the gullies. Marine fossils, often of pelecypods and cephalopods, occur within rust-colored concretions and are locally abundant.

REFERENCES: 12

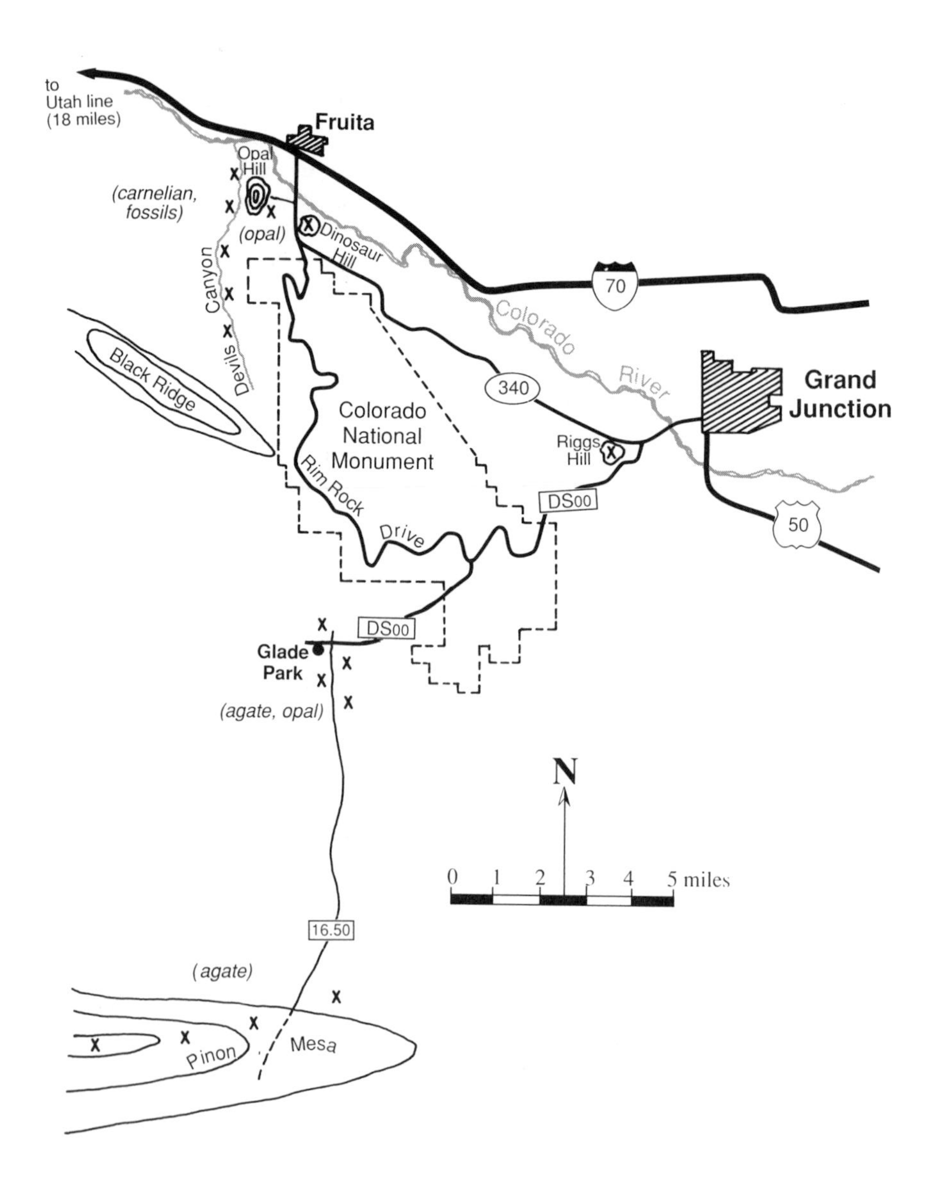
to
Utah line
(18 miles)
Fruita
Opal
Hill
(carnelian,
fossils)
(opal)
Dinosaur
Hill
Devils Canyon
Black Ridge
70
Colorado River
340
Grand
Junction
Colorado
National
Monument
Rim Rock Drive
Riggs
Hill
DS00
50
DS00
Glade
Park
(agate, opal)
N
0 1 2 3 4 5 miles
16.50
(agate)
Pinon Mesa

MESA COUNTY

DINOSAUR QUARRIES AND EXHIBITS

Mesa County bills itself as "Dinosaur Country" for a good reason: the Grand Junction area is one of the world's richest sources of dinosaur fossils. Sheepherders and ranchers discovered huge "dinosaurian" bones in exposures of Jurassic Morrison Sandstone near Grand Junction in the 1880s. Commercial "bone hunters" collected the fossilized bones for shipment to eastern museums and for sale to tourists. Because of the unprofessional manner of collection, the bones had little scientific value.

In 1899, Elmer S. Riggs, assistant curator of paleontology at the Field Columbian Museum (now Chicago's Field Museum of Natural History) sent inquiries to several Colorado and Wyoming towns concerning local fossil occurrences and amateur recoveries. Dr. S. M. Bradbury, president of Grand Junction's Western Colorado Academy of Science, informed Riggs of the local bone hunters and suggested paleontological investigation could be rewarding.

On a shoestring-budget expedition to Grand Junction in 1900, Riggs recovered the shoulder and some vertebrae from a large *Camarasaurus*. Weeks later, three miles west of Grand Junction on a hill that now bears his name, he discovered a complete skeleton of *Brachiosaurus altithorax*, which received worldwide publicity as "the largest land animal that ever lived."

Riggs returned in 1901, but nearly lost the season when a supply raft capsized, dumping food, supplies, and a ton of vital plaster (used to encase excavated bones) into the Colorado River. With generous local assistance, Riggs continued work, recovering the rear two-thirds of a brontosaurus (now *Apatosaurus*). One of the dinosaur's ribs had broken and healed during its lifetime. Study of that rib fossil helped advance a specialized paleontological discipline now known as paleopathology.

Riggs sent six tons of plaster-encased bones to Chicago in 1901. When the reconstructed skeletons were displayed at the prestigious Field Columbian Museum, Grand Junction gained immediate fame as a world-class dinosaur-fossil source.

After observing Riggs's methods, local amateur paleontologists made additional recoveries. In 1937, fossil collector Ed Hansen found in-situ vertebrae in strata forty-two feet above the original

A half-life-sized robotic model of Triceratops *at Dinosaur Valley, Museum of Western Colorado, Grand Junction.*

recovery at Riggs Hill. Edward Holt, a Grand Junction geologist, excavated the sites, recovering partial skeletons of *Stegosaurus*, *Allosaurus*, and *Brachiosaurus*.

Local amateur paleontologist Al Look recognized the educational and tourism potential of the fossils and suggested preserving the historic Riggs Hill quarries. When a commemorative plaque was installed at Riggs Hill in 1938, the guest of honor was Dr. Elmer Riggs, curator of paleontology of the Field Museum and one of world's foremost paleontologists.

But no funds were available to protect Riggs Hill, and the publicity attracted souvenir hunters and "collectors" from as far away as Europe. By 1960, all the visible bones at Riggs Hill, along with the scientific knowledge they could have provided, were gone.

Of the dozens of dinosaur excavation sites near Grand Junction, eight have major scientific importance. They have provided many complete or nearly complete skeletons of Jurassic and Cretaceous dinosaurs, dinosaur eggs, and trackways.

DINOSAUR VALLEY MUSEUM

The tourism potential of dinosaur paleontology in the Grand Junction area first envisioned a half century ago has now been realized. Dinosaur Valley, located at 4th and Main in the downtown district and part of the Museum of Western Colorado, ranks among the world's best "all-dinosaur" museums. Dinosaur Valley, which opened in 1985, has a working paleontological laboratory and exhibits covering many aspects of dinosaur paleontology. Displays include dinosaur-skeleton reconstructions, fossilized dinosaur eggs, and casts of dinosaur trackways. Visitors may touch "hands-on" displays such as gastroliths, coprolite, fossilized bones, and a synthetic recreation of dinosaur skin in its likely coloration, texture, and thickness.

Dinosaur Valley presents dinosaurs not only as reconstructed skeletons, but as living creatures. By combining knowledge of fossilized skeletons and modern comparative anatomy, paleontologists have modeled dinosaur musculature.

Robotic dinosaurs are the most dramatic exhibits. Six one-half-life-sized models of dinosaurs, including *Stegosaurus*, *Tyrannosaurus*, and *Apatosaurus*, have a lifelike exterior appearance and interior electric motors and cams to create a limited but realistic range of motion in jaws, necks, and forearms. The robotic dinosaurs emit loud roars and shrieks that paleontologists believe closely replicate actual dinosaur sounds. Many of Dinosaur Valley's reconstructed dinosaur skeletons originated in nearby quarries, which are open to the public.

RIGGS HILL

Riggs Hill is a 200-foot-high hill of Morrison Formation sediments. The historic 32-acre site was acquired by the Museum of Western Colorado in 1989 for protection and development.

Riggs Hill is three miles from the Dinosaur Valley Museum. From downtown Grand Junction, take Broadway (Colorado Route 340) west across the Colorado River to South Broadway. Signs mark Riggs Hill and an interpretive trail winds through the historic quarries. Collecting is prohibited.

DINOSAUR HILL

Dinosaur Hill, eleven miles west of Grand Junction and just south of Fruita, is another historic dinosaur quarry first excavated by Elmer Riggs in 1901. The Museum of Western Colorado owns the seven-acre tract on the eastern end of Dinosaur Hill.

From Grand Junction, drive west on Colorado Route 340, or take I-70 to the Fruita exit and follow the signs. The Dinosaur Hill interpretive trail is one mile long and takes an hour to complete. Collecting is prohibited.

The historic dinosaur quarries at Dinosaur Hill are open to the public.

Regulations at the Rabbit Valley Research Natural Area are clearly spelled out.

FRUITA PALEONTOLOGICAL RESEARCH NATURAL AREA

This 280-acre research natural area is located three miles southwest of Fruita. It is an active paleontological excavation site of international significance for its wide array of Mesozoic vertebrate fossils. Since its discovery in 1975, paleontologists have recovered fossilized remains of many species of lizardlike reptiles, crocodiles, early mammals, and other terrestrially adapted vertebrates, as well as the dinosaurs *Ceratosaurus*, *Allosaurus*, *Camarasaurus*, and *Stegosaurus*.

The site is known for its tiny adult dinosaurs, one of which, no larger than a chicken, is mentioned in the *Guinness Book of World Records*.

The active research location with the fossil-bearing strata is fenced. For further information contact the Bureau of Land Management District Office in Grand Junction.

RABBIT VALLEY RESEARCH NATURAL AREA

The Rabbit Valley Research Natural Area, with its 1.5-mile-long "Trail Through Time," is Mesa County's premier public dinosaur-quarry site. It is located twenty-eight miles west of Grand Junction, two miles from the Utah line and just north of the I-70 Rabbit Valley exit (Exit 2, no services).

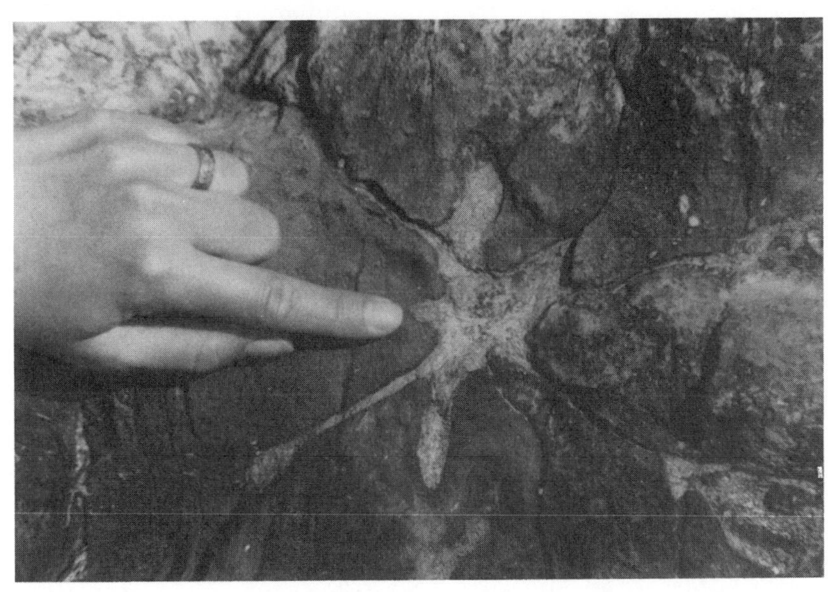

An exposed fossil vertebrae of Camarasaurus *at the Rabbit Valley "Trail Through Time."*

Since quarrying began in 1982, Rabbit Valley has yielded specimens of *Apatosaurus*, *Camarasaurus*, *Diplodocus*, *Allosaurus*, *Stegosaurus*, and the world's oldest *Iguanodon*.

The Trail Through Time leads to quarries with exposed vertebrae and hip sections of *Camarasaurus* and *Diplodocus*, which visitors may study, touch, and photograph. The bones are contained within a durable, in-situ, silicified sandstone. With the difficulty of intact removal obvious, visitors can understand why paleontological excavation demands proper training, tools, and methods, and why anything less will destroy the delicate fossils.

The 280-acre Rabbit Valley Research Natural Area and its Trail Through Time are cooperative projects of the BLM, the Museum of Western Colorado, and state and local governmental agencies. Collecting is prohibited.

Paleontologists are now preparing fossil resource inventories to document the many dinosaur fossils awaiting recovery. Rabbit Valley may soon be developed as an educational dinosaur quarry, where the public may work with paleontologists in actual excavation of 130-million-year-old fossilized dinosaur bones.

REFERENCES: 3, 16

COLORADO NATIONAL MONUMENT

Many of western Colorado's collectible fossils and minerals have weathered from massive Mesozoic sedimentary formations. Although collecting is not permitted, Colorado National Monument is an excellent opportunity to observe and study exposures of Triassic, Jurassic, and Cretaceous sedimentary formations representing the entire 175 million years of the Mesozoic era.

The monument's twenty-eight-mile-long loop drive linking the Grand Junction and Fruita entrances passes many observation points and interpretive signs identifying specific formations. Familiarity with appearance, color, and rock type of these formations can be helpful in fossil collecting outside the monument.

BOOK CLIFFS BARITE AND FOSSILS

The Book Cliffs, capped with Mesaverde Sandstone and sloped with Mancos Shale, are the impressive twenty-mile-long ramparts northeast of Grand Junction. Fossils and barite nodules are locally common at various sites on and at the base of the cliffs.

The Book Cliffs are most accessible north of Palisade near I-70, and also northeast of Walker Field, where unmarked dirt roads and tracks follow the gullies of Indian Wash, Persigo Wash, and Leach Creek.

Cliffs and canyons expose sedimentary strata representing the entire 175 million years of the Mesozoic era.

The Book Cliffs are capped with Mesaverde Sandstone and sloped with Mancos Shale.

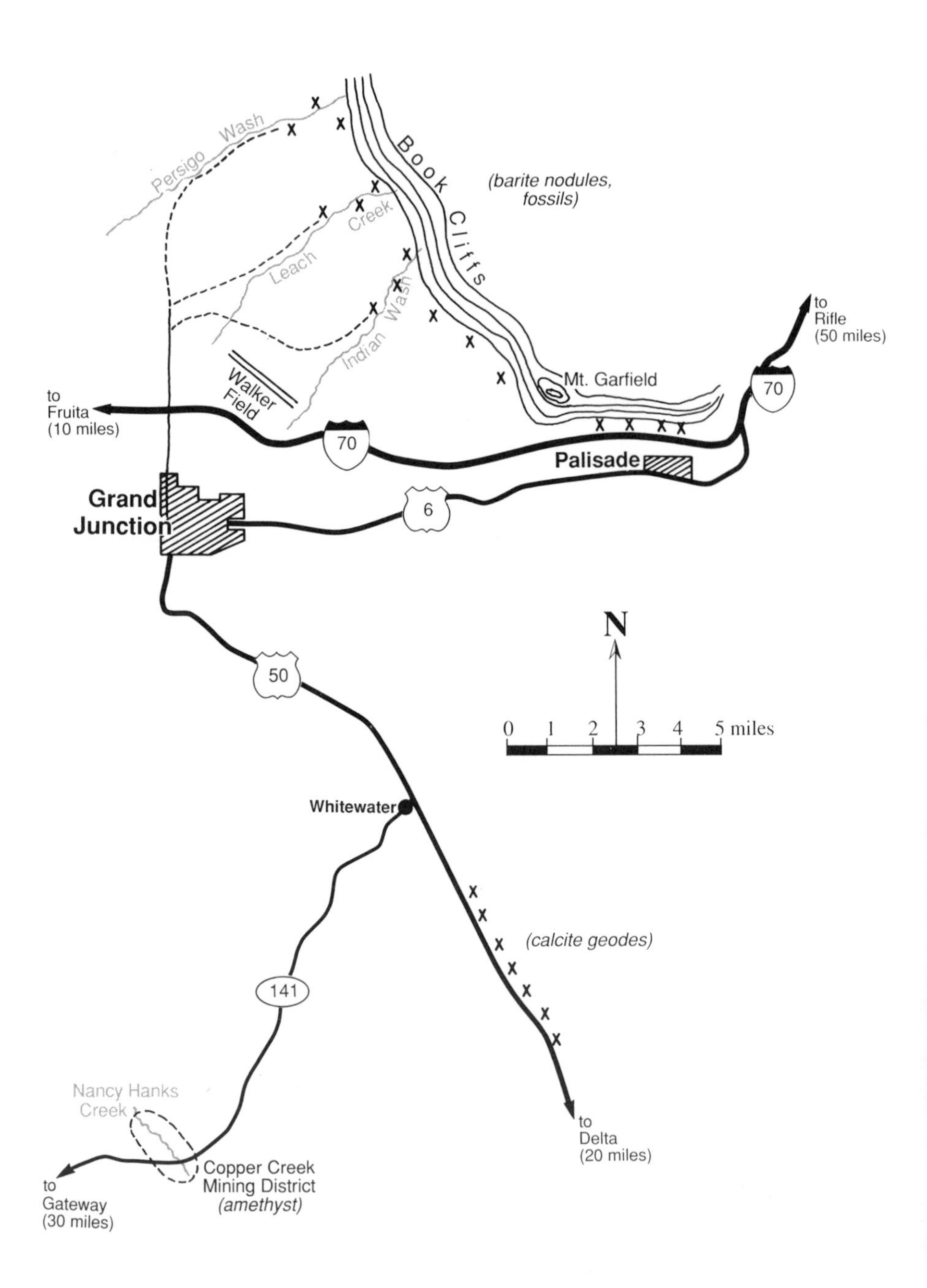

Persigo Wash
Book Cliffs
(barite nodules, fossils)
Leach Creek
Indian Wash
Walker Field
Mt. Garfield
to Rifle (50 miles)
to Fruita (10 miles)
70
Palisade
Grand Junction
6
50
N
0 1 2 3 4 5 miles
Whitewater
(calcite geodes)
141
Nancy Hanks Creek
Copper Creek Mining District
(amethyst)
to Gateway (30 miles)
to Delta (20 miles)

The Book Cliffs near Palisade offer rugged collecting for fossils and barite geodes.

Ammonite fossils occur within certain strata of the Mancos Shale. Fragments of ammonites can be found in talus and in the gullies; intact fossils occur within reddish brown shaley concretions.

Barite nodules, ranging in size from one foot to six feet in diameter, occur as rusty brown concretions within layers of dark gray limestone. Clear, white, and light yellow barite crystals up to two inches long line the interior cavities. Local lapidaries have faceted some transparent barite crystals into gems.

REFERENCES: 12, 14, 54, 67

AGATE, JASPER, OPAL, AND PETRIFIED WOOD

Many of the hills, mesas, and gullies from Grand Junction west into Utah have varying amounts and types of agate, jasper, and petrified and opalized wood. Petrified dinosaur bone, whether in situ or as float, is a vertebrate fossil; collecting is prohibited on public lands.

Unaweep Canyon. Portions of the canyon walls are a good source of copper mineralization and amethyst.

OPAL HILL

Opal Hill (also known as Blue Hill and Agate Hill) is located two miles southeast of Fruita and just south of the Colorado River. The ridge of Morrison Formation sediments was named for its opalized wood, somewhat chalky in appearance, which occurs in various colors. The opal is the common variety rather than the more familiar and striking fire opal, but does have an opalescent surface sheen.

Opal Hill has been well collected for decades. The easily gathered surface material is gone, but specimens can still be found by digging.

From just south of the Colorado Route 340 bridge over the Colorado River, take King's View Road west for a half mile, then turn south on Horse Thief Canyon Road. Take any of the tracks a half mile west to the base of the prominent hill.

DEVILS CANYON

Devils Canyon, just west of Opal Hill, cuts through a hard fossiliferous limestone. Some of the fossilized gastropod shell material has been replaced by a beautiful, deep red carnelian. Similar carnelian replacement of fossil shell material occurs in other western Mesa County fossil localities. Although usually too thin to cut, the carnelian makes attractive fossil specimens.

PIÑON MESA, GLADE PARK, AND BLACK RIDGE

Glade Park and the vast Piñon Mesa area are fifteen miles southwest of Grand Junction. From Grand Junction, follow Colorado Route 340 (Broadway) west, then turn south on Rim Rock Drive (DS 00 Road). Proceed through the eastern entrance to Colorado National Monument and follow the signs to Glade Park, where opalized wood is similar to that found at Opal Hill.

Piñon Mesa, an east-west trending flat-topped ridge south of Glade Park extending west to the Utah line, is a good source of jasper, petrified wood, and banded and moss agate. Search the gullies draining the north slope of the ridge, especially after seasonal runoff has "turned" the creek-bed gravels.

Black Ridge, five miles northeast of Glade Park and just beyond the boundary of Colorado National Monument, is a source of agate and petrified wood that weathers free from fossil-rich Mesozoic sediments. Paleontologists have recently recovered a 120-million-year-old fossilized sycamore, one of the oldest known angiosperms, or flowering plants. Several universities are studying the site, and the BLM has proposed full protection as a research natural area. For further information on the Black Ridge Angiosperm Locality, contact the Grand Junction District Office of the BLM.

WEST SALT CREEK

Agate, jasper, and petrified wood are found along West Salt Creek from its confluence with the Colorado River near Mack north into Garfield County.

***REFERENCES:* 3, 19, 32, 41, 44, 64, 65, 66, 67, 73, 74**

WHITEWATER CALCITE GEODES

Calcite-filled geodes are found along a ten-mile-long section of U.S. 50 south of Whitewater. They occur as lines of highly visible, rusty brown nodules in exposures of gray Mancos Shale in a low ridge immediately east of the highway. The nodules range from one foot to six feet in diameter. Most have hollow centers and radiating seams lined with clear, white, and light yellow calcite crystals, often in well-formed rhombohedrons as long as two inches.

The nodules formed from concentrations of iron and calcium, probably from accumulated shell material, within Cretaceous sea-bottom sediments. Iron oxide accounts for the distinctive rusty brown surface coloration of the nodules. The nodules are crumbly and easily broken. Sections of center cavities and seams make attractive display plates of calcite crystals. The alluvial gravel below the shale exposures contains many abraded, large calcite rhombohedrons.

One productive locality is located 2.5 miles south of Whitewater, where numerous nodules form a thin line in the gray shale exposures.

REFERENCES: 12

UNAWEEP CANYON AMETHYST

Amethyst is commercially collected at the old Copper Creek mining district near Nancy Hanks Gulch. The district is eleven miles southwest of Whitewater on Colorado Route 141 at the eastern end of the spectacular Unaweep Valley.

After prospectors discovered copper mineralization in the late 1890s, miners founded Copper City and Pearl City and built several small smelters during a short-lived copper-mining boom. Today, the district is better known for collectible minerals than for past copper production. Thin and erratic veins of calcite penetrate Precambrian basement rock and overlying Mesozoic sediments. The colorful veins, only a few feet thick, contain a brecciated mix of various green copper minerals, along with purple amethyst, white barite, and green fluorite in a snow white calcite matrix.

The Amethyst Queen Mine is occasionally worked for mineral specimens and rough amethyst through a shaft and drift along a large vein. The amethyst occurs as individual crystals or crystal clusters within the vein matrix. The matrix is dissolved in acid to free the amethyst. Amethyst crystals range in size from one inch to three inches, and in color from light lavender to a valuable deep purple, from which fine gems up to six carats have been cut.

Mineralized veins similar to those at the Amethyst Queen Mine extend up the granite cliffs on both sides of the valley into overlying sedimentary rock. The BLM considers a seven-acre site along Nancy Hanks Gulch as a "significant geologic feature" for its outstanding examples of Tertiary hydrothermal alteration of sandstone. The BLM has not instituted special management regulations because of mixed land control, including patented claims and private ownership. Universities and local colleges conduct geology field trips to the district.

From Whitewater junction, take Colorado Route 141 southwest for 11.3 miles to a roadside parking area just beyond a small bridge. Steep dirt tracks and trails lead a short distance northwest to the mineralized areas.

An inactive pegmatite quarry in Precambrian granite is located a few miles farther west on Colorado Route 141. Easily collected

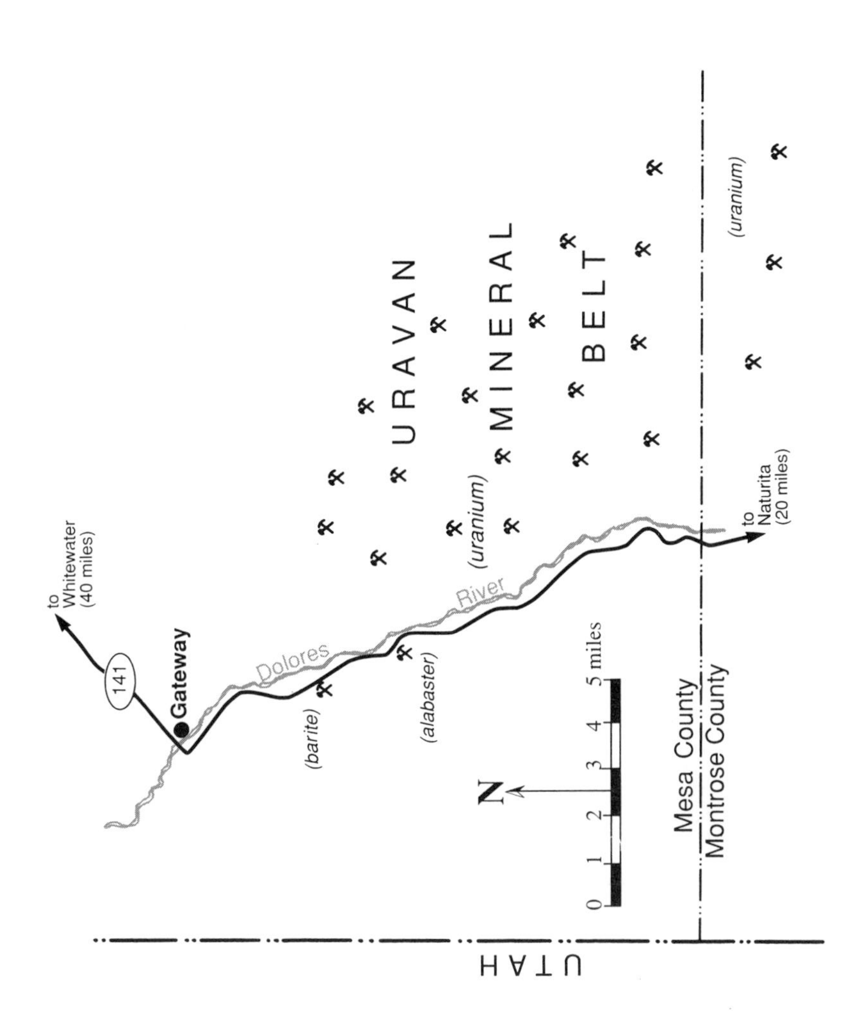
to
Whitewater
(40 miles)
141
Gateway
Dolores
River
(barite)
(alabaster)
(uranium)
URAVAN
MINERAL
BELT
(uranium)
to
Naturita
(20 miles)
0 1 2 3 4 5 miles
N
Mesa County
Montrose County
UTAH

minerals include massive white quartz, large crystals of feldspar and both biotite and muscovite mica, and small crystals of spessartine garnet.

REFERENCES: 3, 32

URAVAN MINERAL BELT

The northern end of the Uravan Mineral Belt extends through southwestern Mesa County. Several mills once operated at Gateway, and Grand Junction was the headquarters for the 1950s uranium rush, with its Atomic Energy Commission field offices, outfitters, radiometric and chemical assayers, and mine supply companies.

About 300 small uranium mines are located between Gateway and the Montrose County line. Carnotite and other radioactive minerals are found on mine dumps and along haulage roads.

REFERENCES: 10, 18, 94

GATEWAY BARITE AND ALABASTER

Small barite and alabaster mines are located south of Gateway. The barite site is located 3.3 miles south of the Dolores River bridge on Colorado Route 141. Several near-vertical barite veins penetrate the sandstone but are difficult to see from the highway. Look for a

Old uranium-vanadium mine near Gateway.

small tunnel at the base of a sandstone cliff 100 yards west of and slightly above the highway. The twenty-foot-long tunnel is driven entirely through solid, snow white barite. The veins are composed of compact crystalline aggregates; small clear barite crystals occasionally occur within seams. Barite specimens, noticeably heavy in the hand, are strewn over a wide area.

An old alabaster mine is located 2 miles farther along Colorado Route 141 (5.3 miles south of the Dolores River bridge at Gateway). The quarry is immediately west of the highway on posted private property with locked gates.

Collecting is possible, however, on adjacent unposted land, for the bright, white alabaster bed can be visually traced in the host sandstone for at least one mile from the quarry on both sides of the Dolores River Canyon. The alabaster is white to very light tan, and sections of the bed contain seams of attractive satin spar.

REFERENCES: 64

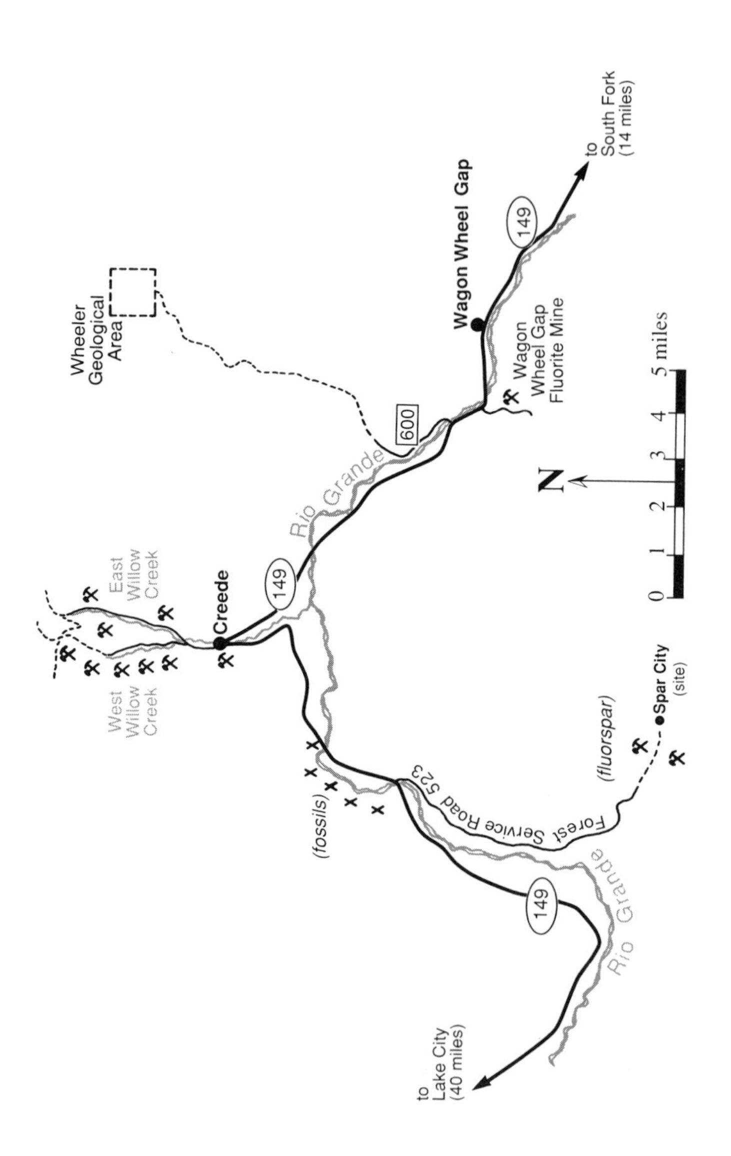
Wheeler Geological Area
Wagon Wheel Gap
to South Fork (14 miles)
149
600
Rio Grande
Wagon Wheel Gap Fluorite Mine
N
0 1 2 3 4 5 miles
East Willow Creek
Creede
West Willow Creek
(fossils)
Forest Service Road 523
(fluorspar)
Spar City (site)
to Lake City (40 miles)

MINERAL COUNTY

CREEDE AREA MINES

Creede, twenty-one miles north of South Fork on Colorado Route 149, is one of the West's great silver camps. Prospectors visited the Creede area in 1860, staked the first lode claims in 1873, and finally made the big strike in 1889, when Nicholas Creede and his partners found outcrops of amethyst quartz and native silver. They named their strike the Holy Moses, after Creede's purported exclamation. Other strikes soon revealed a system of large, rich silver veins, including the bonanza Amethyst Vein, exposed on the surface for nearly two miles.

Creede produced 380,000 troy ounces of silver in 1891. With some ores carrying 2,000 troy ounces of silver per ton, 1893 production soared to 4.8 million troy ounces. Rich mines included the Commodore, Last Chance, and dozens of others with colorful names like the Amethyst, Captive Inca, Happy Thought, Monte Carlo, and Exchequer Tunnel.

Creede was a wild boom town of 10,000 when noted poet and editor Cy Warman described it with these enduring words: "It's day all day in the daytime and there is no night in Creede."

Creede, the site of Colorado's last great silver strike, nestled at the mouth of Willow Creek Canyon.

Collectors search for mineralized amethyst in the bed of Willow Creek near Creede.

Ores were so rich that Creede was Colorado's only silver camp unscathed by the 1893 silver-market crash. By 1912, the Commodore Mine alone had produced 16 million troy ounces—about 550 tons—of silver. Over twenty years, Commodore ores averaged a remarkable 45 troy ounces of silver per ton. When the last mine closed in the early 1980s, Creede's production topped 50 million ounces of silver and 100,000 tons of lead.

Creede, population 800, now survives on seasonal tourism based on mountain scenery, arts, and mining history. To reach the old mines from Creede, follow Main Street north onto the graded gravel road at the base of the 1,000-foot-high rhyolite cliffs of Willow Creek Canyon. After one mile, the right fork heads into East Willow Creek Canyon, the left fork into West Willow Creek Canyon.

A quarter-mile into West Willow Creek, the bleached timbers, tramways, and ore bins of four of the richest Amethyst Vein mines cling precariously to the sides of cliffs. Beyond the Commodore, four-wheel-drive is necessary. The Commodore and other nearby mines are posted. The dumps are very steep, unstable, and dangerous. Some mineral collecting, however, is possible where mine dumps extend to the shoulder of the road. East Willow Creek Canyon, passable for cars, also has many mine ruins and roadside dumps.

Mineralization followed collapse of the ancient Creede Volcano

The Commodore Mine and Mill on West Willow Creek was among Creede's biggest sources of silver.

and formation of a caldera. In Amethyst Vein ores, crystalline and cryptocrystalline quartz are the primary gangue minerals, often composing three-quarters of total ore weight. The so-called "amethyst silver" ore from the Amethyst Vein is among the most distinctive and attractive metal ores mined in the United States.

Amethyst Vein material, whether from sections one inch or ten feet thick, has the appearance and structure of a geode. The outermost layer, which contacts the rhyolite fissure walls, is usually dark quartz heavily laden with sphalerite and argentiferous galena. Next is a delicately banded layer of white or bluish chalcedony, followed by another layer of mineralized quartz. After that is the amethystine quartz, which comprises about half the total weight of the vein. The trace presence of manganese accounts for its delicate violet hues. Next is an intergrowth of white crystalline quartz combs. Finally, in the center of the vein, space permitting, are drusy growths of clear quartz crystals. Because of its attractive appearance and color, lapidaries have slabbed and polished Amethyst Vein material since the 1890s.

Mineral and ore specimens are found in the bed of Willow Creek above Creede and along both the East and West forks. The well-washed rocks facilitate visual identification of mineralized specimens. Thin veins of amethyst, white, and clear quartz lace

Creede ore and gangue minerals displayed at the Denver Museum of Natural History.

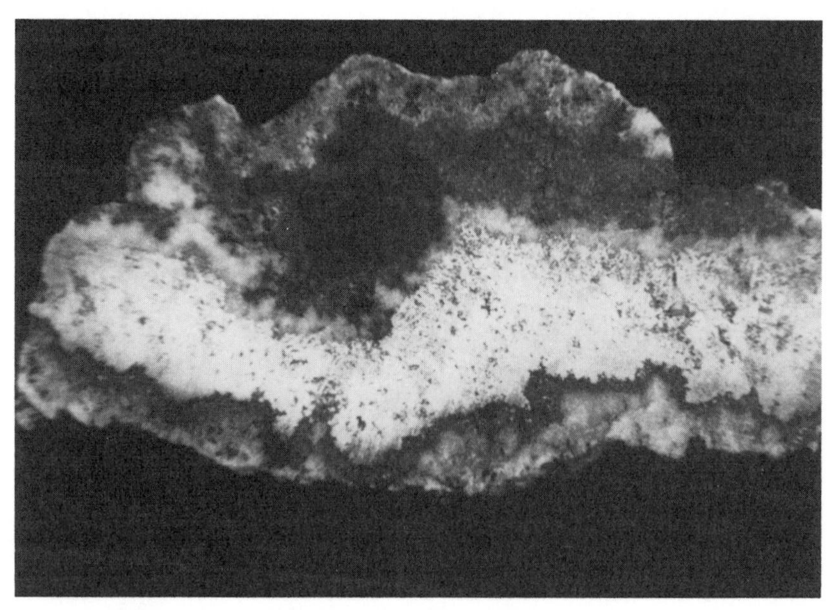

Slabbed ore from Creede contains visible native silver.

many of pieces rhyolite; many carry crystals of galena and pyrite, and even bits of native silver.

Local minerals include mineralized amethyst, galena, native silver, sphalerite, pyrite, siderite, cerussite, malachite, argentite, barite, and calcite. Red jasper and chrysoprase, the apple green variety of chalcedony, also occur. Lapidaries once cut, polished, and sold top-quality, deeply colored chrysoprase as "Creede jade," and even as "turquoise." Miners reported true turquoise at the Last Chance Mine. Float turquoise is sometimes found in Willow Creek and the West Fork.

Creede has an interesting historical museum with local mineral specimens as well as the new Creede Underground Mining Museum. The Underground Museum grew out of the unexpected tourist popularity of Creede's unique Underground Firehouse, located in drifts blasted out of the rhyolite walls of Willow Creek just north of town. The underground workings have mining displays and mineral exhibits.

REFERENCES: 18, 19, 23, 25, 40, 41, 42, 56, 59, 64, 66, 67, 75, 81, 104

WAGON WHEEL GAP FLUORSPAR MINE

The Wagon Wheel Gap fluorspar mine, located at Wagon Wheel Gap, seven miles south of Creede on Colorado Route 149, was Colorado's largest individual fluorspar mine. Prospectors who discovered the huge vein of mineralized purple fluorspar in the early 1900s first thought they had struck an extension of Creede's bonanza Amethyst Vein.

Fluorspar mining began in 1913 and continued until the 1950s, providing flux for Pueblo steel mills. Production topped a half-million tons of fluorspar worth $2 million. The mile-long fluorspar vein averaged 8 feet in thickness and was up to 700 feet wide.

The fluorspar ore has a radiating fibrous structure and ranges from white to purple, lilac, green, and yellow. Fluorite crystals occur in seams as purple octahedrons as large as one-half inch.

The fluorspar mine and mill is located on private posted ranch property one mile west of Wagon Wheel Gap. Permission to visit the site is necessary.

Several small open-cut fluorspar mines are located at the site of Spar City, eight miles southwest of Wagon Wheel Gap. Collectors have found fluorspar specimens and, occasionally, small fluorite crystals. From Creede, follow Colorado Route 149 seven miles southwest, then turn south on Forest Service Road 523 and proceed six miles to the Spar City area.

REFERENCES: 23, 25, 76, 80

CREEDE AREA FOSSILS

Insect and numerous plant fossils occur in large exposures of an easily separated, shalelike volcanic tuff in the upper Rio Grande valley west of Creede. Thirty million years ago, collapse of the Creede Volcano created a caldera, which filled with water, forming a ten-mile-wide lake similar to Oregon's present Crater Lake. Volcanic ash generated by continuing volcanic activity to the west drifted into the lake, accumulating on the bottom in layers containing trapped insects and plant parts. In a fossilization process similar to that at Florissant (Teller County), the sediments compacted into alternating layers of mudstone and shale, the latter providing an excellent environment for fossilization with superb preservation of detail.

Oligocene plant and insect fossils, dated at 25 million years, are found at several tuff exposures along Colorado Route 149 west of Creede. One highway cut exposes the old lake-bed sediments one mile west of Creede.

A more productive site is located about five miles west of Creede, where the Rio Grande flows north of the highway for a distance of one mile. The river makes a broad bend, cutting prominent, steep, 100-foot-high cliffs exposing layers of fossil-rich, buff-colored tuff. The cliffs are very steep, unstable, and dangerous. Outcrops and fragments of the fossiliferous tuff can be collected safely from adjacent areas. Access is easiest from the Rio Grande highway bridges at the ends of the cliffs.

The tuff is composed of alternating layers of a coarse mudstone and fine-grained shale. Fossils occur within the thinly laminated shale layers, which split easily along the planes.

REFERENCES: 35, 63, 64, 68, 104

WHEELER GEOLOGICAL AREA

Located in rugged country northeast of Creede, the Wheeler Geological Area consists of colorful pinnacles, spires, and domes of eroded volcanic tuff. Although not a mineral-collecting locale, this "miniature Bryce Canyon" is a fascinating side trip for anyone interested in geology and unusual natural beauty.

The formations were named for Captain George Wheeler, who surveyed the region for the government in 1874. The area became a national monument under the National Park Service in 1908, but was transferred to the Forest Service in 1950 because it was so remote, inaccessible, and infrequently visited. In 1961, the Forest Service enlarged the original 300-acre area to 640 acres, withdrew it from mineral entry, and later closed it to motorized vehicles.

The Wheeler Geological Area is reached by a rough forty-eight-mile round-trip drive that begins two miles north of Wagon Wheel Gap on Colorado 149. Follow Forest Service Road 600 (Pool Table Road) north along Spring Gulch for ten miles. Beyond Spring Gulch, four-wheel-drive vehicles are necessary for the remaining fourteen miles. From the fenced boundary, a half-mile-long trail leads to the pinnacles and spires.

The road is often impassable in wet spring conditions and may be closed to prevent road damage. The Wheeler Geological Area is at just over 11,000 feet in elevation and is accessible only during the summer months.

West of Creede, the Rio Grande River has cut cliffs through Oligocene sediments rich in fossil leaves and insects.

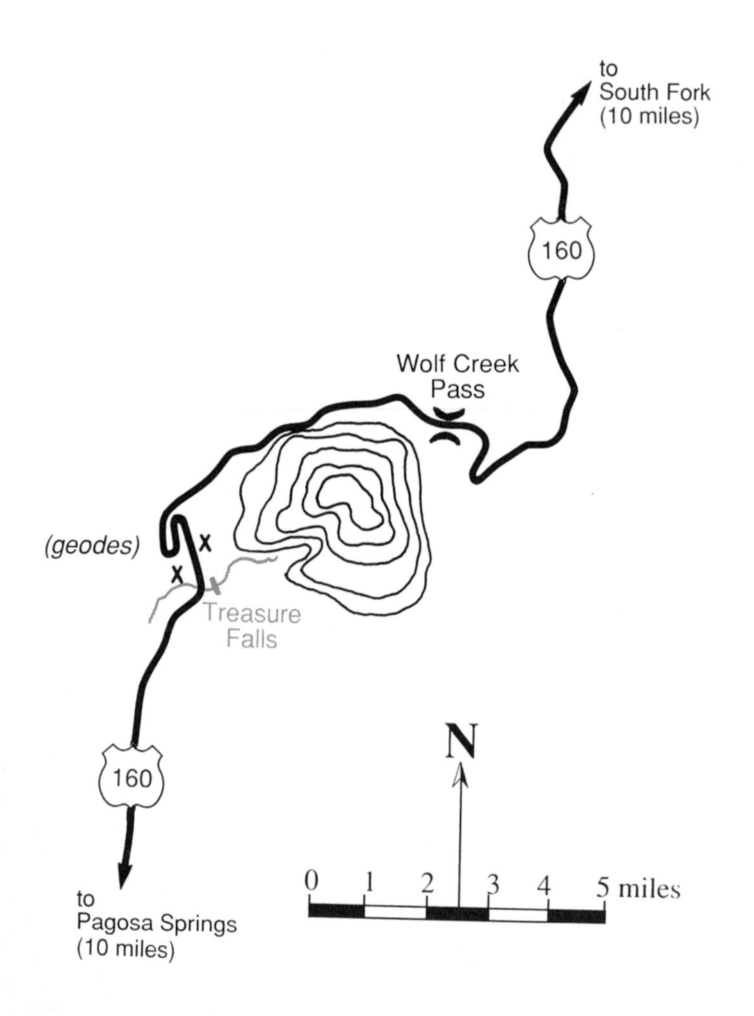
to
South Fork
(10 miles)
160
Wolf Creek
Pass
(geodes)
Treasure
Falls
160
N
0 1 2 3 4 5 miles
to
Pagosa Springs
(10 miles)

The Wolf Creek Pass geode-collecting site is right on the shoulder of U.S. 160.

WOLF CREEK PASS AGATE AND GEODES

One of Colorado's most accessible geode sites is located on U.S. 160 eight miles west of the Wolf Creek Pass summit, or one-half mile east of the Treasure Falls-Fall Creek turnoff. The collecting area is along the base of volcanic cliffs on the eastbound shoulder of the highway and on the opposite talus slopes.

The dark volcanic rock contains countless agate nodules and quartz geodes ranging from pea-sized to as large as five inches in diameter. Nodules and geodes can be found loose in debris and talus as well as imbedded in the host volcanic rock. Collectors have removed most of the larger in-situ geodes within reach, but new material weathers free each year. Most geodes have agate exteriors and center cavities filled with brilliant, beautifully formed, clear quartz crystals and, occasionally, pale amethyst. Others contain snow white natrolite, a zeolite mineral. The delicately banded agate ranges in color from white to light blue.

The toughness of the volcanic rock varies considerably. Some is crumbly and easily broken to free contained geodes. Other sections of rock are quite durable, requiring the use of hammers, chisels,

Small but beautifully formed geode from the Wolf Creek Pass area.

Near Wolf Creek Pass, geodes containing agate and clear crystalline quartz can be collected right from the volcanic rock of a highway road cut.

and suitable eye protection. In-situ nodules and geodes appear nearly black, but geodes that have weathered free usually have rusty colored smooth surfaces.

This is not a site for children, for U.S. 160 is heavily traveled and the shoulders are narrow. The talus slopes below the highway, which also contain geodes, are steep and dangerous.

REFERENCES: 64

MOFFAT COUNTY

PLACER GOLD

Moffat County has produced about 5,000 troy ounces of placer gold from streams along the Iron Springs Divide northwest of Craig. Iron Springs Divide, a 6,000-foot-high ridge extending twenty miles westward from Black Mountain, divides the Yampa River drainage in the south from the Little Snake River drainage in the north.

Gold-bearing northern streams include Timberlake Creek (Moffat County's biggest gold source), Fourmile Creek, and Scandanavian, Dry, and Bighole gulches. On the southern drainage, gold occurs in Lay, Blue Gravel, and Fortification creeks.

Although fine gold can be panned in all these creeks, mining has been limited. Prospectors discovered the placers in the 1880s; when mining began a decade later, it was hindered by low-grade gravels and lack of sufficient water. Two small floating-bucketline dredges operated on Lay and Timberlake creeks in the early 1900s.

Bighole Gulch is also a source of agate, jasper, and leaf fossils.

REFERENCES: 54, 67

MOFFAT COUNTY AGATE, JASPER, AND FOSSILS

Moffat County is a rich source of cryptocrystalline quartz. Agate and jasper, formed within the Pennsylvanian Morgan, Triassic Chinle, and Jurassic Morrison formations of the Uinta Mountains, are common in the alluvial outwash gravels south and east of the Uintas.

The general collecting area covers hundreds of square miles north of U.S. 40 between Cross Mountain and the highway settlement of Blue Mountain. Agate and jasper, smooth and rounded from alluvial wear, is locally abundant in washes and on hillsides. Although early Indians and modern collectors have gathered tons of agate and jasper, erosion continues to expose new material.

There are no services along U.S. 40 for fifty-seven miles between Maybell and Dinosaur, and collectors should be self-sufficient in fuel, water, food, and clothing.

CROSS MOUNTAIN–DOUGLAS MOUNTAIN AREA

From Maybell, follow U.S. 40 west for sixteen miles, then turn north on Twelvemile Gulch Road, which is maintained by the Na-

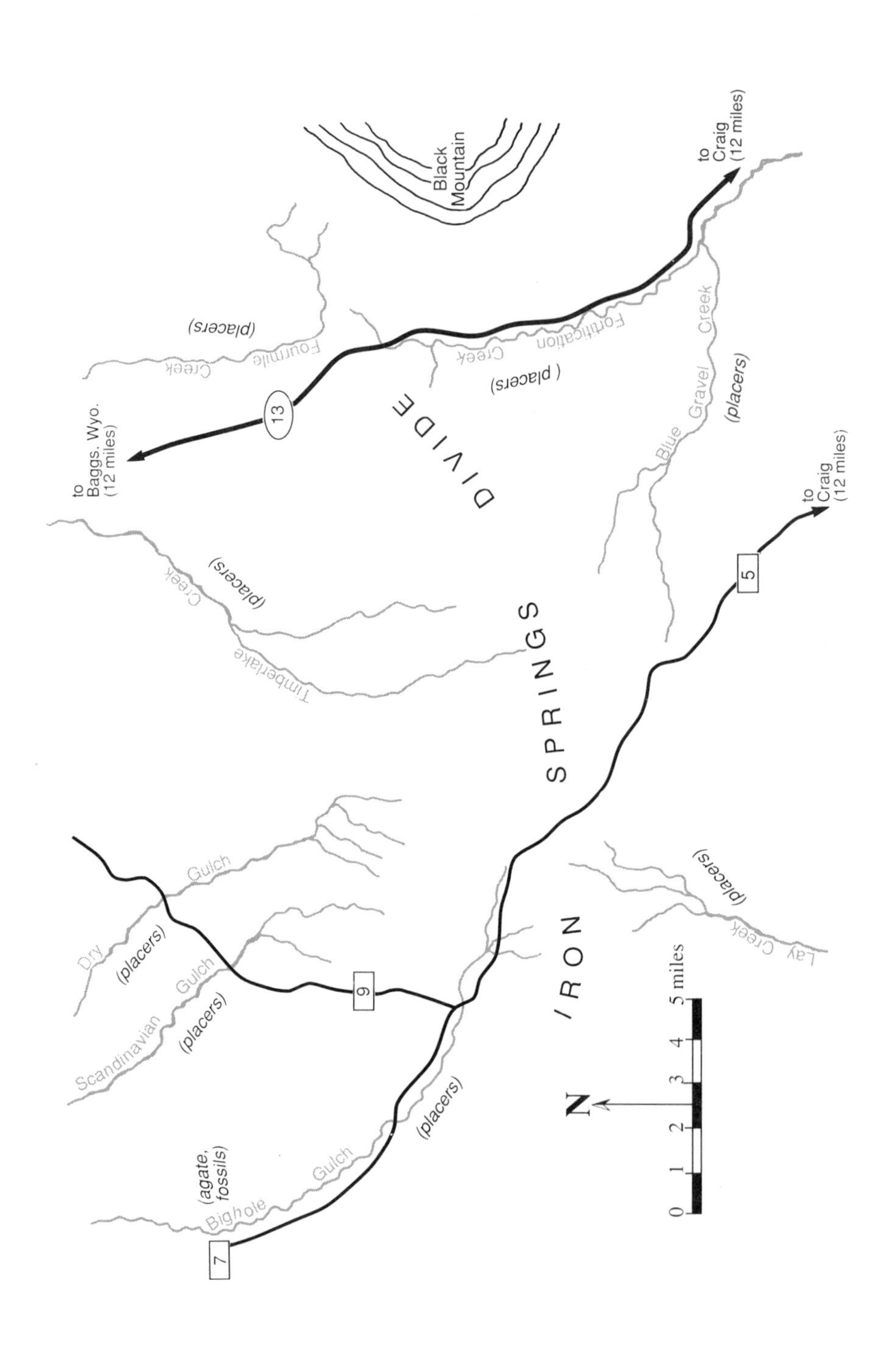
Black Mountain
to Craig (12 miles)
to Baggs, Wyo. (12 miles)
13
Fourmile Creek
(placers)
Fortification Creek
(placers)
Blue Gravel Creek
(placers)
IRON SPRINGS DIVIDE
Timberlake Creek
(placers)
to Craig (12 miles)
5
Dry Gulch
(placers)
Scandinavian Gulch
(placers)
9
Bighole Gulch
(agate, fossils)
(placers)
7
Lay Creek
(placers)
N
0 1 2 3 4 5 miles

The Yampa River cuts through Cross Mountain at Cross Mountain Canyon. Agate is common locally on hillsides and in gullies.

tional Park Service for access to the northeast section of Dinosaur National Monument. The slopes of Cross Mountain, immediately to the east, have red and brown jasper and red-and-white-banded agate. Four miles north of U.S. 40, Cross Mountain is dramatically bisected by the Yampa River and sheer-walled Cross Mountain Canyon.

From Cross Mountain Canyon, continue 1.5 miles farther to the confluence of the Yampa and Little Snake rivers. Take Moffat County Road 25 (four-wheel-drive recommended) along the Little Snake River six miles to Moffat County Road 10. The last two miles parallel a Morgan Formation exposure on the slopes of Douglas Mountain where jasper occurs in several colors.

At the junction of County Roads 25 and 10, cross the Little Snake River bridge and follow Road 10 northwest about one mile. Jasper is locally abundant in the washes and on the slopes of Douglas Mountain.

ELK SPRINGS AREA

From the old highway settlement of Elk Springs, seven miles west of Maybell, take Moffat County Road 14 (Bear Valley Road)

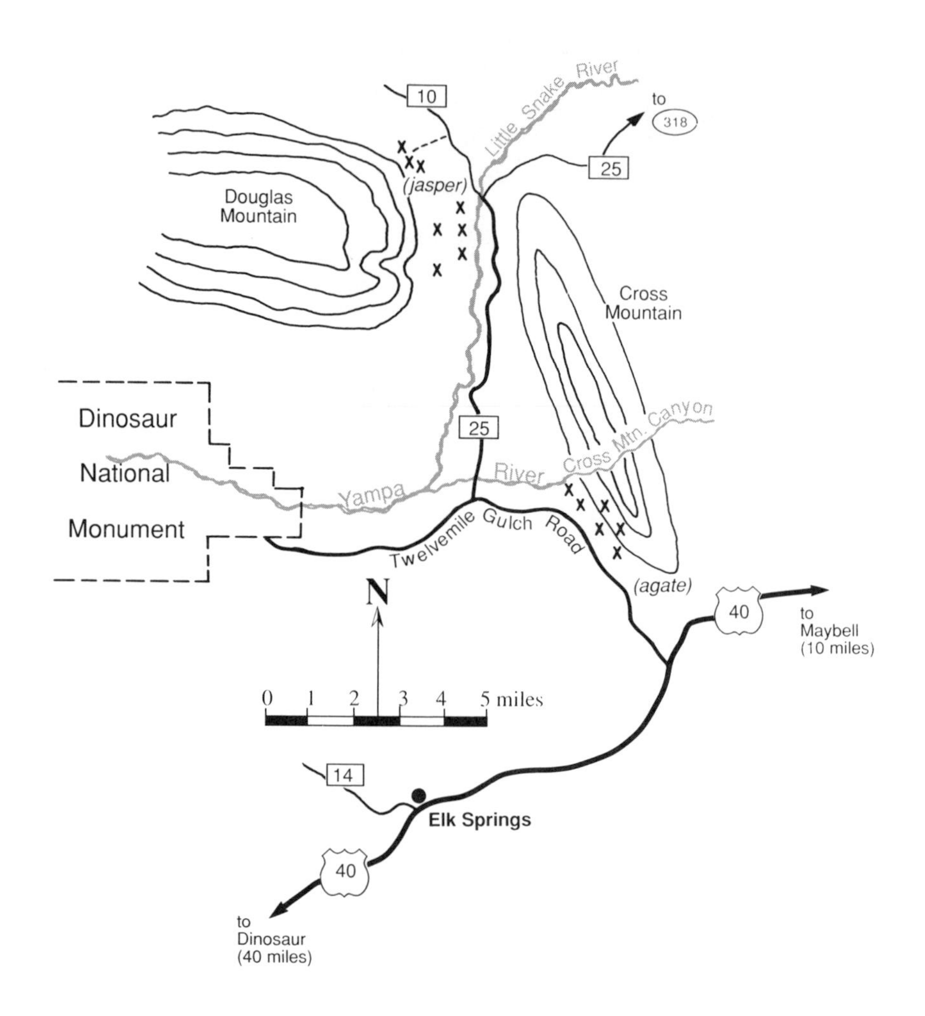

10
Little Snake River
to
318
25
(jasper)
Douglas Mountain
Cross Mountain
Cross Mtn. Canyon
Dinosaur
National
Monument
25
Yampa River
Twelvemile Gulch Road
(agate)
N
40
to
Maybell
(10 miles)
0 1 2 3 4 5 miles
14
Elk Springs
40
to
Dinosaur
(40 miles)

Small marine fossils from Jurassic shoreline sediments near Wolf Creek.

northwest. Agate and jasper are found on hilltops and washes on both sides of the road.

Proceed 4.5 miles to the head of Calico Draw, where several unmarked four-wheel-drive tracks lead north into the draw. Red, brown, and yellow jasper are found in the road gravels. Soil and sagebrush cover is heavy, but gravels and rocks are exposed lower into the draw. Many larger, rounded pieces of agate and jasper, including some beautiful bluish white moss agate with delicate black dendrites, are covered with a white calcareous coating.

Lower Calico Draw is also a paleo locality where Morrison Formation exposures yield dinosaur fossils. Brigham Young University paleontologists recently collected the fossilized bones of an incomplete *Diplodocus*. The BLM has no special management regulations at Calico Draw, but unauthorized collection of vertebrate fossils is not permitted on public land.

WOLF CREEK AREA–COUNTY ROAD 16

The upper Wolf Creek drainage has jasper, agate, and fossils. Wolf Creek crosses U.S. 40 5.5 miles east of Massadona (no services). Follow Moffat County Road 16 (Wolf Creek Road) to the general collecting area.

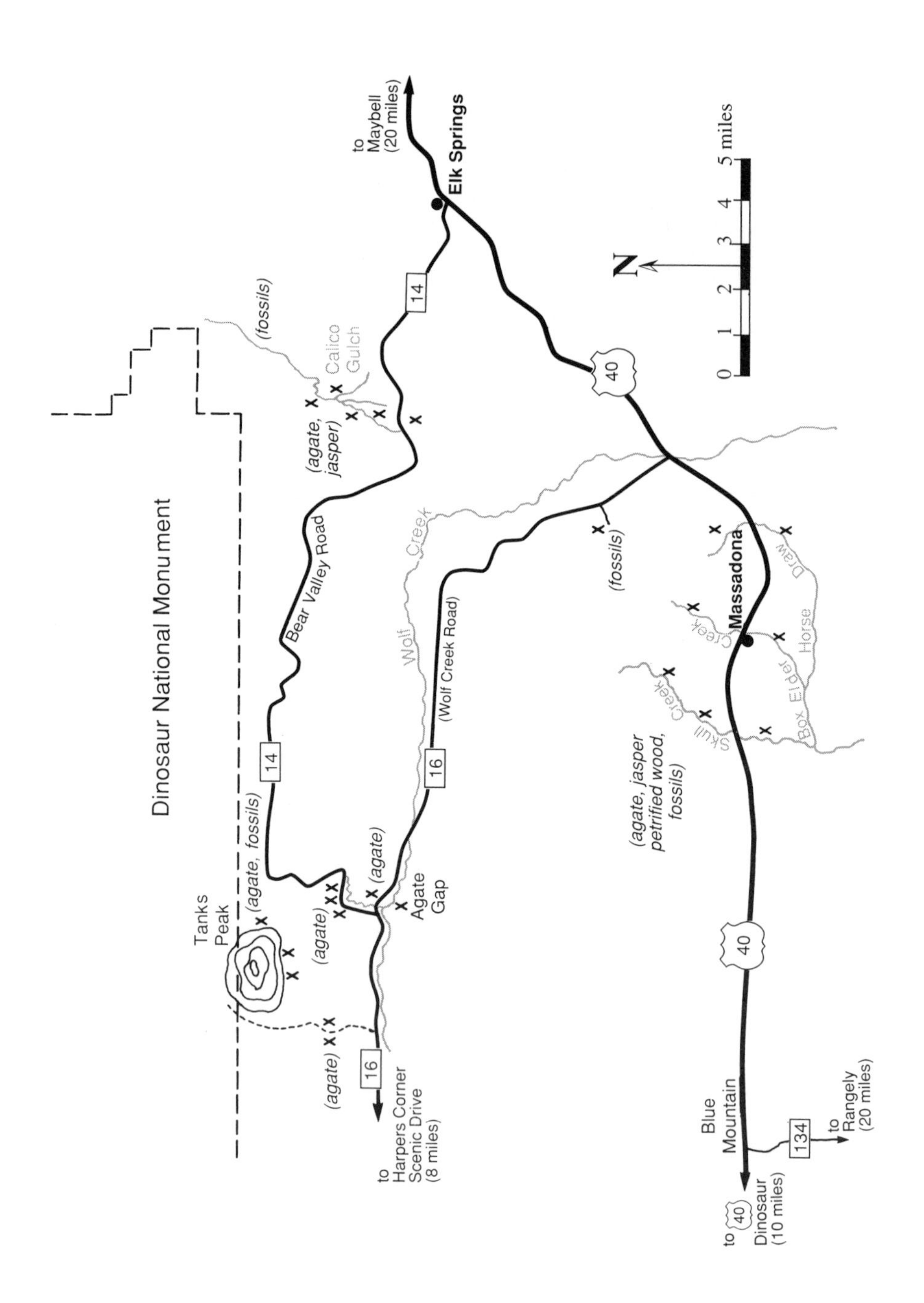
to
Maybell
(20 miles)
Elk Springs
14
(fossils)
Calico
Gulch
(agate,
jasper)
N
0 1 2 3 4 5 miles
40
Dinosaur National Monument
Bear Valley Road
Wolf
Creek
(Wolf Creek Road)
(fossils)
Massadona
Draw
Horse
Box Elder
Creek
Skull
Creek
(agate, jasper
petrified wood,
fossils)
14
16
(agate, fossils)
Tanks
Peak
(agate)
(agate)
Agate
Gap
(agate)
16
to
Harpers Corner
Scenic Drive
(8 miles)
40
Blue
Mountain
134
to
Rangely
(20 miles)
to 40
Dinosaur
(10 miles)

Naturally reliefed crinoid fossils in Jurassic limestone from Wolf Creek.

There are several fossil-rich Morrison Formation exposures along Wolf Creek. Follow County Road 16 north for 2.1 miles, then take the unmarked dirt road west for almost one-half mile and park. Cross the gullies and climb to the prominent high exposures of shales and sandstones 300 yards to the north. Fossils of pelecypods retaining original shell material are common on the slopes. Also present are naturally reliefed fossil crinoids that make nice display specimens. Plant-stem fossils indicate that the sediments within the 100-foot-high exposures are a record of both marine and shoreline environments. Higher on the cliffs, large rust-colored fossilized dinosaur bones protrude from shale layers. Care must be exercised, for the slopes are steep and the cliffs are unstable. County Road 16 continues north and west for over twenty miles, passing many hills and washes where jasper and agate are locally abundant. A topographical map is useful in exploring the area.

Agate Gap and Agate Gulch, about fourteen miles northwest of U.S. 40 along County Road 16, were named for the agate and yellow jasper found nearby on both sides of Wolf Creek. In early literature, settlers referred to Agate Gulch as an "Indian flint mine."

Just beyond Agate Gap and Agate Gulch is the intersection of County Roads 16 and 14. Agate and jasper are common on the hills and in the washes along County Road 14 for one mile north of the intersection.

Three miles farther on County Road 16, Johnson Draw Road leads north for just over a mile to a small flat with scattered agate and jasper. Bear right up Johnson Draw (four-wheel-drive only) for two more miles to Tank Peak. The area immediately to the north is Dinosaur National Monument, where collecting is prohibited.

TANK PEAK

The south and east slopes of Tank Peak have an unusual, banded orange-white agate. Morgan Formation exposures on Tank Peak contain many Pennsylvanian brachiopod and trilobite fossils.

MASSADONA AREA

Massadona is eighteen miles west of Dinosaur on U.S. 40. The drainages of Box Elder Creek, Horse Draw, and Skull Creek north of Massadona have yielded many specimens of agate and jasper, some petrified and opalized wood, and various marine fossils, including belemnites.

REFERENCES: 6, 32, 63, 64, 65, 66, 67, 73, 74, 90

DINOSAUR NATIONAL MONUMENT

Dinosaur National Monument, straddling the border of northwestern Colorado and Utah, is one of the world's most remarkable dinosaur-fossil deposits. Monument attractions include a partially excavated "wall" of in-situ fossilized dinosaur bones, a working paleontological laboratory, and 328 square miles of colorful canyonlands and primitive wilderness created by 550 million years of sedimentary deposition and erosion.

In 1907, steel tycoon Andrew Carnegie sent paleontologist Earl Douglass west to find "something big" to fill the huge new Carnegie Exposition Hall in Pittsburgh. In August 1909, in exposures of Morrison Formation sandstone near the Green River, Douglass discovered eight large fossilized vertebrae from a 35-ton, 145-million-year-old *Apatosaurus*. Douglass took six years to excavate and reconstruct the skeleton. Fifteen feet tall at the hip and eighty feet long, it is still the most complete *Apatosaurus* skeleton ever found.

Douglass worked the site for fifteen years, excavating and cataloging the remains of ten species of adult and rare juvenile Jurassic dinosaurs. By 1924, Douglass had shipped 350 tons of fossilized bones to the Carnegie Museum, the Smithsonian Institution, and the University of Utah. Most were of sauropods, such as *Apatosaurus* and the spined *Stegosaurus*, but the quarry also yielded three types of carnivores, including *Allosaurus*, a three-ton, thirty-foot-

Articulated dinosaur vertebrae exposed in relief at the quarry wall at Dinosaur National Monument.

long predator. Recoveries included fossils of crocodiles, turtles, and small dinosaurs no larger than a modern collie.

Douglass initially filed mineral claims to protect his site from homesteaders and commercial bone hunters. But the government disallowed the claims, declaring that fossils were not "mineral" in the intended sense of the Mining Law. Congress finally protected the site in 1915, designating eighty acres as a national monument. Lack of funds ended Douglass's field work in 1924. In 1933 paleontologists widened and deepened the quarry, but left the main bone layer protected under a ten-foot-thick layer of sandstone. Dinosaur National Monument attained its present size when Congress added 326 square miles of adjacent eastern canyonlands in 1938.

Excavation resumed in 1953 and paleontologists uncovered the fossil-rich layer, but left the bones in place as a permanent exhibit. An exhibition building was completed in 1958, with the steeply tilted bone layer forming one entire wall. The Dinosaur Quarry Building, an imposing structure of glass and steel, encloses the 30-foot-high, 200-foot-long fossil-rich sandstone layer. The cliff face is excavated to "high relief," exposing over 2,200 bones from an estimated 200 different dinosaurs.

Paleontologists call the quarry wall a "snapshot in time" that

Street signs in Dinosaur, Colorado.

provides an extraordinary overview of Jurassic dinosaur life. The "snapshot" was taken about the midpoint of the 180-million-year-long existence of the dinosaurs. The quarry wall was then a sandbar in a winding river that trapped dinosaur carcasses and bones carried by the current. River sand and gravel quickly covered many bones, preventing normal organic decay. Long periods of floodplain deposition buried and compressed the sandbar, and seepage of silica-laden water eventually cemented the sand into sandstone and fossilized the bones.

Tens of millions of years of sedimentary deposition covered the sandbar to a depth of one mile. The uplift of the modern Rockies raised and steeply tilted the sediments. In relatively recent geologic time, erosion uncovered that ancient Jurassic sandbar with its concentration of fossilized dinosaur bones.

The quarry wall contains three distinct fossil layers, each representing brief periods when water level, current, and silting maximized bone deposition, preservation, and eventual fossilization. Paleontologists removed the first, uppermost layer in the 1950s and are now reliefing the third and deepest layer, which contains the greatest number of bones.

The laboratory, open to public view, has preservation and restoration facilities, a paleontological library, such specialized equipment as microscopes for studying microfossils, and a collection of thousands of quarry bones, including a nearly intact, four-foot-long skull of the predator *Allosaurus*.

With only 10 percent of Dinosaur National Monument fully surveyed, paleontologists have found 110 sites with visible, in-situ vertebrate fossils. National Park Service paleontologists ask that visitors leave fossils in place and report new finds immediately. Collecting within the monument boundaries is prohibited.

The Dinosaur Quarry Building is reached from Jensen, Utah. The interior of the national monument is reached from Dinosaur, Colorado, by following thirty-mile-long Harper's Corner Scenic Drive. Harper's Corner, a high promontory 2,500 feet above the Green River and Yampa River canyons, is Pennsylvanian-aged Morgan Limestone studded with thousands of naturally reliefed, clearly visible 300-million-year-old fossil brachiopods and crinoid stems.

In older collecting literature, the town of Dinosaur, Colorado, is noted as Artesia. The name was changed to Dinosaur in 1965 to enhance the local tourism potential. Streets are named after dinosaurs, and three large models of dinosaurs "roam" the local parks.

Further information about Dinosaur National Monument is available from the Headquarters Visitor Center, P.O. Box 210, Dinosaur, Colorado 81610 (303)374-2216.

REFERENCES: 63, 64, 68, 90

THE MOFFAT COUNTY "DIAMOND FIELDS"

Moffat County is the site of "the Great Diamond Hoax," a notorious frontier gemstone-mining fraud that made headlines across the country. Although it happened over a century ago, it's a story that will interest mineral and gemstone collectors.

The 1860s and early 1870s saw many exciting foreign discoveries of diamonds, rubies, and sapphires. It seemed the United States had a precious gemstone bonanza of its own in 1872 when two prospectors arrived at the Bank of California in San Francisco. Philip Arnold and Jack Slack, cousins and native Kentuckians, deposited a canvas bag containing the purported fruits of their latest prospecting venture—uncut diamonds and rubies. A bank officer dutifully alerted bank president William C. Ralston, one of San Francisco's top investment bankers.

Two San Francisco jewelers pronounced the stones genuine and of excellent quality. Ralston, alert for mine speculation opportunities, informed Arnold and Slack that their discovery had great potential and he could arrange the necessary financial backing. But as a prudent financier, he insisted that two of his trusted associates first inspect the gemstone site.

Arnold and Slack took Ralston's men on a thirty-six-hour eastbound train journey, then blindfolded them and set off again on

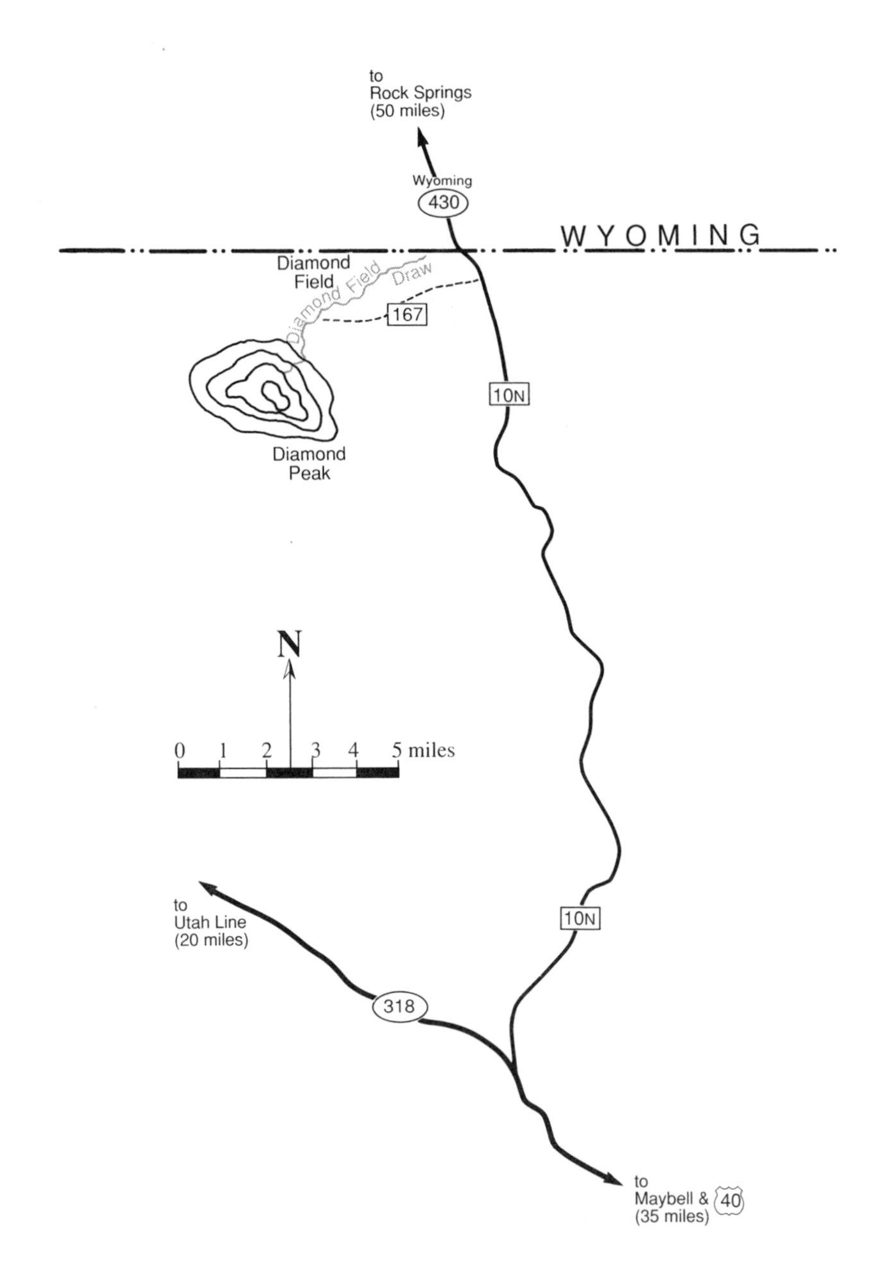
to
Rock Springs
(50 miles)
Wyoming
430
WYOMING
Diamond
Field
Diamond Field Draw
167
10N
Diamond
Peak
N
0 1 2 3 4 5 miles
to
Utah Line
(20 miles)
10N
318
to
Maybell & 40
(35 miles)

mules. Days later they reached the site and Ralston's men greedily scrambled about filling a sack with rough diamonds, rubies, and sapphires. Back in San Francisco, they dumped 7,000 carats of rubies and 1,000 carats of diamonds onto William Ralston's desk.

Ralston's jewelers appraised the stones at "at least $125,000." Wary of salted mine frauds, the cautious Ralston sent the stones to Samuel Barlow, his New York attorney. Barlow personally took them to Charles Lewis Tiffany, founder and president of the nation's most prestigious jewelry firm, who appraised them at $150,000.

Next, Ralston hired respected San Francisco mining engineer Henry Janin to inspect the gemstone site. The engineer proclaimed the gem fields authentic and offered two professional estimates: First, that the "dirt" contained $5,000 in gemstones per ton; second, that a crew of twenty-five men could wash out a million dollars' worth every month.

Convinced, William Ralston founded the San Francisco and New York Mining and Commercial Company, funding it with $2 million raised from twenty-five personal friends and wealthy contacts. Notable investors included General George McClellan, the Civil War general and unsuccessful presidential candidate, and Senator Ben Butler, who immediately began cutting through federal red tape to speed acquisition of the site.

The story hit the newspapers from San Francisco to London. In Wyoming Territory, the *Laramie Daily Independent* called the discovery site "The Great Diamond Fields of America." A dozen other companies quickly incorporated and raised another $2 million to get in on the action.

Meanwhile, Ralston's elite circle of investors, eager "to get rid of those two Kentucky bumpkins," bought out Philip Arnold and Jack Slack for $600,000 in cash and a percentage of future profits. Arnold and Slack took the money and promptly vanished.

A thirty-year-old government geologist named Clarence King was next to appear in the unfolding drama. As a United States geologist, King had headed the Fortieth Parallel Survey in 1867, compiling cartographic records and detailed reports on everything from mineralization to vegetation within a 100-mile-wide, 1,000-mile-long survey tract along the proposed transcontinental rail route.

In 1872, King was deeply troubled by the purported discovery of "The Great Diamond Fields of America." King knew that simultaneous natural occurrence of diamonds and rubies was improbable, and wondered why relatively dense gemstones would concentrate on the surface, and not in deep placers. King even suspected the site was possibly within the limits of his own Fortieth Parallel Survey. That was particularly disturbing, for King's published survey report indicated no precious gemstone occurrences.

King tracked down mining engineer Henry Janin, who told of his train journey to southwestern Wyoming Territory, his long blindfolded mule trek, and how he had found the gemstones below a prominent flat-topped mountain. Most interesting were Janin's descriptions of topography, vegetation, and geology. "From my knowledge of the country," King later wrote, "there was only one place that answered that description . . . that place lay within the limits of the Fortieth Parallel Survey."

King traveled by train to Rawlings Springs, near Green River, Wyoming, in October 1872, then followed a circuitous 150-mile-long route into northwestern Colorado Territory. Finally, King knew he had found "The Great Diamond Fields of America" when he saw a sign nailed to a tree. Claiming water rights from a nearby wash, it was signed "Henry Janin."

Gemstones were everywhere, King noted with a smile, especially in crevices that had apparently been made with sharp instruments. Later investigation revealed the diamonds were African rejects called "niggerheads," which Arnold and Slack had purchased in London and Amsterdam for practically nothing. King even found one "rough" diamond partially faceted. Some of the rubies were garnets; others were real—Burmese rejects. The sapphires were Ceylonese rejects.

As newspapers headlined the "hoax," Henry Janin trekked back to the "diamond fields" to confirm his own folly. A red-faced Charles Lewis Tiffany admitted he was inexperienced with uncut diamonds and had grossly overestimated their value. William C. Ralston, the butt of many jokes, repaid his investors and absorbed the entire loss himself. By 1875, the diamond hoax, along with other bad investments, helped send his Bank of California into temporary insolvency. The day after the board of directors demanded his resignation, Ralston's body was pulled from San Francisco Bay.

Private detectives tracked down Philip Arnold in Elizabethtown, Kentucky, where he had used his share of the profits to buy his own bank. Although Kentucky refused to extradite, Arnold returned $150,000 to California in return for immunity from further prosecution.

Jack Slack was never found. He took up a quiet career as an undertaker in White Oaks, New Mexico, where he died in 1896.

Clarence King emerged from the Great Diamond Hoax as nearly a folk hero, earning the everlasting gratitude of the San Francisco millionaires he had saved from further swindle and embarrassment. When Congress established the United States Geological Survey in 1879, Clarence King, a mere field geologist, was hardly considered for the post of director. But a host of prominent San Franciscans were eager to repay their old debt with powerful political clout. They succeeded, and Clarence King was appointed the first director of the United States Geological Survey.

The location of "The Great Diamond Fields of America" is in northwestern Moffat County, just one mile south of the Wyoming line and ten miles east of the Utah line. The Great Diamond Hoax is memorialized on topographical maps with place names like Diamond Peak, Diamond Wash Draw, and Diamond Field.

From Maybell and U.S. 40, follow Colorado Route 318 northwest for thirty-four miles. Take Moffat County Road 10N north for eighteen miles to a point one mile south of the Wyoming line. Follow Road 167 three miles west to Diamond Wash Draw. The flat-topped mountain to the south is Diamond Peak, and Diamond Field, a one-square-mile plain, is just to the north.

If you do find happen to find rough diamonds, rubies, and sapphires, add them to your collection, but only as mementos of the West's greatest gemstone fraud.

REFERENCES: 88

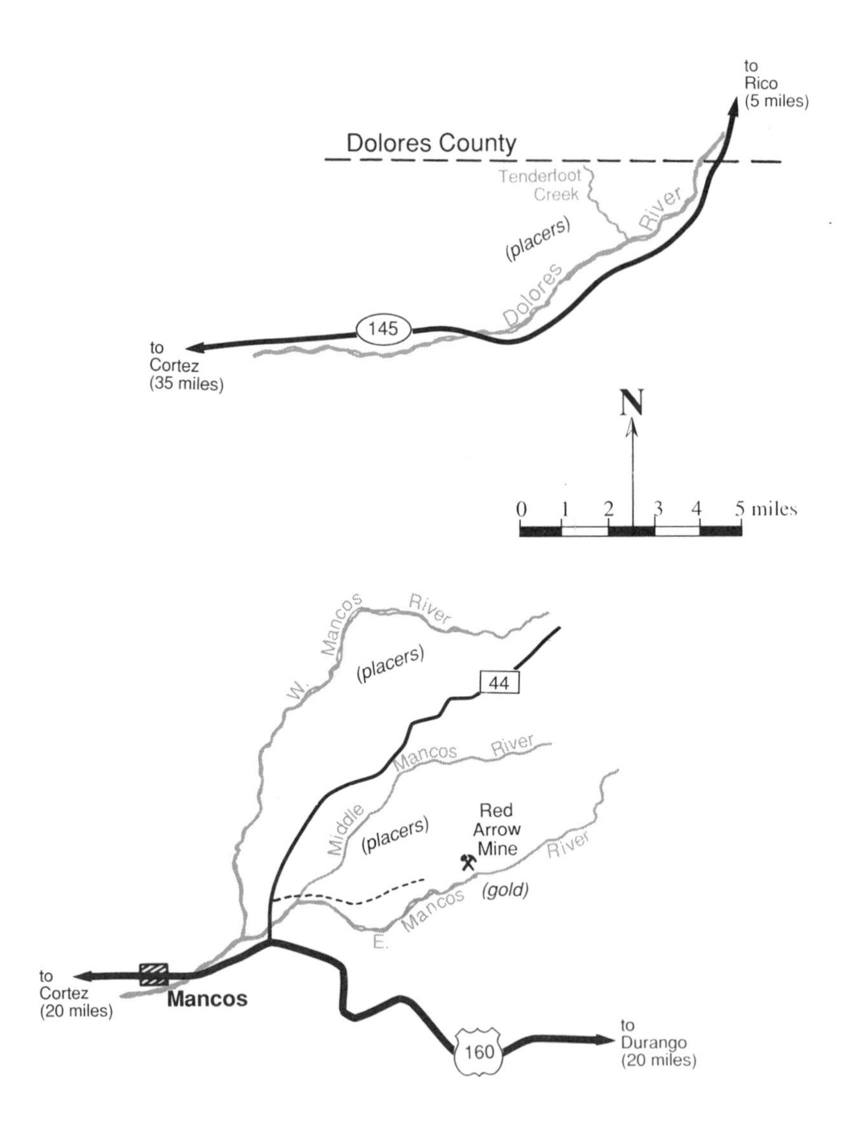

to
Rico
(5 miles)
Dolores County
Tenderfoot
Creek
River
(placers)
Dolores
145
to
Cortez
(35 miles)
N
0 1 2 3 4 5 miles
W. Mancos River
(placers)
44
Middle Mancos River
(placers)
Red
Arrow
Mine
River
(gold)
E. Mancos
to
Cortez
(20 miles)
Mancos
160
to
Durango
(20 miles)

MONTEZUMA COUNTY

PLACER GOLD

Montezuma County has produced about 100 troy ounces of placer gold, most from the upper East, Middle, and West Mancos rivers, which head in the La Plata Mountains on the Montezuma-La Plata county line. County Road 44 leads into the Mancos River placer districts. Miners recovered lesser amounts of placer gold near the confluence of Tenderfoot Creek and the Dolores River, just south of the Montezuma-Dolores county line.

REFERENCES: 54

THE RED ARROW LODE GOLD STRIKE

Although Montezuma County produced little placer gold, the search for it led to Colorado's last and perhaps most unexpected bonanza lode strike. During the Depression, the imminent revaluation of gold sent otherwise unemployed prospectors back into the gold fields. The Starr brothers, Raymond and Charles, prospected Gold Run, a small tributary of the East Mancos River, where miners had recovered small amounts of coarse placer gold decades earlier. No one had ever found nor expected to find lode gold in the Entrada Sandstone country rock.

But on June 3 1933, near the mouth of Gold Run, the Starrs found an encouraging pocket of placer gold. Following the trail of coarse gold flakes upstream, they discovered a reddish fissure that "shot" across the gully like a "red arrow." Along the fissure, the Starrs found a mineralized vein outcrop heavily stained with malachite and azurite. They crushed and washed samples, and found their pan concentrates glittering with gold.

The Starrs staked claims and recovered 200 troy ounces of gold from the first three tons of loose ore scraped from the surface. Extravagant newspaper coverage, aimed largely at encouraging other Depression-era prospectors, made the Starrs folk heroes. Within six years, thanks to ores worth as much as $30 per pound, their Red Arrow Mine produced $400,000 in gold. Even after the oxidation-enriched bonanza surface ores were gone, overall ore grades averaged seven troy ounces of gold per ton.

The Red Arrow, with a half mile of workings on three levels, operated sporadically until 1950. There is hope that the mine may produce again.

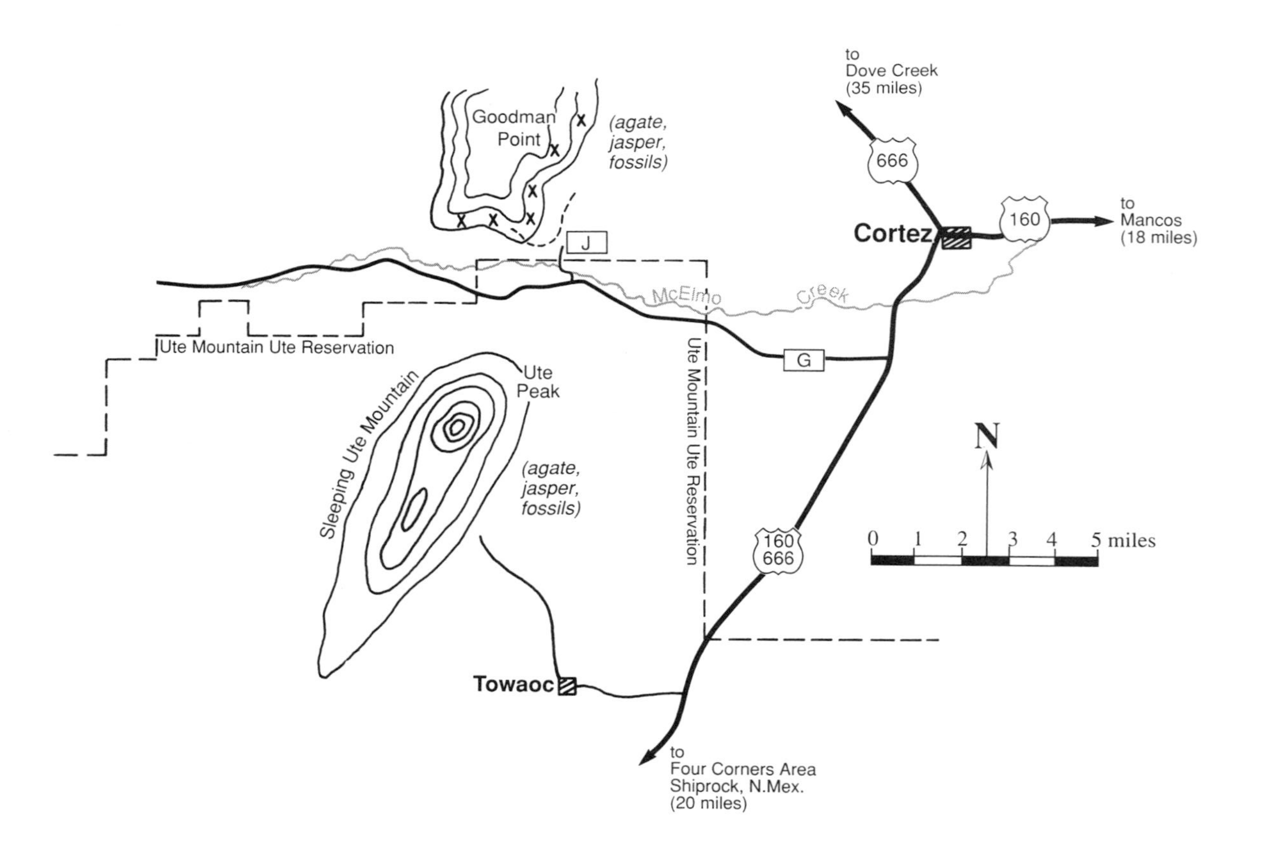
Goodman
Point
(agate,
jasper,
fossils)
to
Dove Creek
(35 miles)
666
Cortez
160
to
Mancos
(18 miles)
J
McElmo
Creek
Ute Mountain Ute Reservation
G
Ute
Peak
Sleeping Ute Mountain
(agate,
jasper,
fossils)
Ute Mountain Ute Reservation
N
0 1 2 3 4 5 miles
160
666
Towaoc
to
Four Corners Area
Shiprock, N.Mex.
(20 miles)

Although the Red Arrow is not a collecting site, it is notable for its spectacular specimens of native gold and silver, many of which were sold to collectors. Vein pockets contained "sheets" of gold weighing up to 80 troy ounces, or about one-half pound. Many native gold specimens were distinctively "antiqued" with encrustations of dark chalcocite and green malachite.

The Red Arrow Mine is located about nine miles northeast of Mancos.

REFERENCES: 25, 54, 108

MONTEZUMA COUNTY AGATE AND FOSSILS

McELMO CANYON AND GOODMAN POINT

Agate, jasper, fossils, and petrified wood are locally abundant in Montezuma County. The slopes of Goodman Point, a prominent sandstone mesa eight miles west of Cortez, are known for delicately banded light blue agate, yellow and brown jasper, and fossils. The fossils are ammonites and pelecypods that weather free from underlying exposures of Mancos Shale.

From Cortez, follow U.S. 160-666 south for three miles, then take G Road (McElmo Road) west for seven miles into McElmo

Goodman Point towers over McElmo Canyon west of Cortez. The slopes of Goodman Point are a source of agate, jasper, and fossils.

Canyon. Turn north on J Road, cross McElmo Creek, and proceed toward the steep Dakota Sandstone slopes of Goodman Point. Permission should be obtained to cross the posted private land to reach the slopes.

Complex land ownership should be checked before collecting, for private land separates the irregular boundary of the Ute Mountain Ute Reservation to the south and BLM land to the north. The Goodman Point area also has protected archaeological ruins and, not far to the north, noncontiguous tracts of Hovenweep National Monument where all collecting is prohibited.

Collectors will enjoy a visit to San Juan Gems, the region's largest and most complete lapidary, jewelry, and rock shop, located two miles north of Cortez on Colorado Route 145. San Juan Gems has an extensive stock of Montezuma County agate, jasper, geodes, minerals, petrified wood, and dinosaur bones.

UTE PEAK

Agate, jasper, and fossils also occur on the slopes of Ute Peak, the "folded arms" portion of a prominent regional landmark known as Sleeping Ute Mountain. Sleeping Ute Mountain is a Tertiary laccolith of resistant volcanic rock containing thin seams of a light blue, banded agate once mined by Indians for tool and projectile points. The once-overlying Mancos Shale has completely eroded away, freeing many marine fossils.

Sleeping Ute Mountain is within the Ute Mountain Ute Reservation and is reached by roads from Towaoc, the Ute Mountain Agency, ten miles south of Cortez.

Nontribal members may not collect on the reservation without express permission of the Ute Mountain Ute Tribal Council in Towaoc. In the past, the requests of hopeful collectors have been denied.

REFERENCES: 12, 65, 66, 67, 68

MONTROSE COUNTY

PLACER GOLD

Montrose County has produced 3,000 troy ounces of placer gold from the Dolores and San Miguel rivers. The San Miguel River contains placer gold for 75 miles downstream from its source near Telluride (San Miguel County).

Montrose County's two commercial placer-mining districts are on the San Miguel River from the confluence of Cottonwood Creek downstream to Naturita, and on the Dolores River from the confluence of the San Miguel River downstream to Roc Creek.

Determination to mine the mouth of Mesa Creek inspired Colorado's most ambitious placer-mining water-diversion project. In 1890, engineers constructed the famed $100,000 "Hanging Flume" through the Dolores Canyon. Six miles of the flume is suspended by brackets on the side of a sheer cliff 150 feet above the river. The flume carried 80 million gallons of water daily on a perfect grade of six feet to the mile. Although the Hanging Flume was rightfully acclaimed an engineering marvel, placer mining never returned construction costs. The Hanging Flume may still be seen along Colorado Route 141 five miles west of Uravan.

REFERENCES: 54, 108

THE URAVAN MINERAL BELT

The Uravan Mineral Belt crosses western Montrose County. Over 400 uranium and vanadium mines are grouped in three general areas: west and south of Uravan, south of Paradox Valley on the San Miguel County line, and along the Utah border.

The radium used by French physicist Madame Marie Curie in her landmark radiation experiments was obtained from high-grade carnotite mined from a site near Roc Creek. Carnotite and other radioactive minerals can be found on mine dumps and along ore haulage roads. Western Montrose County's many reminders of the 1950s uranium rush include Naturita's Yellow Rock Cafe, the old Uranium Drive-In outdoor movie, and the Rimrock Historical Museum, which has displays of radioactive ores, relics from the uranium rush, as well as local fossils.

REFERENCES: 10, 18, 81, 94

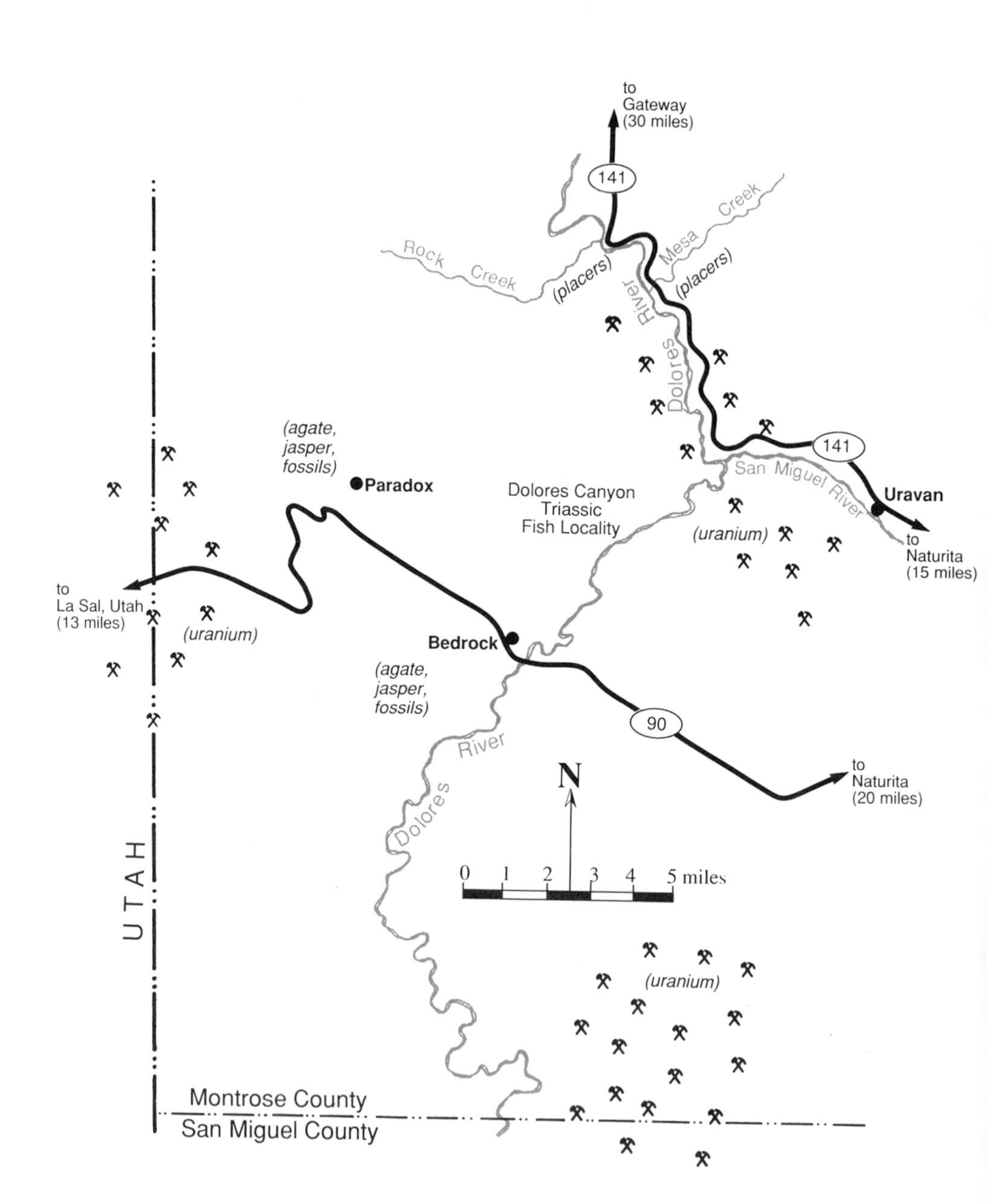

to
Gateway
(30 miles)
141
Rock Creek
Mesa Creek
(placers)
(placers)
River
Dolores
(agate, jasper, fossils)
Paradox
Dolores Canyon
Triassic
Fish Locality
141
San Miguel River
Uravan
(uranium)
to
Naturita
(15 miles)
to
La Sal, Utah
(13 miles)
(uranium)
Bedrock
(agate, jasper, fossils)
90
River
Dolores
to
Naturita
(20 miles)
N
0 1 2 3 4 5 miles
UTAH
(uranium)
Montrose County
San Miguel County

PARADOX VALLEY AGATE AND FOSSILS

Twenty-five-mile-long Paradox Valley, an inverted salt anticline, is a true geologic paradox, for the Dolores River flows across the valley rather than through it. Banded agate and jasper is locally common near Bedrock and northwest of Paradox.

The Paradox Valley is also known for superb fish fossils, some of which are displayed at the Rimrock Historical Museum in Naturita. The Bureau of Land Management has recently designated the Dolores Canyon Triassic Fish Locality, one of North America's most important freshwater fish fossil sources, a research natural area. Articulated fish fossils occur within a 150-foot-thick section of the Dolores Formation. Shifting Triassic stream channels deposited sediments that trapped fish in backwaters, then covered them with mud. Deposits yield reptilian fossils, including phytosaurs, or armored crocodiles. The dune sands of the overlying Wingate Formation contain trackways of Jurassic bipedal dinosaurs.

Paleontologists from the University of Colorado and the USGS are currently studying the Dolores Canyon Triassic Fish Locality. For further information about the site, contact the Montrose District Office of the Bureau of Land Management.

REFERENCES: 3, 67

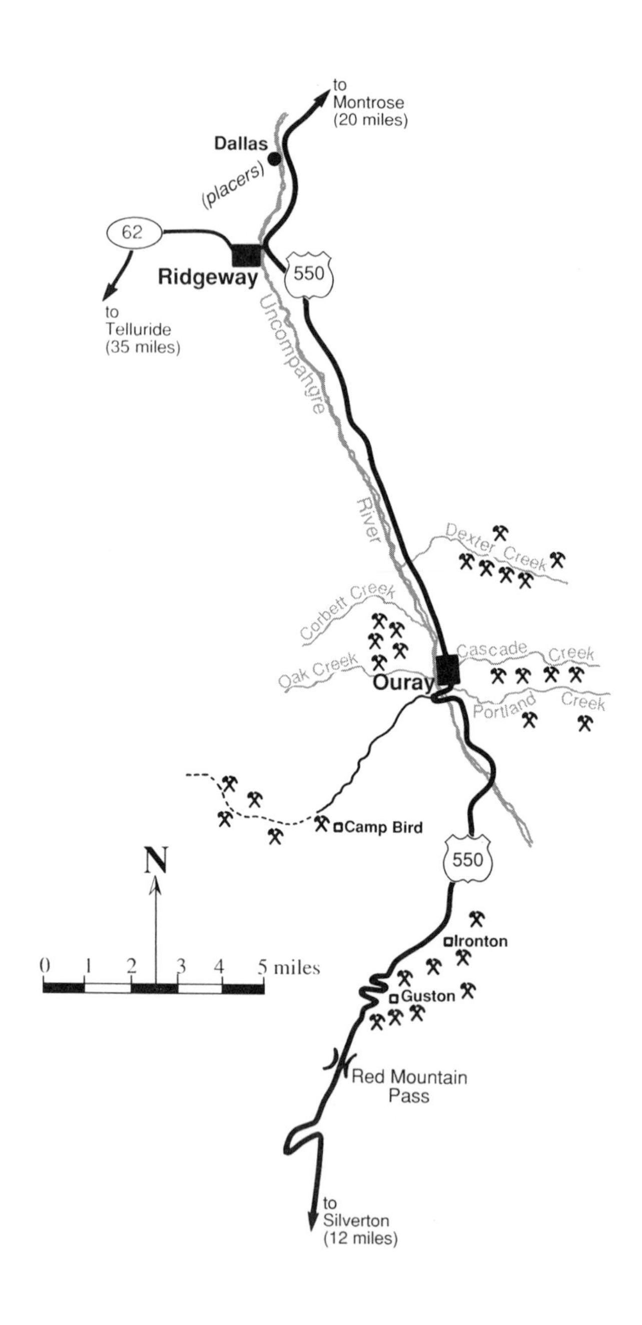
to
Montrose
(20 miles)
Dallas
(placers)
62
Ridgeway
550
to
Telluride
(35 miles)
Uncompahgre
River
Dexter Creek
Corbett Creek
Cascade Creek
Oak Creek
Ouray
Portland Creek
Camp Bird
N
550
0 1 2 3 4 5 miles
Ironton
Guston
Red Mountain
Pass
to
Silverton
(12 miles)

OURAY COUNTY

PLACER GOLD

Although Ouray County has extensive lode mineral deposits, it is only a minor source of placer gold. Total production, less than 200 troy ounces, has come from the Uncompahgre River near Dallas and Ridgway.

REFERENCES: 54

OURAY COUNTY MINES

Prospectors discovered mineralized vein outcrops near the present site of Ouray in 1874, a year after the Brunot Treaty legalized mineral entry. Inability to identify and treat complex ores, inaccessibility, and high elevations delayed mine development for nearly a decade. The camps of Ouray, Ironton, and Red Mountain finally boomed in the 1880s, thanks largely to big producers like the Virginius, Yankee Boy, Governor, National Belle, and Revenue mines. Many mines closed after the 1893 silver-market crash, but reopened after 1900 to produce copper and lead.

Ouray, one of Colorado's most picturesque towns, is surrounded by about 1,000 mines and mineral prospects.

By 1924, Ouray County had produced $100 million in gold, silver, copper, lead, and zinc. Today, with few mines operating, total production stands at a quarter-billion dollars.

There are over a thousand old mines and prospects in southern Ouray County. Ervan Kushner, in his 1972 *Guide to Mineral Collecting at Ouray, Colorado*, divided the mining areas of Ouray County into regional groups: Dexter Creek northeast of Ouray; the Cascade and Portland Creek areas east of Ouray; the Red Mountain-Guston-Ironton area on U.S 550 south of Ouray; the heights between Corbett and Oak Creeks west of Ouray; and the large, Sneffels-Camp Bird timberline area southwest of Ouray along County Road 361 (Camp Bird Road). Many roads lead to the mining districts; four-wheel-drive is often necessary.

Mineralization varies in individual veins, but primary ore minerals are galena, argentite, sphalerite, chalcopyrite, pyrite, arsenopyrite, tetrahedrite-tennantite, native gold and some native silver, azurite, and malachite. The most abundant gangue mineral is quartz, and Ouray County mines are known for museum-grade specimens of clear and white quartz crystals. Rhodochrosite and rhodonite also occur, along with fluorite, barite, calcite, garnet, and some epidote.

Ouray has a fine local museum with exhibits of mining equipment and local minerals and ores. The Bachelor-Syracuse Mine on Dexter Creek offers underground tours into a turn-of-the-century silver and gold producer. The Columbine Rock Shop and The Sandman, both on Main Street in Ouray, display fine collections of Colorado and Ouray County minerals.

THE CAMP BIRD MINE

The Camp Bird Mine is a classic rags-to-riches mining story. Two English prospectors discovered the Camp Bird Vein in 11,500-foot-high Imogene Basin, six miles southwest of Ouray, in 1877. The vein seemed to contain only low-grade copper-silver-lead mineralization, but they managed to sell the two discovery claims, the Gertrude and the Una, to the Allied Mines Company in 1881 for $40,000.

Allied drove a short exploratory crosscut, then fell into bankruptcy. Miners drove the last twelve feet in 1884, apparently to fulfill patent requirements. In one of the great mistakes of Colorado mining, those last twelve feet were never properly sampled, and the claims lay idle for twelve years.

In 1896, part-time prospector Thomas F. Walsh picked up the claims for back taxes. He sampled the old dumps below the exploration tunnel and found traces of gold. Walsh then scraped up enough money to hire a miner to sample the old crosscut. Assays revealed only low-grade lead-copper-zinc, leaving Walsh puzzled over the low-grade gold in the dump samples.

Although sick with fever, Walsh climbed to Imogene Basin to personally sample the crosscut. Alongside the copper-lead-zinc vein he found a quartz vein, seemingly barren except for tiny specks of a blackish crystalline mineral. Again, Walsh took samples.

Walsh lay sick in bed when the results came back. Walsh studied them quietly, then summoned his six-year-old daughter, Evalyn, to his side.

"You must keep a secret I'm going to tell you, promise?"

"Yes, Papa," the little girl answered.

"I have the reports on the samples I took up at the Gertrude," Walsh confided, "and they run as high as three thousand dollars of gold per ton. Daughter, I've struck it rich."

Walsh quietly bought or staked all the ground near the vein, then announced his discovery. He took no partners and sold no shares, instead borrowing development capital based on uncontested property control. The Camp Bird Mine, named for the gray jay, the familiar high-country "camp robber," reached full production in 1899 with a $5,000 daily profit.

By 1902, the Camp Bird had produced 200,000 troy ounces of gold worth $4 million, a record 60 percent of which was clear profit. The Camp Bird was the biggest, richest gold mine ever owned by one individual. In 1902, Walsh sold out for shares, future production royalties, and $3.5 million in cash. His daughter, Evalyn, went on to achieve international notoriety as the wealthy but troubled heiress who used part of the Camp Bird fortune to purchase the Hope Diamond.

By 1916, the Camp Bird had produced over 1 million troy ounces—sixteen tons—of gold, earning a clear profit of $15 million. Cumulative production now exceeds 1.5 million troy ounces of gold, along with large amounts of silver, lead, copper, and zinc.

The Camp Bird Mine has yielded many fine mineral specimens, most notably gold-in-quartz and superb milky quartz crystals. The mine is currently inactive and on caretaker status, and the dumps are not open to collecting. Nevertheless, a visit to the mine along Camp Bird Road—for those who don't mind heights and sheer drop-offs—is worth it for the drive alone. Four-wheel-drive is required beyond the Camp Bird Mill.

REFERENCES: 18, 19, 23, 25, 32, 38, 64, 66, 67, 74, 81, 108

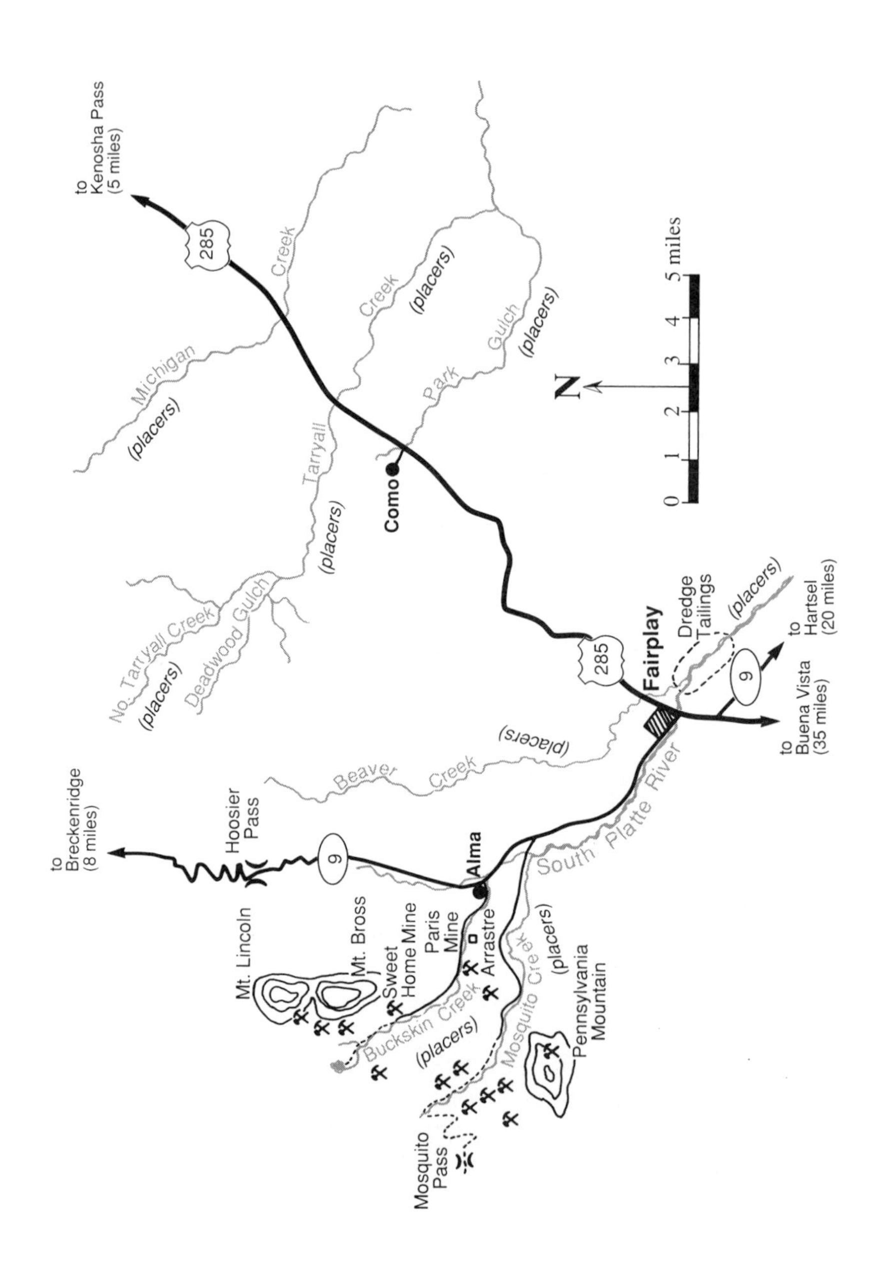

to Kenosha Pass (5 miles)
285
Michigan
(placers)
Creek
Tarryall
(placers)
Creek
(placers)
Park
Gulch
(placers)
Como
N
0 1 2 3 4 5 miles
No. Tarryall Creek
(placers)
Deadwood Gulch
285
Fairplay
Dredge Tailings
(placers)
to Hartsel (20 miles)
9
to Buena Vista (35 miles)
Beaver Creek
(placers)
South Platte River
to Breckenridge (8 miles)
Hoosier Pass
9
Alma
Mt. Lincoln
Mt. Bross
Sweet Home Mine
Paris Mine
Arrastre
Buckskin Creek
(placers)
Mosquito Creek
(placers)
Pennsylvania Mountain
Mosquito Pass

PARK COUNTY

PLACER GOLD

Park County has produced 350,000 troy ounces—fourteen tons— of placer gold, third among Colorado counties. The upper South Platte River drainage has over 100 miles of gold-bearing creeks.

Prospectors discovered placer gold on Tarryall Creek in the summer of 1859 and within weeks founded the boom camps of Tarryall, Fairplay, Buckskin Joe, Sterling City, and Montgomery. By 1867, Park County placers had yielded over $2 million in gold. Mining eventually employed every recovery method from panning and sluicing to ground sluicing, hydraulicking, draglining, and, most importantly, dredging.

Colorado's largest and last floating-bucketline dredge, the South Platte Dredging Company's Dredge No. 1, known as the "Fairplay Dredge," began working the South Platte River gravels south of Fairplay in 1941. Its 150-foot-long steel hull drew 9 feet of water, and 105 13-cubic-foot buckets took down 35-foot-high banks and excavated 70 feet below waterline. In its first year, the dredge recovered a quarter-million dollars in gold.

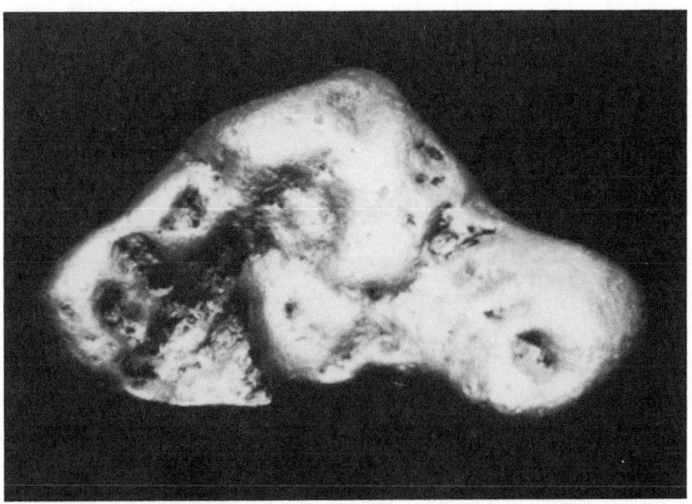

The Penn Hill Nugget, at 11.95 troy ounces, is Colorado's largest known gold nugget. It was mined on Pennsylvania Mountain in 1937.

When the Fairplay Dredge shut down in 1951 after seven years of operation, it had recovered 120,000 troy ounces of gold worth over $3 million. The Fairplay Dredge was dismantled in 1980, but miles of conical tailing heaps still cover the South Platte River's bottom.

Pennsylvania Mountain, four miles southwest of Alma, has North America's highest commercial gold placers. The placers are eluvial, formed from quartz-gold outcrops that weathered away in place without alluvial movement.

The 12,500-foot-high placers have produced Colorado's two largest known gold nuggets. Miners found the 11.95-troy ounce "Penn Hill Nugget," Colorado's largest, in 1937. A small-scale commercial placer miner recovered Colorado's second largest nugget, the 8.75-troy ounce, distinctively shaped "Turtle Nugget," in 1990. Both nuggets are displayed at the Denver Museum of Natural History.

The Pennsylvania Mountain placers are under claim and worked during the summer. Trespassing is prohibited. Recreational gold panning and mining is popular along many other Park County creeks.

REFERENCES: 23, 53, 108

ALMA AREA MINES

In 1860, prospectors discovered lode gold outcrops above Buckskin Gulch and built *arrastres* to crush the high-grade quartz ores. By 1870, miners found silver mineralization, but high elevations, inaccessibility, and unsuitable smelting methods hindered development. In 1873 Alma emerged as the district supply and transportation center.

Major mineralized zones include the London Fault, a highly mineralized geologic fault in upper Mosquito Gulch, and the Buckskin Gulch stockwork, with both vein and replacement-type ores in porphyry and limestone.

Alma boomed in the 1880s. Mines above 11,500 feet used aerial tramways or burro pack trains to transport ore. By 1924, the London, Moose, Fanny Barrett, Hilltop, and dozens of smaller mines had produced $25 million in gold, silver, lead, copper, and zinc. Many mines reopened in the 1930s when gold was revalued, and boomed again during World War II when base metal prices soared.

Alma's cumulative production now exceeds $50 million. Geologists re-explored the deposits in the early 1980s when gold prices were high, but the mines remain inactive.

From Alma, take Forest Service Road 416 (Buckskin Gulch Road) west to the site of Buckskin Joe, a gold rush-era *arrastre*, and the old Paris mill. The Paris Mine, high on the steep cliffs to the south, delivered ore to the mill with an aerial tramway that is still in place. A

Deep red rhodochrosite "rhomb" from the Sweet Home Mine.

mile and a half beyond the Paris mill, the road winds through an 11,500-foot-high timberline amphitheater and the Sweet Home Mine property.

Developed in 1872 to mine argentiferous galena, the Sweet Home is now known for its rhodochrosite. In 1965, a rich vein provided many spectacular, deep-red, museum-grade specimens. Sweet Home rhodochrosite typically has a deep red color, well-developed crystals, and occurs with tetrahedrite and white and clear quartz. In 1965, a rich vein provided specimen miners with many deep red, museum-grade specimens.

Specimen miners have again leased the Sweet Home. In selective underground mining operations in 1991 and 1992, miners extracted a wealth of spectacular rhodochrosite specimens that rank high among the most notable U.S. gemstone-mineral specimen recoveries in recent years. The new rhodochrosite specimens stole the spotlight at the 1992 and 1993 Denver Gem and Mineral Shows.

The Sweet Home property remains under lease for selective rhodochrosite mining and is posted. Several other old mines, prospect holes, and dumps follow the same trend of mineralization, extending from the south side of Buckskin Creek northward up the steep side of Mt. Bross.

Several mines are located on the high peaks immediately to the north. The Moose Mine, at 13,700 feet just below the summit of Mt. Bross, was North America's highest major precious-metal mine, yielding $5 million in silver from argentite and argentiferous galena ores. In the 1880s, selected ore graded 700 troy ounces of silver per ton. The Russia, Sovereign, and several other mines below the summit of nearby Mt. Lincoln produced similar ores.

Specimens of these historically and mineralogically interesting ores make attractive display pieces. Argentite occurs as masses or coatings of soft, dark, lead-gray cubic crystals on galena. Galena has a somewhat similar appearance, but its lead-colored cubic crystals are larger with bright, silvery fracture surfaces. As silver content increases, crystals of argentiferous galena become smaller and appear as sugary masses of tiny, bright, silver-colored crystals.

Beyond the Sweet Home Mine, the road becomes rough for the final 1.5 miles to 12,000-foot-high Kite Lake. A well-marked trail leads to the high mines and summits of nearby 14,169-foot Mt. Bross and 14,284-foot Mt. Lincoln.

To reach the Mosquito Gulch mines from Alma, take Colorado Route 9 one mile south, then turn west on Park County Road 12 toward the site of old Park City, the Mosquito Creek placers, and Pennsylvania Mountain. The London Group Mines, including the London, North London, London Extension, London Butte, and American mines, are located six miles west into Mosquito Gulch. London Mountain, bisected by the mineralized London Fault, has 100 miles of underground workings that have produced $25 million in gold, silver, lead, and zinc.

Mine dumps are heavily oxidized, but yield small specimens of pyrite, chalcopyrite, argentite, galena, tetrahedrite, and sphalerite. Gangue minerals include rhodochrosite, green and purple fluorite, barite, selenite, and clear and white crystalline quartz.

Four-wheel-drive is necessary beyond the London Group mines. County Road 12, steep, rough and no longer maintained, climbs to 13,118-foot Mosquito Pass—North America's highest road pass—then descends into the Leadville mining district in Lake County.

Collectors planning field trips should consider elevation and weather. Alma, elevation 10,355 feet, is North America's highest active town, and the mines are considerably higher. The district covers seventy square miles of rugged mountain terrain, mostly above 11,000 feet. Heavy snow is possible in late May and early October, and light snow squalls and severe electrical storms occur in July and August.

The alpine tundra is fragile and susceptible to permanent damage. The old pack trails high on Mt. Bross and Mt. Lincoln look much as they did a hundred years ago—and will look a hundred years into the future. When on alpine tundra, stay on established trails and tracks, and limit collecting to dumps and already-disturbed ground.

Mining-related points of interest in Fairplay include the South Park Historical Museum; South Park City, a historically authentic recreation of a mining boom camp; and the Prunes Monument, named after a miner's burro and which honors all the burros that packed down high-grade ores from the high mountain mines a century ago.

REFERENCES: 12, 18, 19, 23, 25, 41, 74, 81, 106

HARTSEL BARITE

An old open-cut barite mine near Hartsel is a source of clear, white, and green-blue barite crystals. Barite veins are emplaced within clay derived from limestone and stained brown by iron oxide. The veins, over one foot thick, contain massive barite and seams lined with clusters of tabular barite crystals. Collectors have found crystals as long as four inches, but most are smaller.

The workings include several open cuts, dumps, and a long-collapsed small tunnel. Collectors have dug "coyote holes" into the clay, which is unstable and dangerous. Specimens may be collected safely by digging the surface dumps.

The in situ crystals are colored a delicate blue from contact with naturally radioactive hot springs. The blue color in the dump material is intensified by exposure to solar ultraviolet radiation.

From the junction of U.S. 24 and Colorado Route 9 just west of Hartsel, follow U.S. 24 west for just over half a mile. Turn south on the graded road and go a half mile, then follow the track for the final half mile up a grassy hill to the workings. This site is currently under claim and permission is required before collecting.

REFERENCES: 5, 8, 64, 82

PARK COUNTY PEGMATITES

Eastern Park County north of Lake George on U.S. 24 has numerous granite pegmatites near Tarryall Creek providing amazonite, smoky quartz, and topaz (see Teller County). Other Park County pegmatites occur at Harris Park and near Guffy.

HARRIS PARK

Harris Park, a summer residential settlement, is located near the head of Elk Creek in northeastern Park County. From Parkview on U.S, 285, take Deer Creek Road six miles north to the adjacent residential area of Highland Park, then follow the signs to Harris Park.

Since prospectors discovered local pegmatites in 1867, Harris

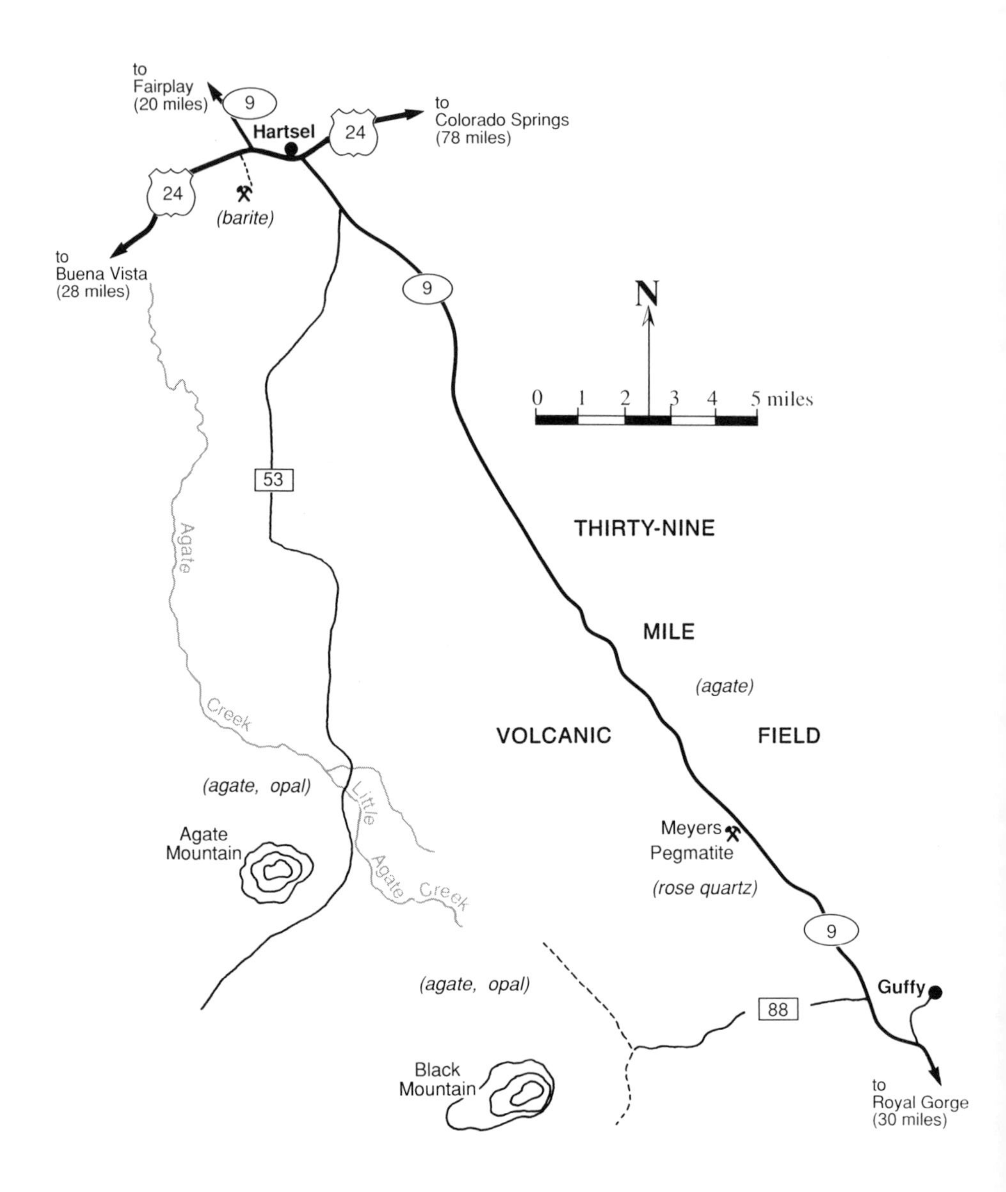
to
Fairplay
(20 miles)
9
Hartsel
24
to
Colorado Springs
(78 miles)
24
(barite)
to
Buena Vista
(28 miles)
9
N
0 1 2 3 4 5 miles
53
THIRTY-NINE
Agate
MILE
(agate)
Creek
VOLCANIC
FIELD
(agate, opal)
Little
Agate
Mountain
Agate Creek
Meyers
Pegmatite
(rose quartz)
9
(agate, opal)
Guffy
88
Black
Mountain
to
Royal Gorge
(30 miles)

Park has yielded many fine specimens of smoky quartz, light brown-yellow topaz, and microcline feldspar. Collectors have found quartz crystals to seventeen inches and topaz crystals to three inches in length.

Most Harris Park pegmatites are claimed or on private land. Permission is required before collecting.

MEYERS RANCH PEGMATITE

The Meyers Ranch Pegmatite, a former commercial source of rose quartz, is located twenty miles south of Hartsel on Colorado Route 9, or six and one-half miles north of the Guffy turnoff. The open cut is visible west of the highway near a wooded hilltop. Obtain permission before crossing the private land along Current Creek.

The pegmatite mine produced rose quartz, columbite-tantalite, microcline feldspar crystals up to six feet long, muscovite and biotite mica, and beryl.

Opposite the Meyers Ranch Pegmatite, a Colorado Route 9 road cut exposes a small pegmatite. Specimens of white quartz, pink feldspar, and muscovite mica four inches in length are found on the east highway shoulder.

REFERENCES: 64

THIRTY-NINE MILE VOLCANIC FIELD AGATE

Agate and jasper are present in surface gravels in many parts of South Park. An attractive blue agate occurs in the Thirty-Nine Mile Volcanic Field. Colorado Route 9 cuts through the lava flows ten miles south of Hartsel, and scattered agate may be collected from that point south to the Guffy turnoff.

Agate and white moss opal have been reported along Agate Creek and near Agate Mountain, Black Mountain, and Thirty-One Mile Mountain in southwestern Park County. Follow Agate Creek Road (Park County Road 53) twenty miles south to Agate Mountain. The large area between Agate Mountain and Colorado Route 9 is accessible by several four-wheel-drive roads and offers good prospecting for agate and moss opal.

REFERENCES: 5, 66

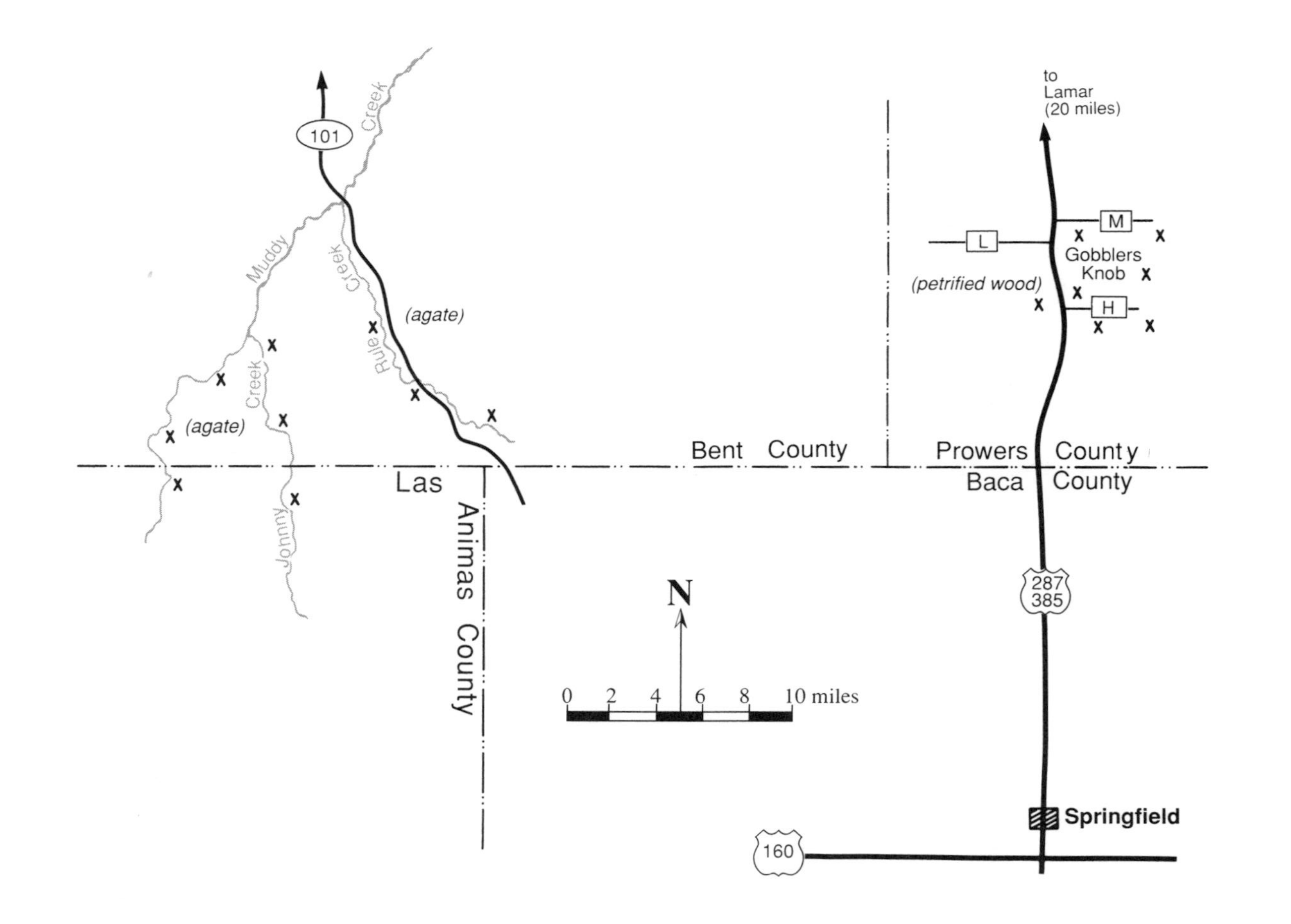
101
Creek
Muddy
Creek
Rule
(agate)
Creek
(agate)
Johnny
Las
Animas County
Bent County
N
0 2 4 6 8 10 miles
to
Lamar
(20 miles)
M
L
Gobblers
Knob
(petrified wood)
H
Prowers County
Baca County
287
385
Springfield
160

PROWERS COUNTY BENT COUNTY

PETRIFIED WOOD

Early collecting literature notes abundant petrified wood in Prowers County near U.S. 287-385 eighteen miles south of Lamar. The general location is near M Road and the low bluffs and gullies of Gobblers Knob, which expose Dakota Sandstone and small sections of the underlying Morrison Formation. Fragments of well-silicified petrified wood are found in many of the local washes.

Local residents collected commercial quantities of large petrified tree trunks in the 1930s. Much was slabbed and polished for the Lamar tourist trade. While less brightly colored than the better-known material from Arizona's Petrified Forest, the petrified wood has pleasing multicolor patterns of soft reds, yellows, browns, and whites.

In 1932 Lamar businessman W. G. Brown collected forty tons of petrified logs to build a gas station, which doubled as an attraction

The "Petrified Service Station" in Lamar as it appears today.

to bring this "interesting phenomena of southeastern Colorado to the attention of the stream of tourists hurrying to see more highly advertised regions." Brown promoted his Petrified Service Station as "an all-wood building with scarcely a piece of wood in it." The biggest petrified logs used in construction were eight feet long, three feet in diameter, and weighed 3,200 pounds. Even the "hardwood" floors were slabbed and polished trunk sections. The building later appeared in *Ripley's Believe It Or Not.*

The Lamar Tire Service, the current owner, took over the building in 1962. Today, the building is occupied by a real estate office. Sixty years of continuous termite-free service testifies to the durability of Prowers County "hardwood." The petrified-wood building is located in Lamar at 501 Main Street.

REFERENCES: 32, 33, 66, 67

AGATE AND JASPER

Agate and jasper are fairly common in southern Prowers and Bent counties. Exposures of Morrison Formation sediments in the extreme southern part of Bent County along the upper tributaries of Rule, Johnny, and Muddy creeks contain colorful patterned agates. Most notable is a "flower" agate found near gully exposures of the Morrison Sandstones, where agate seams are several inches thick and up to 100 feet long. The agate is marked by distinctive yellow, orange, and red fortification patterns.

REFERENCES: 32, 66, 67

PUEBLO COUNTY
OTERO COUNTY

PUEBLO AREA FOSSILS

The gullies and mesas surrounding Pueblo have numerous exposures of fossiliferous Cretaceous sediments.

BACULITE MESA

Baculite Mesa is a large, prominent mesa east of I-25 and six miles north of Pueblo. Fossils of the straight-shelled ammonite *Baculites* are locally abundant in Pierre Shale exposures. The soil and alluvial gravel surrounding the mesa are derived largely from Pierre Shale and contain countless casts and fragmented fossils of clams and baculites.

DRY CREEK

Just northwest of Pueblo, Dry Creek cuts through Pierre Shale and into the underlying fossiliferous Niobrara Formation. Fossils found in the bed and banks of Dry Creek include *Inoceramus deformis*, a small, round clam, and inch-long fish scales.

Baculite Mesa, near Pueblo, has massive exposures of fossiliferous Pierre Shale.

Fossil casts of clams and ammonites are abundant at Baculite Mesa.

Section of Baculite *fossil from Baculite Mesa.*

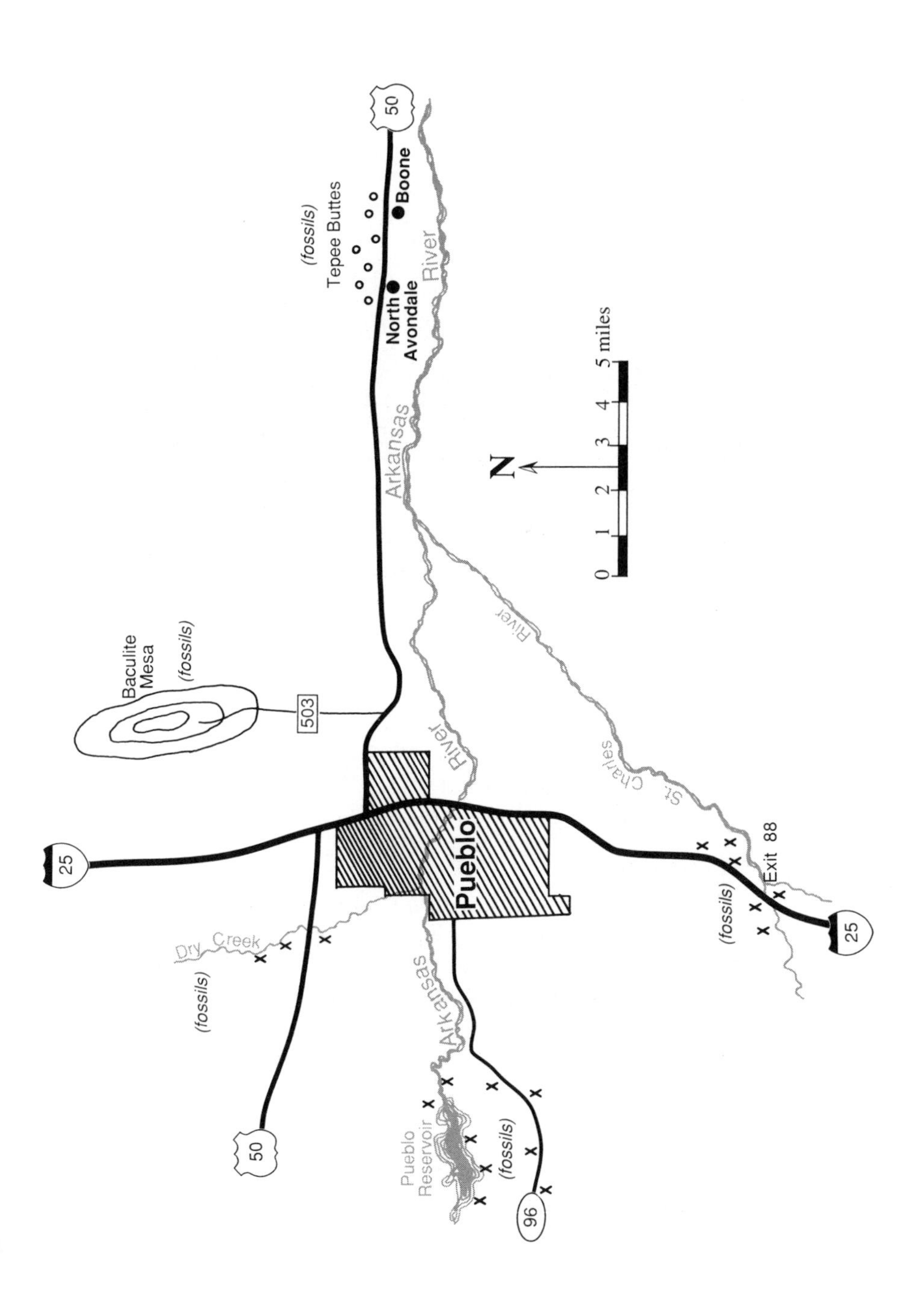
Tepee Buttes
(fossils)
Boone
North
Avondale
Arkansas
River
Baculite
Mesa
(fossils)
503
5 miles
0 1 2 3 4 5
N
River
St. Charles
Pueblo
25
50
Dry Creek
(fossils)
Exit 88
(fossils)
25
Arkansas
Pueblo
Reservoir
(fossils)
96
50

Low, conical hills north of Avondale and Boone were once reefs on the floor of a Cretaceous sea.

WEST OF PUEBLO

Marine fossils weather free from Pierre Shale and are locally common in a broad area west of Pueblo, including the hills around Pueblo Reservoir and the gullies along Colorado Route 96.

SOUTH OF PUEBLO

Exposures of fossiliferous Pierre Shale, Niobrara Sandstone, and Greenhorn Limestone occur near I-25 south of Pueblo. Cretaceous marine fossils, such as the clams and oysters *Inoceramus deformis* and *Ostrea congesta*, are locally abundant in gullies near and north of I-25 Exit 88. Specimens recently weathered free from sedimentary exposures retain much of the lustrous shell material.

EAST OF PUEBLO

Numerous "tepee buttes" are located ten miles east of Pueblo and just north of U.S. 50 near the towns of Avondale and Boone. The low, inconspicuous buttes, originally reefs on a shallow Cretaceous sea floor, contain seams of crystalline calcite and abundant small fossil clams.

REFERENCES: 12, 22, 63, 68

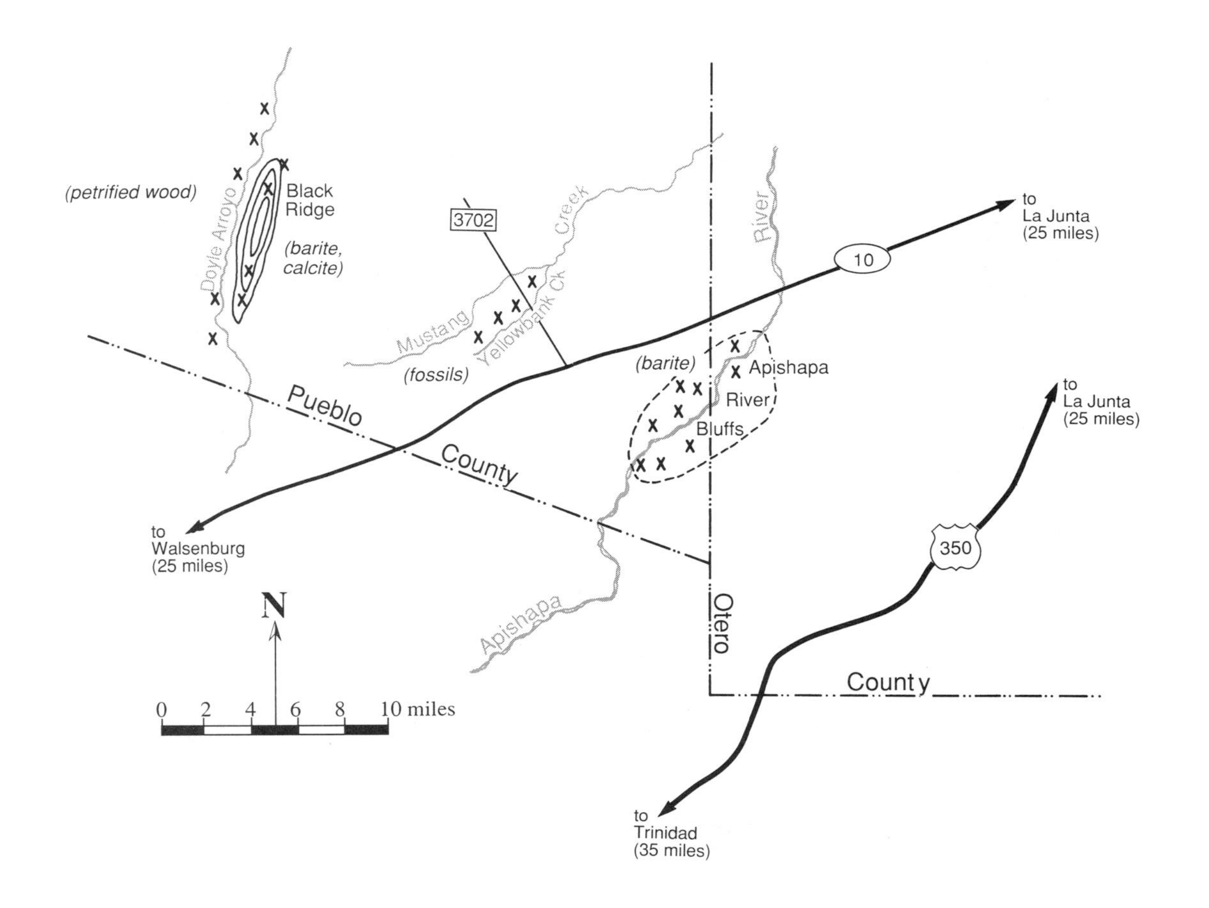

(petrified wood)
Black Ridge
(barite, calcite)
Doyle Arroyo
3702
Creek
Mustang
Yellowbank Ck
(fossils)
River
to La Junta (25 miles)
10
(barite)
Apishapa River Bluffs
Pueblo
County
to Walsenburg (25 miles)
to La Junta (25 miles)
350
Apishapa
Otero
County
N
0 2 4 6 8 10 miles
to Trinidad (35 miles)

THE APISHAPA RIVER BLUFFS AREA

Doyle Arroyo, Black Ridge, Yellowbank Creek, and the Apishapa River Bluffs, located near the Pueblo-Otero county line midway between Walsenburg and La Junta, are reached by county roads and tracks from Colorado Route 10.

Doyle Arroyo and Black Ridge are located in southeastern Pueblo County twenty-five miles northeast of Walsenburg. Petrified wood is abundant where Doyle Arroyo cuts through Dakota Sandstone. In the 1930s, collectors reported sections of petrified wood six feet long and one foot thick.

Black Ridge is a low north-south trending ridge immediately east of Doyle Arroyo. The ridge exposes Carlisle Shale containing thin veins and concretions of calcite and barite. White to dark brown calcite occurs in botryoidal form and as half-inch-long rhombohedrons lining the hollow centers of septaria and concretions. Barite seams within the shale occasionally open into small pockets filled with white to light blue barite crystals up to one inch long.

APISHAPA RIVER BLUFFS

The low bluffs of the Apishapa River on the Pueblo-Otero county line were a popular 1930s collecting locale for calcite and light blue barite. The bluffs expose Apishapa Shale and Carlisle Shale within a thirty-square-mile area. Veins of white calcite and light blue barite occur in the crumbling shale exposures. Crystals weather free from shale exposures and are found in the arroyo beds. Barite crystals are well formed and up to one inch long, but lack the aquamarine-blue color of barite from the better-known Stoneham locality in Weld County. A portion of the Apishapa River Bluffs is within Comanche National Grassland.

YELLOWBANK CREEK

Located a short distance northeast of the Apishapa River Bluffs, Yellowbank Creek is a source of marine fossils from exposures of Timpas Limestone. Concretions containing pyrite and marcasite, along with clam and fish-scale fossils, are found in the creek bed.

REFERENCES: 78

RIO GRANDE COUNTY

PLACER GOLD

Rio Grande County, although a major source of lode gold, has produced only 1,000 troy ounces of placer gold from the Summitville district. Prospectors discovered placer gold in Wightman Fork in June 1870.

REFERENCES: 54

THE SUMMITVILLE MINING DISTRICT

In 1873 placer miners in Wightman Fork discovered the Little Annie Vein, a rich lode gold outcrop on nearby South Mountain. Summitville, elevation 11,300 feet, was at times North America's highest town. During the 1880s, Summitville was Colorado's leading gold district.

Depletion of the rich discovery veins turned Summitville into a ghost town by the turn of the century, but it revived during the 1920s, when miners found an overlooked bonanza vein. Although small, the ore from that one vein averaged twenty-eight troy ounces of gold per ton and was worth a half-million dollars.

Summitville boomed again in the 1930s when gold was revalued to $35 per troy ounce. One celebrated 139-pound piece of ore contained an estimated 420 troy ounces of native gold. By the 1950s, Summitville's cumulative production reached 250,000 ounces of gold worth $5 million.

In the 1970s, geologists delineated a large, very low-grade gold resource on South Mountain. In the mid-1980s, miners developed a huge open-pit heap-leach mine, recovering 270,000 troy ounces of gold in just six years—more than the old-timers had mined in a century. Summitville's total production now stands at more than a half-million troy ounces of gold.

Along with native gold, Summitville has provided specimens of galena, sphalerite, pyrite, a variety of copper sulfide and oxide minerals, barite, calcite, and quartz. The district is especially known for enargite and Colorado's best covellite crystals. Covellite occurs in bladed crusts up to six inches thick and ranges in color from blue-purple to an unusual gold.

Summitville is twenty miles north of Del Norte and is reached by Forest Service Roads 380 and 330 from South Fork and Del

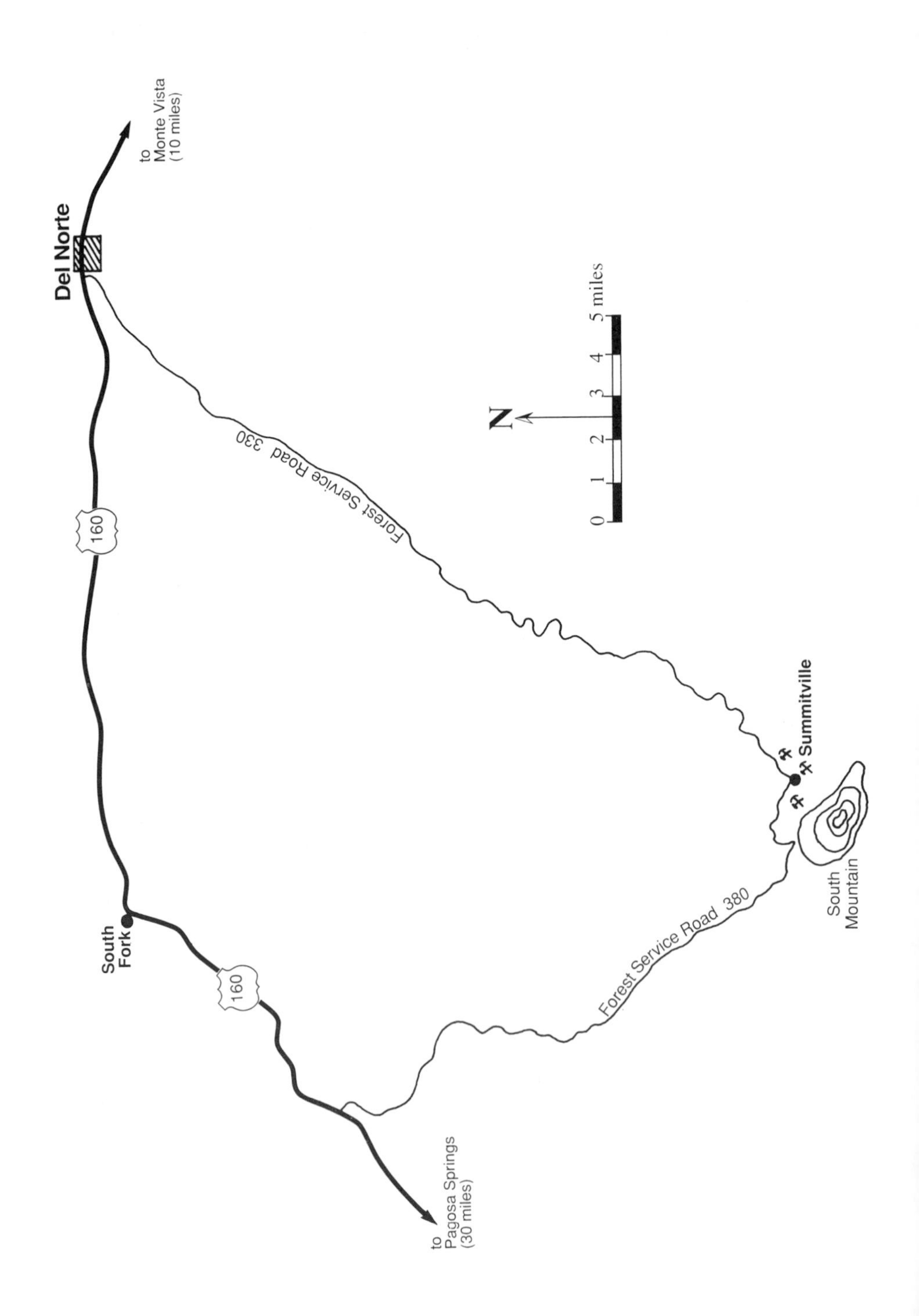
Del Norte
to
Monte Vista
(10 miles)
160
Forest Service Road 330
N
0 1 2 3 4 5 miles
Summitville
South
Mountain
Forest Service Road 380
South
Fork
160
to
Pagosa Springs
(30 miles)

Norte, respectively. Most mine properties are consolidated under the Summitville Mining Company, which prohibits collecting.

THE SUMMITVILLE GOLD BOULDER

The Summitville gold boulder is Colorado's most remarkable chance discovery of gold in recent times. In 1975, ASARCO geologists surveyed Summitville's low-grade gold resource on South Mountain. The only free gold specimens found contained particles barely visible, and no one expected to find much more.

But on the afternoon of October 3, equipment operator Bob Ellithorpe noticed a large rock lying in full view just off the shoulder of a public road below the early mine ruins. Ellithorpe was shocked to find it laced with native gold. Returning excitedly to camp, he asked the project geologist if he wanted to see some gold. The geologist casually agreed, but when Ellithorpe said it would take two of them to put it in the pickup, he suspected a joke.

The specimen turned out to be, literally, a "gold boulder." The 141-pound float boulder of breccia consisted of silicified vuggy quartz latite fragments cemented together with a matrix of fine-grained crystalline quartz and barite, and crystallized gold. The boulder had apparently moved downhill by natural processes from the Little Annie Vein to the point where it was discovered. The geologist believed the boulder had been fully exposed by the roadside for at least thirty years. Specific-gravity tests determined that the boulder contained about 350 troy ounces of gold.

Ownership of the "Summitville gold boulder" became a confusing issue. When Ellithorpe first reported the boulder, the company geologist verbally agreed to his request to "keep half of it." Traditionally, prospectors providing mineral-location information received a 10 percent finder's fee, if the find was worthy of claiming or development. But the gold boulder was alluvial float found on patented ground. Ellithorpe was technically not a prospector, but a paid private contractor. Another question concerned the geologist's legal right to negotiate on behalf of the company. The matter was closed when Ellithorpe accepted ASARCO's offer of a $21,000 "finder's fee."

ASARCO kept the discovery secret for a year to avert a "gold rush." The company finally announced the find in November 1976, just as Summitville was snowed in for the long winter. With the agreement of the landowner, the Reynolds Mining Company, ASARCO donated the Summitville gold boulder to the Denver Museum of Natural History, where it is currently displayed. Because of higher gold prices and added worth as a rare and spectacular speci-

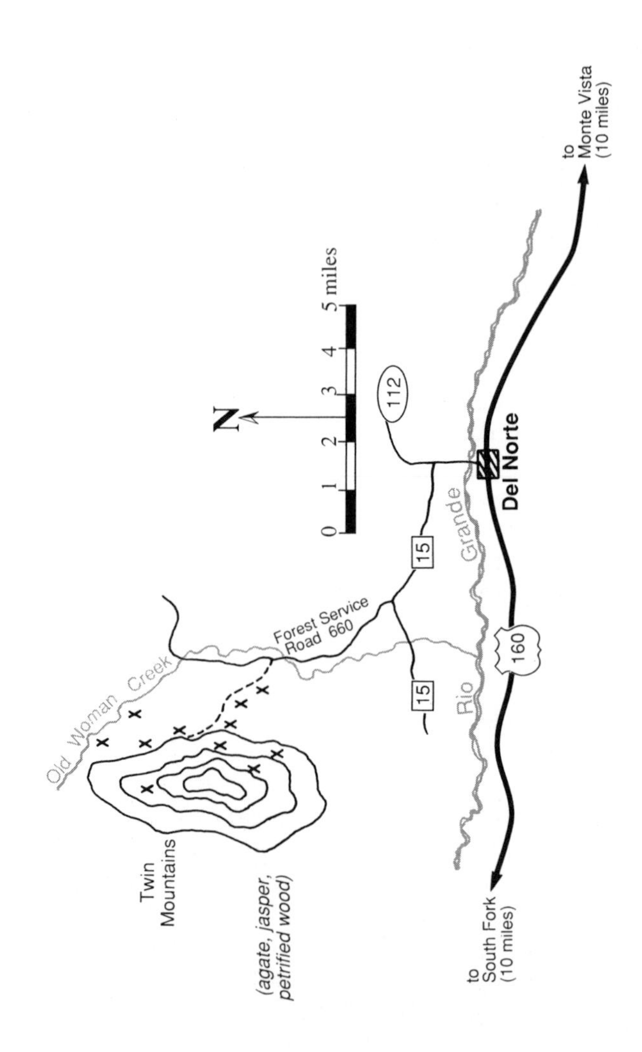
N
0
1
2
3
4
5 miles
to
Monte Vista
(10 miles)
112
Del Norte
15
Forest Service
Road 660
Old Woman Creek
Rio
Grande
160
15
Twin
Mountains
(agate, jasper,
petrified wood)
to
South Fork
(10 miles)

men, the Summitville gold boulder is now valued at $350,000.

Rockhounds and treasure hunters armed with picks, shovels, gold pans, and metal detectors rushed to Summitville the following spring. ASARCO closed several roads when gold seekers interfered with exploration activities.

If prospectors did find more gold, they never reported it. Nevertheless, the Summitville gold boulder will always be a reminder of the bonanza ores of the nineteenth century—and also that the old-timers didn't find it all.

REFERENCES: 18, 23, 81, 104, 108

DEL NORTE AGATE

A productive agate locality is located six miles northwest of Del Norte along the Rio Grande-Saguache county line. Present are a variety of banded agates, colorful jaspers, geodes, some opal, abundant and well-silicified petrified wood, and some of Colorado's best dendritic, or "plume," agate.

The geodes, agate, and jasper formed in the cliffs and slopes of Twin Mountains, a distinctive twin-peaked formation of dark volcanic rock containing nodules and seams of cryptocrystalline quartz. Geodes, known locally as "Monte Vista eggs" and contain-

The area surrounding Twin Mountains, north of Del Norte, is rich in agate, jasper, and petrified wood.

ing both banded and botryoidal agate and clear crystalline quartz, weather free from the volcanic rock. Agate and jasper is collected over a large area between Old Woman Creek and Twin Mountains. Chrysoprase and bloodstone are also present.

From Del Norte and U.S. 160, take Colorado Route 112 north for one mile, then follow 15 Road north for three and one-half miles. Take Forest Service Road 660 north for three miles along Old Woman Creek to an unmarked track leading west toward Twin Mountains. Fragments of geodes and agate are found along the roadside, and gullies have numerous pieces of petrified wood as long as two feet, some retaining detailed original bark structure. The track is passable in dry weather for passenger cars nearly to the base of Twin Mountains.

The massif of Twin Mountains, nearly three miles long, is a fine prospecting area, especially on the rugged higher slopes. Numerous stripped pockets and seams appear in the lava where collectors have extracted geodes and agate. Commercial collectors have left behind several deep trenches.

The land between Old Woman Creek and Twin Mountains is private, but not posted. Permission should be obtained before collecting on private land.

REFERENCES: 64, 65, 67

ROUTT COUNTY

PLACER GOLD

Routt County has produced 10,000 troy ounces of placer gold, mostly from Deep Creek and Ways Gulch on the southeast side of Hahns Peak. Prospectors discovered the placers in 1864, and production peaked in 1910 when a small floating-bucketline dredge worked the confluence of Ways Gulch and Willow Creek.

Because of the proximity to its lode sources on Hahns Peak, the placer gold is coarse. Steamboat Lake, just west of Hahns Peak, now covers some of the early placer workings.

Hahns Peak and the summer settlements of Hahns Peak and Columbine are located twenty-five miles north of Steamboat Springs on Routt County Road 129.

REFERENCES: 54

HAHNS PEAK

Hahns Peak, a prominent 10,820-foot-high pyramidal peak, produced $200,000 in lead, copper, silver, and gold from a dozen small hardrock mines, including the Tom Thumb, Minnie D.,

Hahns Peak is a source of metal sulfide minerals and clear quartz crystals.

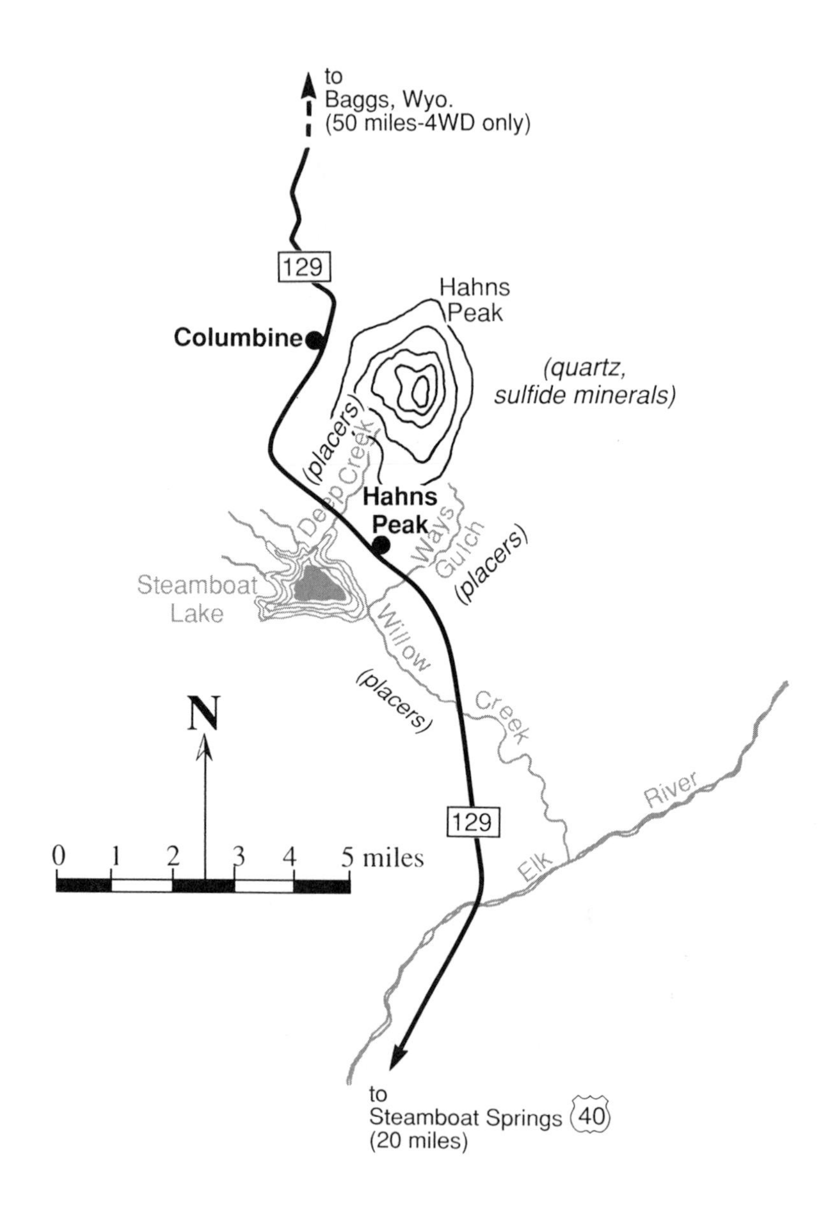

to
Baggs, Wyo.
(50 miles-4WD only)
129
Hahns
Peak
Columbine
(quartz,
sulfide minerals)
(placers)
Deep Creek
Hahns
Peak
Ways
Gulch
(placers)
Steamboat
Lake
Willow
(placers)
Creek
N
0 1 2 3 4 5 miles
129
River
Elk
to
Steamboat Springs 40
(20 miles)

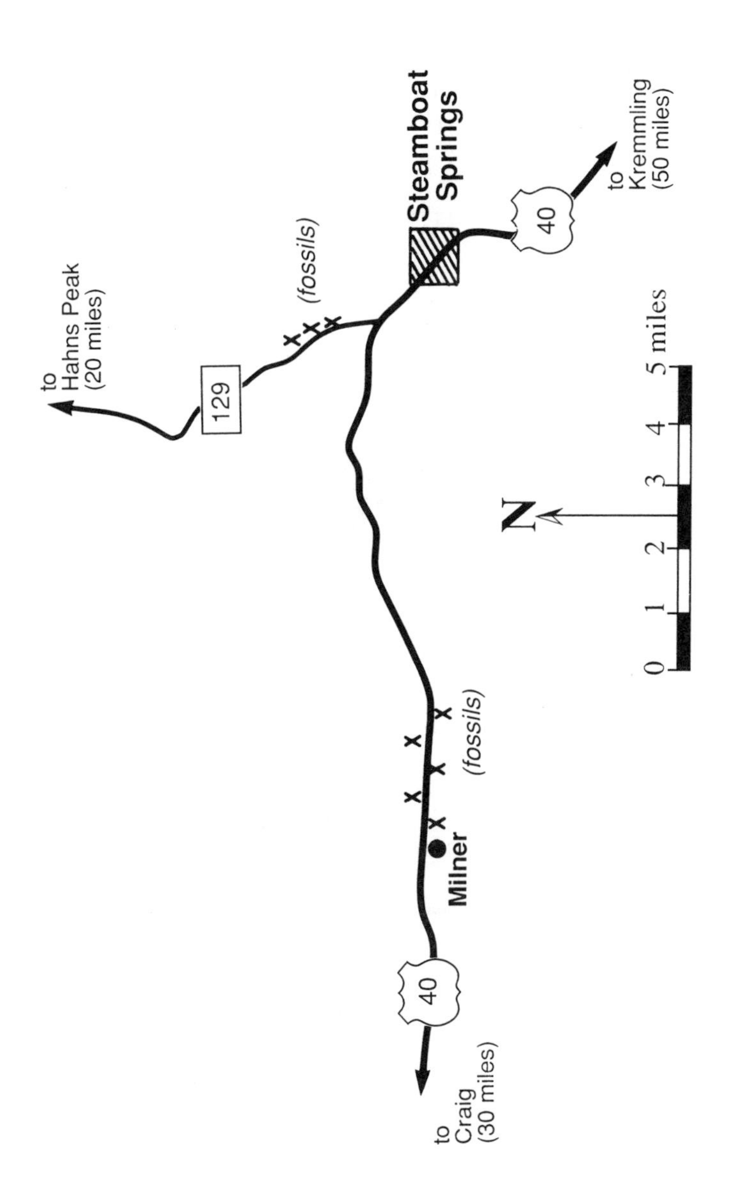
to
Hahns Peak
(20 miles)
129
(fossils)
Steamboat
Springs
40
to
Kremmling
(50 miles)
N
0
1
2
3
4
5 miles
(fossils)
Milner
40
to
Craig
(30 miles)

Southern Cross, and Master Key. The upper half of the peak is barren of timber, crisscrossed with rough mine roads, and dotted by mines and dumps.

Dumps yield small specimens of metal sulfide minerals, but Hahns Peak is best known for clear quartz crystals. The crystals, up to five inches long and coated with "rusty" limonite, formed in cavities in the porphyrytic country rock. Crystals, usually abraded and fragmented, are dug from the steep upper talus slopes.

Several Forest Service and unmarked four-wheel-drive jeep roads from the towns of Hahns Peak and Columbine lead to the mines on the peak. The summit may also be reached by a two-mile, fairly steep hike.

REFERENCES: 18, 23, 64

FOSSILS

Road cuts and gullies east of Milner along U.S. 40 expose fossiliferous Mancos Shale. In western Colorado, Mancos Shale is generally equivalent in age and stratigraphic position to the fossil-rich Pierre Shale of eastern Colorado.

Mancos Shale is also exposed in road cuts along Routt County Road 129, one mile north of the U.S. 40 junction. The roadside exposures contain many marine fossils, such as the clam *Inoceramus*, and seams of white calcite up to two inches thick.

REFERENCES: 12, 13, 68

SAGUACHE COUNTY

BONANZA AREA MINES

Bonanza is located eighteen miles west of Villa Grove and U.S. 285 on County Road LL56.

In 1879, prospectors discovered rich silver-lead-manganese veins along upper Kerber Creek and 4,000 people rushed into the boom camp of Bonanza. The Rawley, developed in 1880, was the biggest producer. When the district shut down in the 1950s, total production topped $10 million in silver, lead, and zinc and lesser amounts of copper and gold.

Two types of mineralization include high-sulfide quartz veins (containing lead, zinc, copper, silver, and some gold) and quartz-rhodochrosite-fluorite veins with minor metal sulfide content. Ores were usually high-sulfide quartz, but exploratory drifting into the quartz-rhodochrosite-barite veins provided many interesting specimens. The primary ore minerals are galena, sphalerite, pyrite, chalcopyrite, bornite, enargite, and argentiferous tennantite.

Mine dumps and ruins in the Bonanza district.

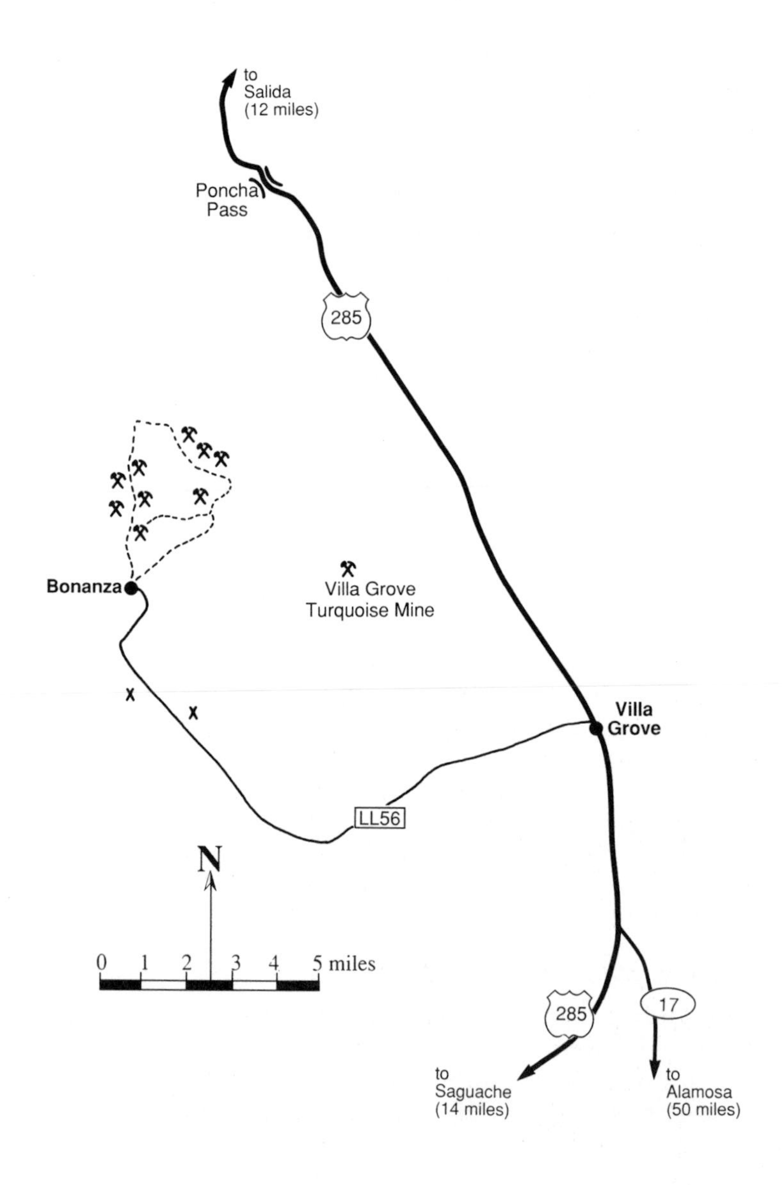
to
Salida
(12 miles)
Poncha
Pass
285
Bonanza
Villa Grove
Turquoise Mine
Villa
Grove
LL56
N
0 1 2 3 4 5 miles
285
17
to
Saguache
(14 miles)
to
Alamosa
(50 miles)

Gangue minerals are quartz, rhodochrosite, barite, and calcite. The ten-square-mile district has twenty-five major mines and at least fifty others.

Bonanza has only a few summer residents and no services. Located well off the main highways, the Bonanza dumps have not been heavily collected and have interesting specimens of ore and gangue minerals.

REFERENCES: 9, 18, 19, 23, 25, 41, 81

THE VILLA GROVE TURQUOISE MINE

The Villa Grove Turquoise Mine is located seven miles northwest of Villa Grove. Prospectors discovered copper mineralization in the 1880s. Miners identified turquoise "nuggets" (float) in 1893, then dug exploratory trenches into turquoise veins.

Mining peaked in the 1940s, when miners recovered turquoise from both open cuts and underground drifts, hand-cobbed the rough, then "turned" it with water in a cement mixer to separate the matrix rock. Trimmed rough then sold for $15 to $45 per pound.

The Colorado Bureau of Mines valued the 1950s turquoise production at $80,000. In 1957, 450 pounds of trimmed, high-quality rough brought $15,600. Turquoise from the Villa Grove Mine (then known as the Hall Mine) was Saguache County's second most valuable mine product.

Villa Grove turquoise, remarkably free of veining, is a clean sky blue, with little green coloration. Some experts consider it the finest turquoise now mined in the United States.

Commercial mining resumed in 1992. The Villa Grove Turquoise Mine, owned by a Villa Grove resident, is not open to collectors.

REFERENCES: 56, 59, 60, 73, 74

CRYSTAL HILL AMETHYST

Crystal Hill, eight miles northwest of the little town of La Garita, is a popular collecting site for amethyst and clear quartz crystals.

The Crystal Hill area, originally known as Beidell, was a minor source of lode gold from small underground mines pursuing erratic veins in brecciated volcanic rock. Miners found little gold, but an abundance of clear, smoky, and amethyst quartz crystals.

Quartz prisms exhibit delicate violet color zoning toward the termination and distinctive "water bubble" inclusions. Beidell amethyst and the name Crystal Hill first appeared in mineralogical literature in 1890.

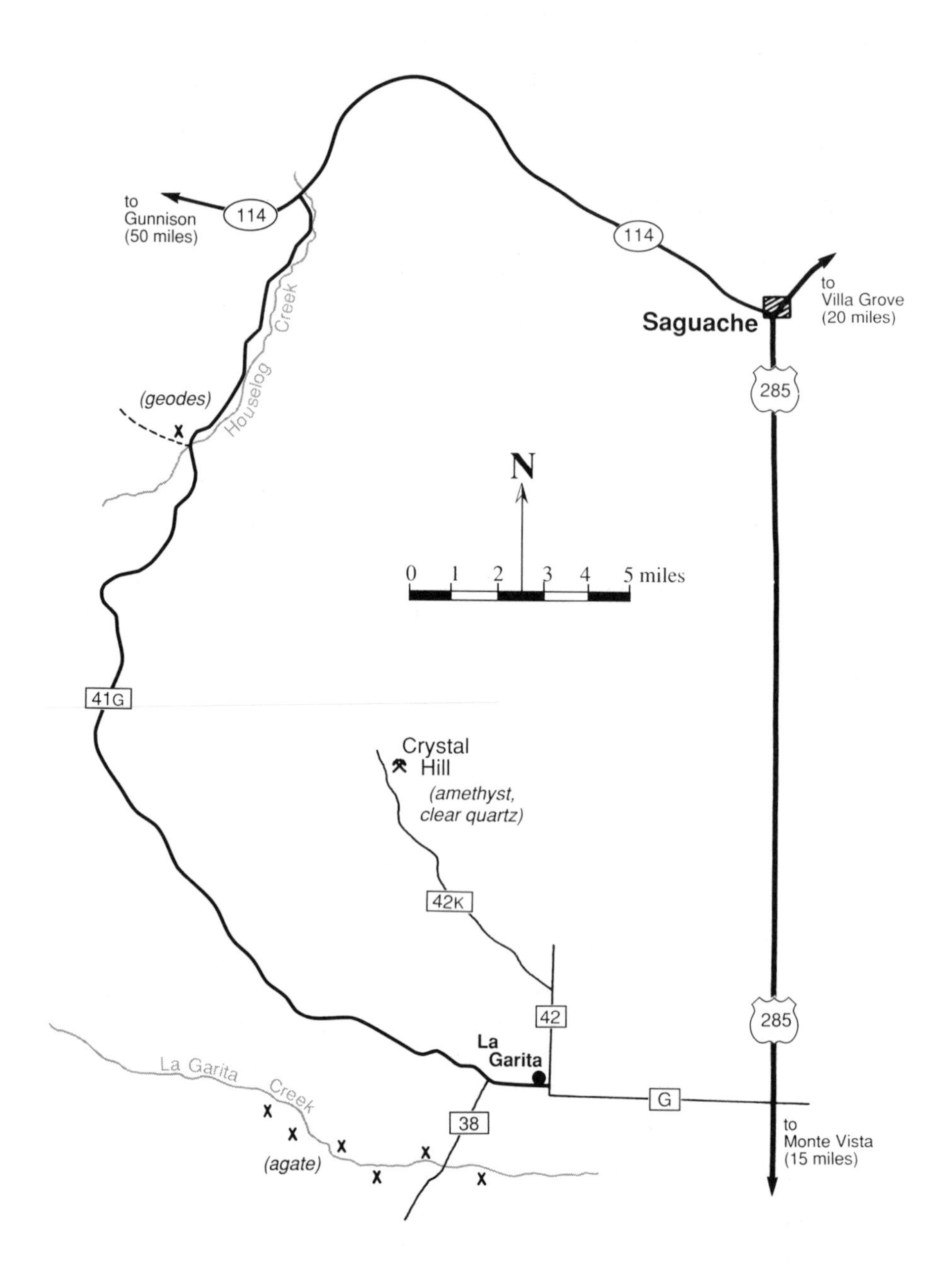
to
Gunnison
(50 miles)
114
114
to
Villa Grove
(20 miles)
Saguache
285
Houselog Creek
(geodes)
N
0 1 2 3 4 5 miles
41G
Crystal
Hill
(amethyst,
clear quartz)
42K
42
285
La
Garita
La Garita Creek
G
38
(agate)
to
Monte Vista
(15 miles)

Exploration geologists returned to Crystal Hill in the late 1970s, delineating a large, low-grade zone of disseminated gold near the top of the hill. The Crystal Hill Mining Company developed an open-cut heap-leach mine, recovering 30,000 troy ounces of gold four years.

Crystal Hill's complexly mineralized breccias formed when mineral-bearing solutions penetrated fragmented volcanic rock. A gold-bearing high-silica solution preceded solutions rich in two common brecciating agents, iron and manganese. The last solution, which leached downward from the surface, carried calcium carbonate from dissolved limestone.

The silica crystallized as drusy quartz and, in vugs, as well-developed crystals of clear, smoky, and amethyst quartz. Iron and manganese remain as yellow and black coatings of their respective oxides. The dissolved limestone recrystallized as white needles of drusy calcite.

Drusy quartz and calcite fill most cavities within the breccia; many contain well-developed clear, smoky, and amethyst quartz crystals, including some doubly terminated crystals and penetration twins. The mineralized breccia fills the mine dumps and is even found on roads and in nearby Beidell Creek. Typically, clusters of quartz crystals with amethyst coloration tend to "pyramid" with one long prism attractively rising above a base of shorter prisms. A loupe will reveal the tiny water-bubble inclusions common in Crystal Hill amethyst.

Collectors break pieces of breccia with heavy hammers and cold chisels to expose crystal-lined cavities. Old clothes are recommended because of the abundance of black, hard-to-remove manganese oxide.

To reach Crystal Hill from La Garita and County Road G, take County Road 42 north for two miles, cross the cattle guard, then follow County Road 42K west into the foothills of the La Garita Mountains at Beidell Gulch. Eight miles from La Garita, the open cut of the Crystal Hill mine appears ahead. Proceed one mile farther and park at the base of Crystal Hill near the lower mine dumps and the road leading to the upper mine workings.

REFERENCES: 5, 19, 25, 32, 60, 64

SAGUACHE COUNTY AGATE, GEODES & APACHE TEARS

HOUSELOG CREEK GEODE LOCALITY

From Saguache, follow Colorado Route 114 west for 15.5 miles. Then take County Road 41G south for 6.8 miles to the Houselog

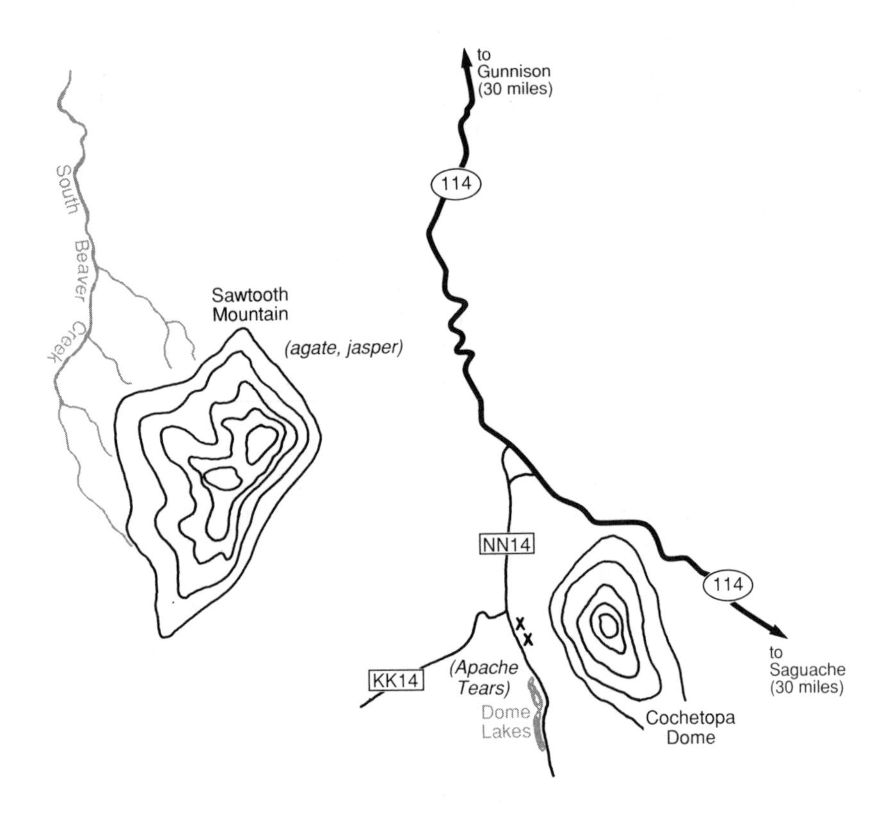
to
Gunnison
(30 miles)
114
South Beaver Creek
Sawtooth
Mountain
(agate, jasper)
NN14
114
to
Saguache
(30 miles)
KK14
(Apache
Tears)
Dome
Lakes
Cochetopa
Dome

If, indeed, the story is true, then the source of the jasper seems to have been lost again. For interested rockhounds, Sawtooth Mountain could be an interesting prospecting opportunity.

LA GARITA CREEK

Three miles southwest of La Garita, La Garita Creek is a source of agate, including the moss variety. Upper La Garita Creek is about five miles north of the Twin Mountains-Old Woman Creek agate locality (see Rio Grande County).

REFERENCES: 5, 19, 26, 41, 67

Creek culvert at a small aspen park and a marked Forest Service road junction. Collect on the hill immediately to the north where trenches expose weathered lava fifty yards up the hillside.

Chalcedony nodules, seams, and geodes occur in situ in the weathering volcanic rock. Geodes are lined or filled with botryoidal jasper and banded agate. Fragments of jasper and banded agate are also found by digging and screening the rocky soil at the base of the hill. Color ranges from gray to red, with green-yellow predominant. Also present is an attractive opaque white jasper with a slight opalescent sheen.

COCHETOPA DOME APACHE TEAR LOCALITY

Apache tears, rounded pieces of natural volcanic glass, are found along the western base of 11,132-foot Cochetopa Dome. From Saguache, follow Colorado Route 114 west for forty-one miles, then take County Road NN14 south for almost four miles to a 200-yard-long road cut. Black Apache tears can be collected on the roadside or dug and screened from the alluvial gravels in the wall of the cut.

The Apache tears, ranging from one-quarter to two inches in size, weather free from the volcanic rock of Cochetopa Dome and

Cochetopa Dome, a volcanic formation rich in obsidian nodules, is the source of local "Apache tears."

"Apache tears" from Cochetopa Dome are obsidian, or natural volcanic glass.

are locally abundant. Although they appear black, most are translucent with tiny mossy inclusions. They take a nice tumble polish.

SAWTOOTH MOUNTAIN AGATE AND JASPER

Sawtooth Mountain rises west of Colorado Route 114 near Cochetopa Dome in northwestern Saguache County. It is forty-five miles west of Saguache and fifteen miles south of Gunnison. The volcanic mountain, topped by 11,719-foot-high Agency Peak, is the source of the agate and jasper found in creek beds along its western and northern drainages.

An interesting tale about the Sawtooth Mountain jasper was published in 1913 in *Early Days on the Western Slope of Colorado*, and reprinted in 1935 in *Southwestern Lore*, the journal of the Colorado Archaeological Society. It tells of a Spanish adventurer, Juan Rivera, who purportedly visited the Gunnison River headwaters in 1761, finding gold in several streams near Taylor Park (Gunnison County).

> A tradition exists that the beautiful seven-riven-jasper sparingly used on interior decoration in the Cathedral of the City of Mexico, was mined somewhere on Saw Tooth Mountain, on the west side of Cochetopa Pass, one hundred and forty

Sawtooth Mountain is the legendary source of agate mined by the Spanish and transported to Mexico City in the 1700s.

> years ago, probably by Rivera's men, and transported over Cochetopa Pass and down the Rio Grande to Santa Fe.
>
> If a fact, which there is no reason to doubt, they must have carefully covered up or obliterated all traces of the vein it came from. This, reasoning the Spanish nature, they would likely do to prevent its use on any other edifice in the New World but this one cherished masterpiece of their architecture.

The editor of *Southwestern Lore* added the following comments:

> For quite a period of years, Mr. H. W. Endner of Gunnison, has tried to trace the location of this deposit of jasper. He has been in the habit of asking every old-timer, prospector, or cowpuncher if anything was known about it. No success attended these efforts. Many of those approached knew about the tradition but no one knew where the deposit was to be found. Persistent effort, however, brought success. The deposit has now been located. It shows evidence of early Spanish workings. Much discarded material was found, indicating that only selected blocks were brought to Mexico.
>
> A very beautiful specimen of this jasper was once exhibited in the old museum of Western State College [Gunnison].

SAN JUAN COUNTY

PLACER GOLD

San Juan County has produced about 500 troy ounces of placer gold. Charles Baker and other prospectors discovered placer gold on the upper Animas River above the site of Silverton in 1860. Exaggerated reports of the strike attracted 300 men in April 1861. Indian trouble discouraged them, and prospectors did not return until 1869.

Cement Creek, a tributary of the upper Animas above Silverton, has limited, but fairly rich placers formed in historic times from tailings of early stamp mills and the Gold King Mill, testimony to the inefficiency of early milling methods.

REFERENCES: 23, 54

SILVERTON AREA MINES

Daring prospectors, defying both the Utes and federal orders to keep out of the San Juan Mountains, discovered a rich gold outcrop in Arrastra Gulch near the future site of Silverton in 1870. They developed the Little Giant Mine and crushed the high-grade gold ore in *arrastres*.

Prospectors rushed in after the 1873 Brunot Treaty legalized mineral entry, and discovered many important veins, but lack of railroad transportation, short working seasons, high elevations, avalanches, and an inability to recognize and treat complex ores delayed development. The rich Sunny Side Vein, discovered in 1873, wasn't developed until 1882, when the Denver & Rio Grande narrow-gauge railroad arrived. The district peaked after 1900, when the flotation separation process made lower-grade ores profitable.

By 1924, 200 mines had produced $80 million in metals. Today, total silver-gold-lead-zinc-copper production from San Juan County mines has exceeded $120 million. In recent decades the Sunnyside Mine, working sections of the original Sunny Side Vein, dominated production. In the 1970s, the Sunnyside, San Juan County's last big mine, produced 150,000 ounces of gold along with substantial quantities of silver and base metals.

Mineralization in the Silverton area was emplaced within faults and fissures of a caldera, or collapsed volcanic system. Numerous mines and mine dumps have provided countless fine mineral specimens. Primary ore minerals are native gold, argentite, argentifer-

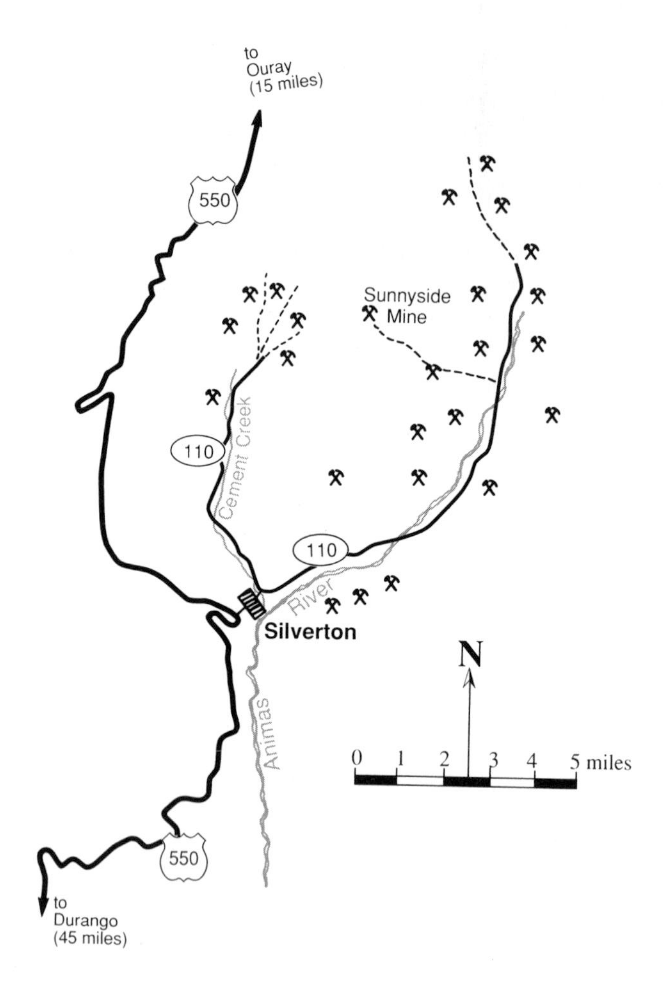
to
Ouray
(15 miles)
550
110
Cement Creek
Sunnyside
Mine
110
River
Silverton
Animas
N
0 1 2 3 4 5 miles
550
to
Durango
(45 miles)

Silverton area mines have produced over $120 million in metals. Hundreds of mine dumps still yield specimens.

ous galena, pyrite, sphalerite, chalcopyrite, and argentiferous tetrahedrite. Gangue minerals include quartz, rhodochrosite, rhodonite, fluorite, and calcite.

Quartz, in thick, massive veins, is the most abundant gangue mineral. Vein seams and pockets are often filled with crystalline quartz in combination with other minerals, forming such colorful composite specimens as green fluorite octahedrons and pink rhodochrosite rhombohedrons on white quartz.

San Juan County has produced most of Colorado's rhodochrosite, but is best known for rhodonite, which occurs in large pink masses in association with quartz and metal sulfides. Collectors have recovered tons of massive, pink rhodonite from the Sunnyside and other mine dumps. Lapidaries fashioned much of it into bookends and other ornamental objects.

Over 200 minerals occur in San Juan County, and mine dumps yield many specimens in miniature and micromount sizes. The mines are easily reached from Silverton by following Colorado Route 110 north along Cement Creek and northeast along the Animas River.

The Old Hundred Mine offers an underground tour of a turn-of-the-century gold mine. The San Juan County Historical Museum displays local minerals and mining artifacts.

REFERENCES: 18, 19, 23, 32, 49, 64, 71, 81

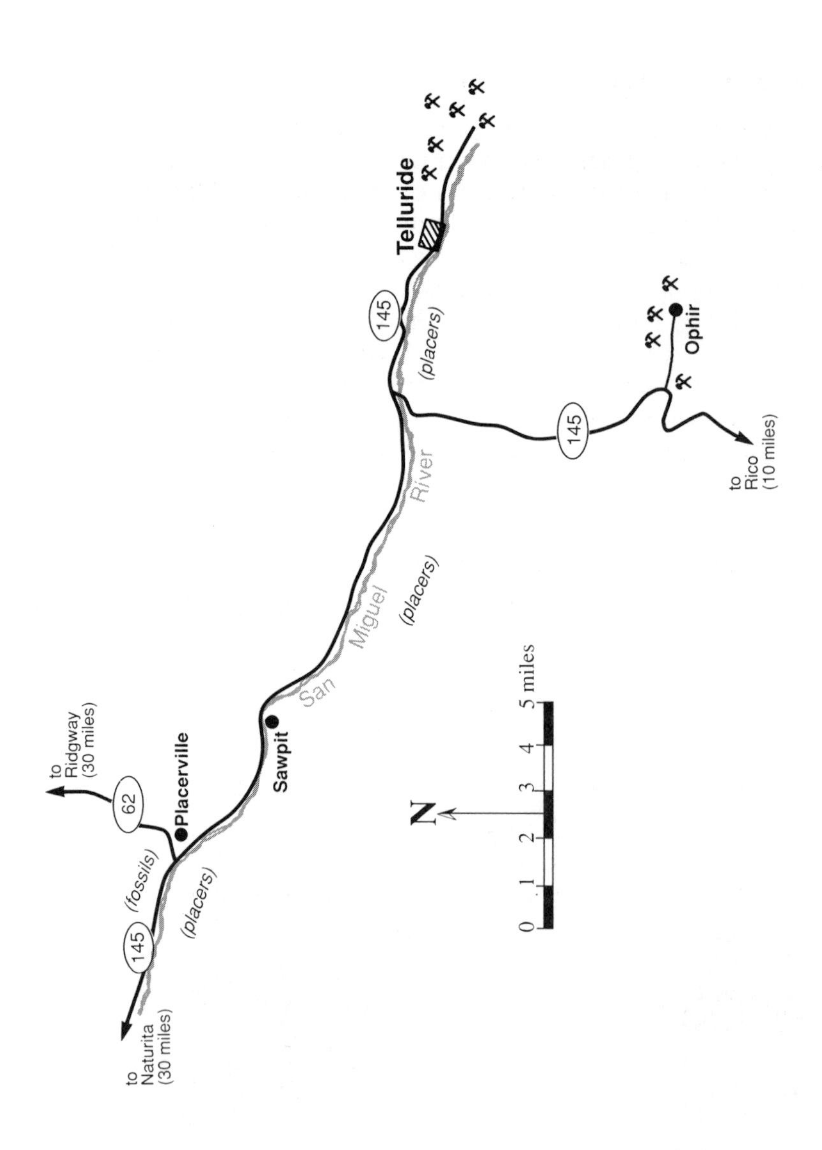

to
Ridgway
(30 miles)
62
Placerville
(fossils)
(placers)
145
to
Naturita
(30 miles)
Sawpit
San
Miguel
River
(placers)
(placers)
145
Telluride
145
Ophir
to
Rico
(10 miles)
N
0
1
2
3
4
5 miles

SAN MIGUEL COUNTY

PLACER GOLD

San Miguel County has produced 10,000 ounces of placer gold, highest among San Juan Mountain counties. Fine placer gold occurs in the gravels and terraces of the upper San Miguel River from its headwaters above Telluride downstream through the towns of Sawpit, Placerville, and Keystone into Montrose County. Gold panning is a popular summer activity.

Telluride Outside Goldpanning (303-725-3895) offers two-hour outings to pan gold in the upper San Miguel River placers. Expert guides provide pans, shovels, tweezers, magnifying glasses, and vials, along with background information on local mining history and geology. The basic fee is $20 for adults and $15 for children and seniors.

REFERENCES: 54

THE TELLURIDE AREA MINES

In 1875 prospectors discovered lode mineralization in Marshall Basin, high above the site of Telluride. Isolation, timberline elevations, and inability to identify and treat complex ores delayed development until the 1880s. Production increased sharply with con-

One-inch quartz crystals from Telluride.

solidation of the Mendota, Smuggler, and Union properties into the famed Smuggler-Union Mine. Selected shipments of Smuggler Vein ore graded as high as 800 ounces of silver and 18 ounces of gold per ton. Other big producers included the Cleveland, Bullion, Hidden Treasure, Liberty Bell, and Tomboy mines.

Mining on the steep mountain slopes of Marshall Basin demanded aerial tramways. Most high mines had full living facilities, for miners were often stranded by heavy snow for weeks. Avalanches sometimes destroyed mills and tramways and claimed lives. Nevertheless, the Marshall Basin mines yielded more than $1 million in gold annually for thirty consecutive years, along with large quantities of silver, lead, zinc, and some copper.

Telluride Mines, Inc., consolidated many properties in 1942 and produced lead, zinc, and copper. The Idarado Mining Company took over in 1953, connecting many individual mines in a sprawling underground network with 300 miles of workings. When mining operations finally halted in 1978, Telluride area mines had produced a quarter-billion dollars in metals.

Ore minerals include pyrite, chalcopyrite, galena, sphalerite, tetrahedrite, and several tellurides, typically in a matrix of white

Telluride area mines have provided a quarter-billion dollars in metals. The richest mines are high in Marshall Basin (center).

quartz. Primary gangue minerals were calcite, barite, siderite, and some fluorite.

Many mineral specimens are found, both at Marshall Basin and the roadside dumps at Pandora, two miles east of Telluride. Several spectacular four-wheel-drive roads lead to the higher mines in Marshall Basin. Commercial jeep tours are available in Telluride.

Marcie Ryan (303-728-4405), a professional geologist, offers personalized tours and mineral-collecting expeditions. Three-hour trips, costing $35 for adults and $20 for children, are scheduled for Monday and Wednesday afternoons; longer collecting trips are by advance arrangement only.

The San Miguel County Museum in Telluride displays local minerals, ores, and mining equipment. Gem Dandy, a Telluride jewelry and rock shop, has specimens of local ores and minerals.

REFERENCES: 18, 19, 23, 41, 67, 81

OPHIR AREA MINES

The silver camp of Ophir, ten miles south of Telluride on Colorado Route 145, offers interesting mine dump collecting. Ophir, and the surrounding Iron Springs Mining District, boomed in the 1880s, then collapsed with the 1893 silver-market crash. Production remained sporadic until 1940, when properties were consolidated for base metal mining.

Several veins contained huebnerite, an ore of tungsten. Hopes for economic survival hung on tungsten through the 1950s, but the last Ophir mines closed in the early 1960s.

Several dozen mines are located near Ophir, which is nestled below the sheer, 2,000-foot-high Ophir Valley walls. Primary ore minerals are argentiferous galena, chalcopyrite, sphalerite, tetrahedrite, and huebnerite; gangue minerals include pyrite, quartz, siderite, fluorite, and calcite.

REFERENCES: 18, 23, 25

PLACERVILLE FOSSILS

West of Telluride, the San Miguel River cuts through sediments deposited over 100 million years of geologic time. Exposures of Cretaceous Mancos Shale near Placerville, at the junction of Colorado Routes 145 and 62, contain numerous fossilized pelecypods, fish teeth, and fish scales.

Near Placerville, lower exposures of earlier Permian sandstones have many important vertebrate fossils. One locale, designated the Placerville Permian Vertebrate Site by the BLM, provides the only well-preserved Permian bone material found in Colorado. Paleon-

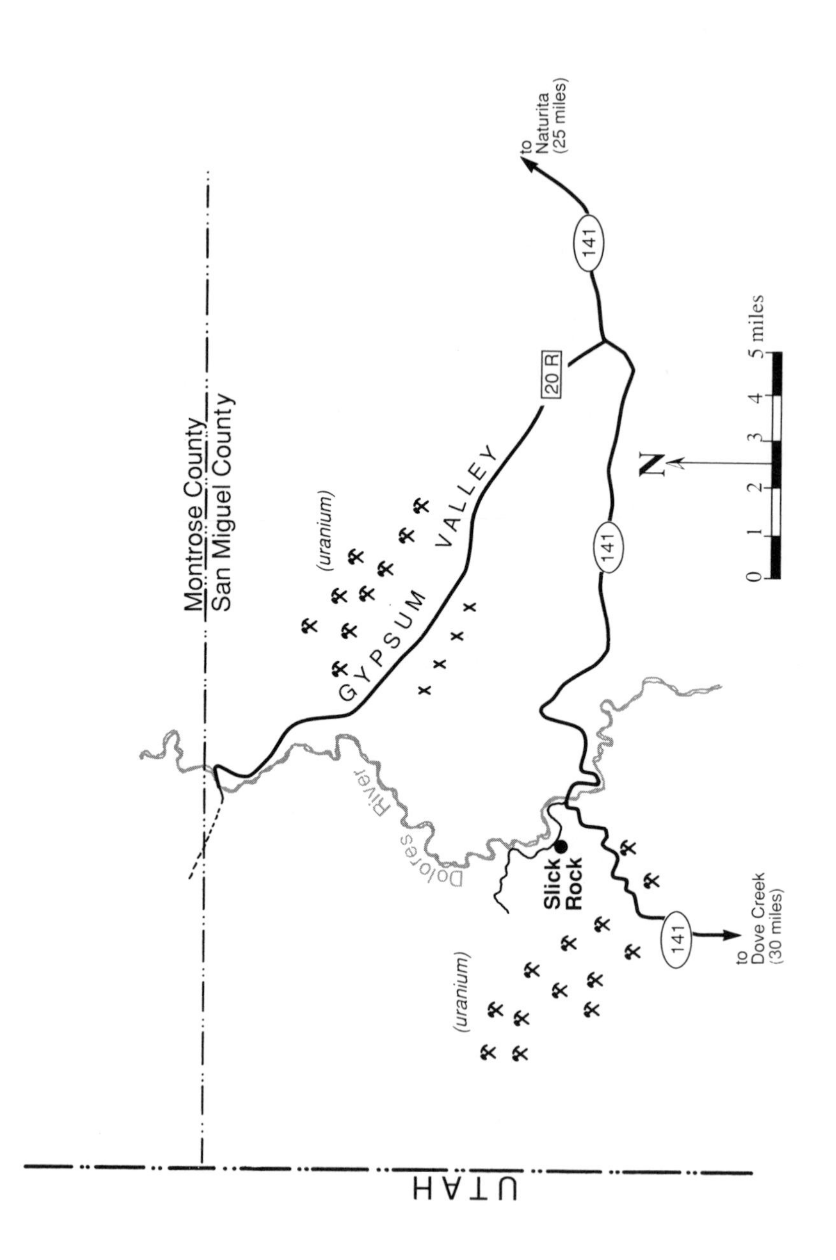
Montrose County
San Miguel County
to Naturita (25 miles)
141
20 R
GYPSUM VALLEY
(uranium)
Dolores River
Slick Rock
to Dove Creek (30 miles)
0 1 2 3 4 5 miles
N
UTAH

tologists have recovered five new vertebrate species at Placerville, including the sail-backed reptile *Platyhistrix*.

Because of inaccessibility, very difficult terrain, steepness of the fossiliferous exposures, and fragmented land ownership, the site is not specially protected. However, it is heavily collected, and since the durable sandstones weather very slowly, few fossils are exposed at any one time.

Unauthorized collecting of vertebrate fossils on public lands is prohibited. For further information on the Placerville Permian Vertebrate Site, contact the Montrose District Office of the Bureau of Land Management.

REFERENCES: 3, 68

THE URAVAN MINERAL BELT

The southern end of the Uravan Mineral Belt extends across western San Miguel County and into adjacent Utah. More than 300 individual uranium mines are grouped near Gypsum Valley and west and south of Slick Rock, a former uranium milling center. Bright yellow carnotite and other radioactive minerals are present on mine dumps.

REFERENCES: 10, 81, 94

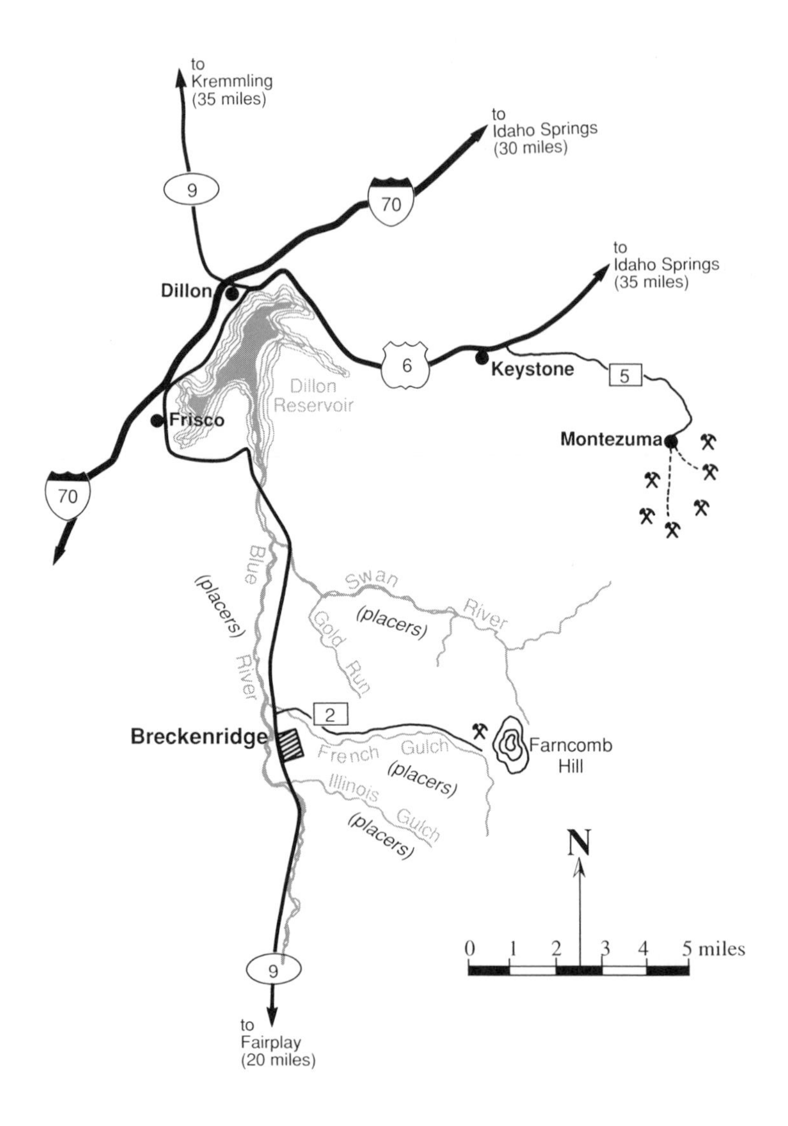
to
Kremmling
(35 miles)
to
Idaho Springs
(30 miles)
9
70
to
Idaho Springs
(35 miles)
Dillon
6
Keystone
5
Dillon
Reservoir
Frisco
Montezuma
70
Blue
(placers)
Swan
(placers)
River
Gold
Run
River
2
Breckenridge
French
Gulch
(placers)
Farncomb
Hill
Illinois
Gulch
(placers)
N
0 1 2 3 4 5 miles
9
to
Fairplay
(20 miles)

SUMMIT COUNTY

PLACER GOLD

Summit County has produced 750,000 troy ounces (over twenty-five standard tons) of placer gold, highest among Colorado's counties, all from the upper Blue River and its tributaries.

In August 1859, gold rush prospectors from South Park crossed the Continental Divide at Georgia Pass to strike gold at Gold Run; in Georgia, American, French and Humbug gulches; and along the Blue and Swan rivers. Some placers were "pound diggings," where a miner could wash out a pound of gold every day with a simple sluice box. The Weaver Brothers, who discovered Gold Run, recovered ninety-six pounds of gold in the first six weeks.

Placer mining was a major industry in Summit County. Miners constructed over 150 miles of ditches and flumes to provide supplemental water for large ground-sluicing and hydraulicking operations.

In 1898 the nation's second floating-bucketline gold dredge began work near Breckenridge. The early "gold boats" were too light for the deep channel gravels, but by 1906 engineers designed heavier and

Ruins of floating-bucketline dredge in French Gulch.

very successful steam- and electric-powered dredges. In 1912 the eight big dredges of the "Breckenridge Navy" each recovered more than 100 troy ounces of gold per day. The last Summit County gold dredge shut down in 1939.

Summit County had coarser gold and more nuggets than other Colorado placer districts. Dredge operators reported many nine- and ten-troy ounce nuggets, and occasionally nuggets of twenty and thirty troy ounces, which were neither documented nor preserved. Nuggets of silver and bismuth sometimes turned up in the sluices.

Today, reclamation is restoring miles of dredge tailing heaps along the Blue and Swan rivers and French Gulch for residential, commercial, and park use. The ruins of two old dredges can still be seen on the Swan River and in French Gulch. Recreational gold panning is popular throughout the upper Blue River drainage.

REFERENCES: 53, 108

FARNCOMB HILL CRYSTALLIZED GOLD

Farncomb Hill, five miles east of Breckenridge in upper French Gulch, has provided Colorado's most spectacular specimens of crystallized gold, many of which are now in top private and museum collections across the country. Harry Farncomb discovered

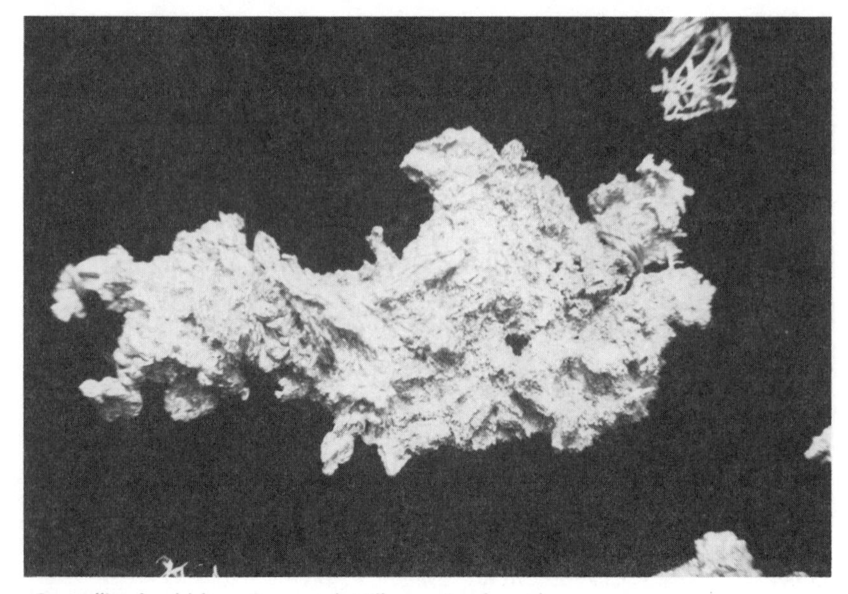

Crystallized gold from Farncomb Hill near Breckenridge.

the veins of crystallized gold in 1879. Farncomb, a French Gulch placer miner for nineteen years, was always curious about the lode source of the coarse French Gulch placer gold. Like others before him, he traced the placers upstream to a low, shale-covered hill which, with no visible vein outcrops, seemed an unlikely source of lode gold.

Nevertheless, Farncomb dug a few shallow holes into the weathered shale to strike a rich eluvial gold deposit where a high-grade vein had weathered away in place. The gold, which Farncomb recovered by panning, occurred as large, delicate leaves and wires.

Farncomb named his strike the Wire Patch and quietly began sluicing the large, delicate leaves and wires of gold. He also acquired the surrounding ground, never by claiming, but by leasing or purchase so as not to attract attention.

In 1880, Farncomb revealed his discovery by depositing 300 ounces of distinctively shaped gold in a Denver bank. Prospectors and promoters rushed up French Gulch to get in on the action, only to learn that Harry Farncomb already controlled everything in sight. Farncomb was accused of "cheating" by depriving others of "fair chance." The subsequent French Gulch shoot-outs and lawsuits went down in local history as the "Ten Years War."

Farncomb soon discovered the tips of several in-situ veins,

"Tom's Baby," mined from Farncomb Hill in 1887, is the largest piece of native gold found in Colorado. As displayed in the Denver Museum of Natural History, it weighs 102 troy ounces.

which he named the Wire Patch Mine. Within two years, Farncomb recovered 7,000 troy ounces of gold worth $140,000, sold out, and retired as one of Colorado's wealthiest men.

Miners found six major vein systems in a small section of Farncomb Hill. The shallow veins were thin and erratic, tending to either pinch out or blossom into fabulously rich pockets of "picture rock." In 1887, two miners blasted into such a pocket, recovering 243 troy ounces of gold in four hours. Miners named the largest piece "Tom's Baby." At 136 troy ounces, it is still the largest single piece of gold ever recovered in Colorado.

"Tom's Baby" was among thousands of beautiful specimens of Farncomb Hill crystallized gold. The gold occurred in three habits: leaf, wire, and spongelike arborescent growths. Leaf gold was most common, appearing as parallel intergrowths of flattened octahedral crystals as long as six inches.

Wire gold occurred as both individual wires and masses of wires, usually less than an inch long and about one millimeter in diameter. Exceptional specimens were two inches long and a quarter-inch in diameter. Miners called dense tangled wire masses "bird's nest" gold. The delicate arborescent growths were flattened, up to three inches long, and slightly flexible.

The Denver Museum of Natural History has a spectacular display of Farncomb Hill crystallized gold specimens, including "Tom's Baby" and the 600-piece John F. Campion collection.

From Breckenridge, follow Colorado Route 9 south for one mile, then take Summit County Road 2 (French Gulch Road) east for five miles through historic French Gulch to the Farncomb Hill area. Farncomb Hill is privately owned and posted.

REFERENCES: 81, 108

MONTEZUMA AREA MINES

Soon after the 1864 discovery of silver in the Argentine District (Clear Creek County), prospectors crossed the Continental Divide and made new strikes near the site of Montezuma. Silver-lead mines prospered through the 1880s, then again in the 1930s and 1940s, producing about 200,000 troy ounces of silver, along with lead, zinc, copper, and small amounts of gold.

Ore minerals are sphalerite, pyrite, tetrahedrite-tennantite, chalcopyrite, and several silver sulfides. Gangue minerals include clear quartz, calcite, and barite. Because of remoteness and rugged country, the mine dumps yield many composite specimens of metal sulfides and quartz. Some dumps also have nice specimens of golden barite.

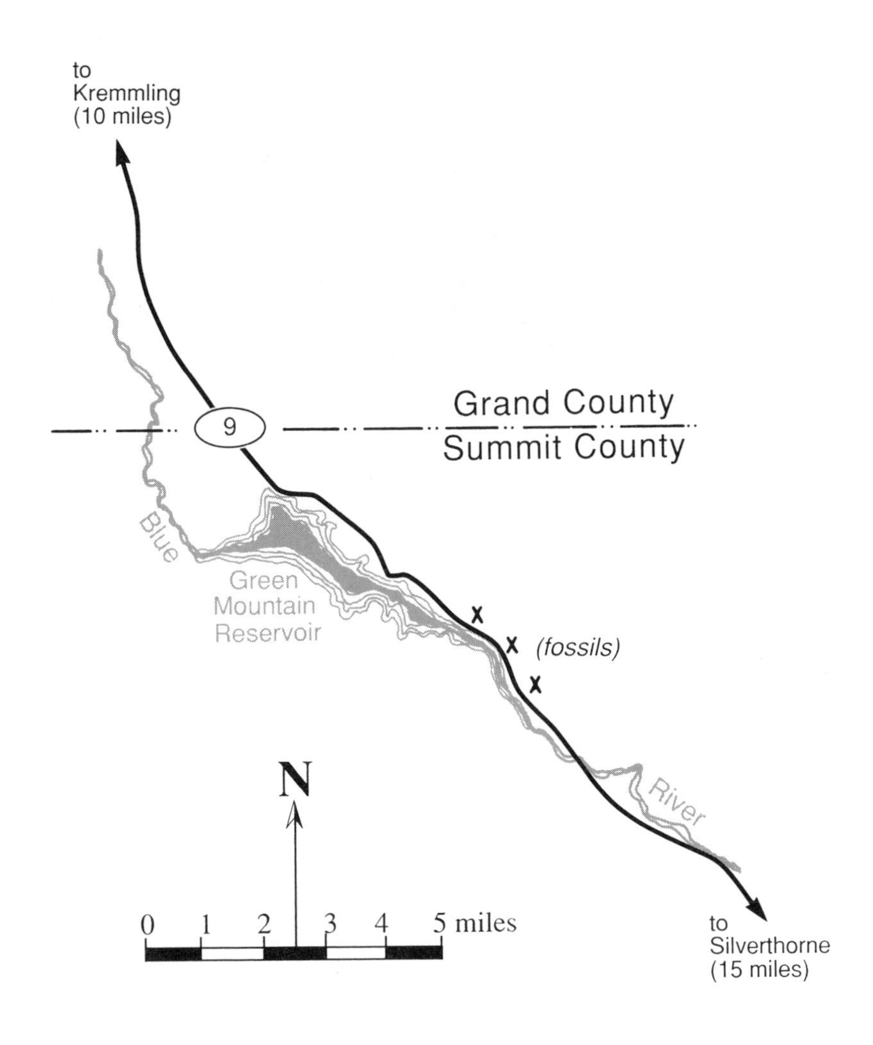
to
Kremmling
(10 miles)
9
Grand County
Summit County
Blue
Green
Mountain
Reservoir
(fossils)
River
N
0 1 2 3 4 5 miles
to
Silverthorne
(15 miles)

An old metal sulfide mine dump awaits collectors near Montezuma.

From the Keystone Ski Area on U.S. 6, take County Road 5 six miles south to Montezuma. Several dozen mines are located south and east of Montezuma.

REFERENCES: 67, 81

SUMMIT COUNTY FOSSILS

There are numerous exposures of Cretaceous Pierre Shale in northern Summit County. Several massive road cuts along Colorado Route 9 near Green Mountain Reservoir and the Grand County line expose large sections of fossiliferous strata.

Pelecypod and cephalopod fossils occur within rusty concretions in the dark gray, papery shale. Many of the fossil-bearing concretions crumble easily. Fossils from concretions occurring in the highest exposed strata seem to form more durable casts that retain parts of the original lustrous shell material.

REFERENCES: 12

TELLER COUNTY

CRIPPLE CREEK AREA GOLD MINES

Cripple Creek, a twenty-square-mile district located 9,600 feet high on the southwest shoulder of Pikes Peak, is among North America's richest gold sources.

Bob Womack, a cowboy and part-time prospector, sank a shallow shaft into the El Paso Lode in 1890, after geologists had dismissed Cripple Creek as an unlikely area for significant mineralization. Word of Womack's strike attracted other prospectors, among them Winfield Scott Stratton, a Colorado Springs carpenter, who struck gold at nearby Victor in 1892. When Stratton dug out $200,000 in rich ore, 10,000 miners, prospectors, speculators, and adventurers rushed to Cripple Creek. In 1893, the mines produced $2 million in gold.

Cripple Creek's gold occurred in a brecciated volcanic plug. Thirty-five million years ago, an ancient volcano collapsed and filled with water and sediments. Repetitive fracturing and cement-

The old gold-mining town of Victor nestles at the southern end of the Cripple Creek mining district.

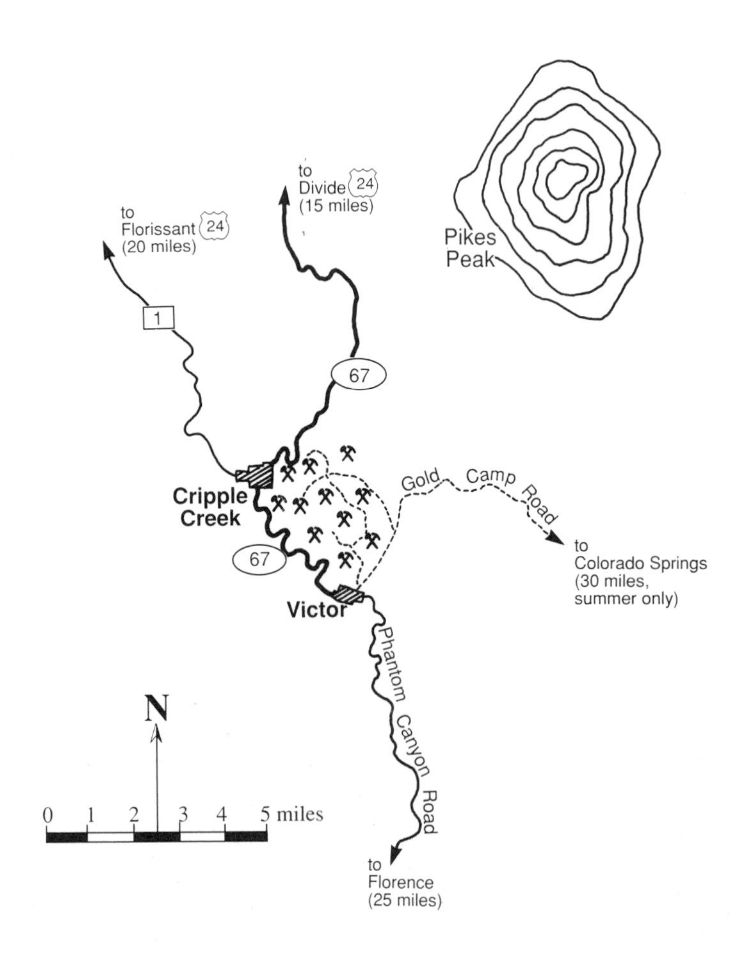

to
Divide 24
(15 miles)
to
Florissant 24
(20 miles)
Pikes
Peak
1
67
Gold Camp Road
to
Colorado Springs
(30 miles,
summer only)
Cripple
Creek
67
Victor
Phantom Canyon Road
N
0 1 2 3 4 5 miles
to
Florence
(25 miles)

ing eventually formed a breccia "neck" 3,000 feet deep. Renewed volcanic surges forced magma through the fractured breccia, creating dikes and pipes of volcanic rock. Associated hot mineral-bearing solutions impregnated the breccia with gold, primarily as the telluride minerals sylvanite and calaverite, and sometimes as extraordinarily rich veins and pockets.

In 1895, Cripple Creek produced 300,000 troy ounces of gold, then 500,000 troy ounces in 1897. Production peaked in 1900, when 475 mines poured out 910,000 troy ounces—thirty standard tons—of gold worth $18.2 million. Cripple Creek remained the leading U.S. gold district for the next consecutive fifteen years. Select ores contained thousands of ounces of gold per ton, and assayers sometimes expressed grades not in the conventional troy ounces per ton, but in dollars per pound.

The most celebrated high-grade pocket was the "Cresson vug." In 1914, miners in the Cresson Mine blasted into a thirty-six-foot-long geode, or vug, filled with sylvanite and calaverite. The richest ore graded $16,000 per ton. In one month, miners extracted 60,000 troy ounces of gold worth $1.2 million.

By 1920, 475 Cripple Creek mines had produced 500 tons of gold worth a third of a billion dollars, and Cripple Creek ranked second among the world's gold-producing districts. Gold mining faded in the 1950s, but is now making a comeback as miners turn to open-pit heap-leach methods to exploit low-grade ores.

Cripple Creek is located at the north end of the district; Victor, a one-time rival for supremacy and site of the richest mines, is at the south end. State and county roads make the district easily accessible.

Although most properties are now consolidated, corporate-owned, and posted, some collecting is possible. District auto tour routes pass huge mine dumps that spill onto road shoulders. Dump material contains specimens with small crystals of sylvanite and calaverite. Purple fluorite is a common gangue mineral. Inexperienced collectors may confuse sylvanite and calaverite with pyrite.

THE CRIPPLE CREEK DISTRICT MUSEUM

The Cripple Creek District Museum, acclaimed as one of the nation's finest small museums, has a superb collection of mining and assaying equipment, claim maps, models of major local mines, and specimens of calaverite, sylvanite, and free gold ores.

MOLLIE KATHLEEN GOLD MINE

Just east of Cripple Creek on Colorado Route 67, the Mollie Kathleen Mine is Colorado's only shaft-type underground tour mine.

One thousand feet below the surface, miner-guides explain underground equipment and workings, then provide visitors with samples of gold ore.

REFERENCES: 18, 19, 23, 67, 81, 91, 108

FLORISSANT FOSSILS

Florissant Fossil Beds National Monument, located five miles south of Florissant on Teller County Route 1, is North America's premier insect-fossil site. Insect paleontologists (paleoentomologists) rate Florissant behind only two other sites in the world: Germany's Solnhofen quarry and the Baltic Coast amber sources.

The Florissant fossil deposits are a record of life in an early Oligocene forest. Paleontologists believe the local paleoenvironment 35.5 million years ago was wet, semitropical, and dominated by 300-foot-tall sequoia trees. Oligocene forest life was interrupted by intense volcanic activity fifteen miles to the west in the Thirty-Nine Mile Volcanic Field (Park County). Geologists compare the subsequent series of destructive events to the recent eruptions of Washington's Mount St. Helens. Fifteen-foot-deep mudflows, rich in volcanic matter and dissolved silica, swept through the forests,

Florissant has always been popular with mineral and fossil collectors. Crystal Peak rises in the distance.

burying all but the massive sequoias. The upper exposed sections of the dead trees decayed normally; the low trunk sections, buried in an oxygen-devoid environment, began the slow process of cellular petrifaction.

The mudflows dammed a drainage, creating an irregular lake twelve miles long and two miles wide that supported a profusion of plant, insect, and fish life. Along its shores, recovering forests of graceful palms and broad-leafed trees became habitat for Oligocene birds and mammals similar to modern species.

Meanwhile, diminishing eruptions in the nearby volcanic field periodically generated clouds of powderlike volcanic ash. The ash fell onto the ancient lake, carrying millions of insects and leaves to the bottom, where successive layers of fine volcanic mud provided near-perfect conditions for fossilization.

Seasonal floods washed accumulated volcanic ash from adjacent hills into the lake as a coarse mud, adding to the layered, deepening sediments. A final volcanic eruption generated a mudflow that covered the ancient lake. The deeply buried, layered lake bottom eventually became a fine-grained, thinly laminated gray shale; the seasonal runoff of hillside ash and sediments formed intermittent

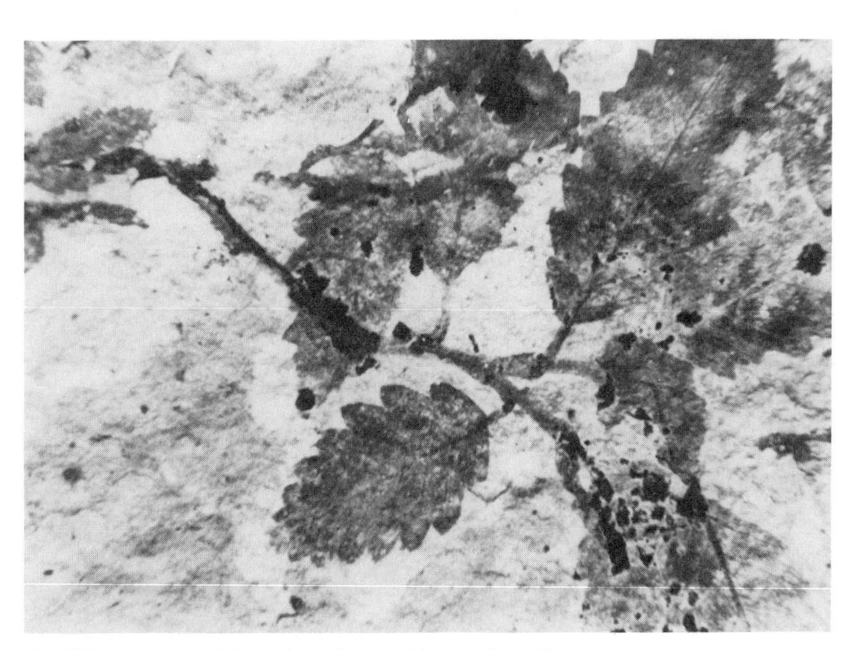

Fossil impressions of branch, twigs, and leaves from Florissant.

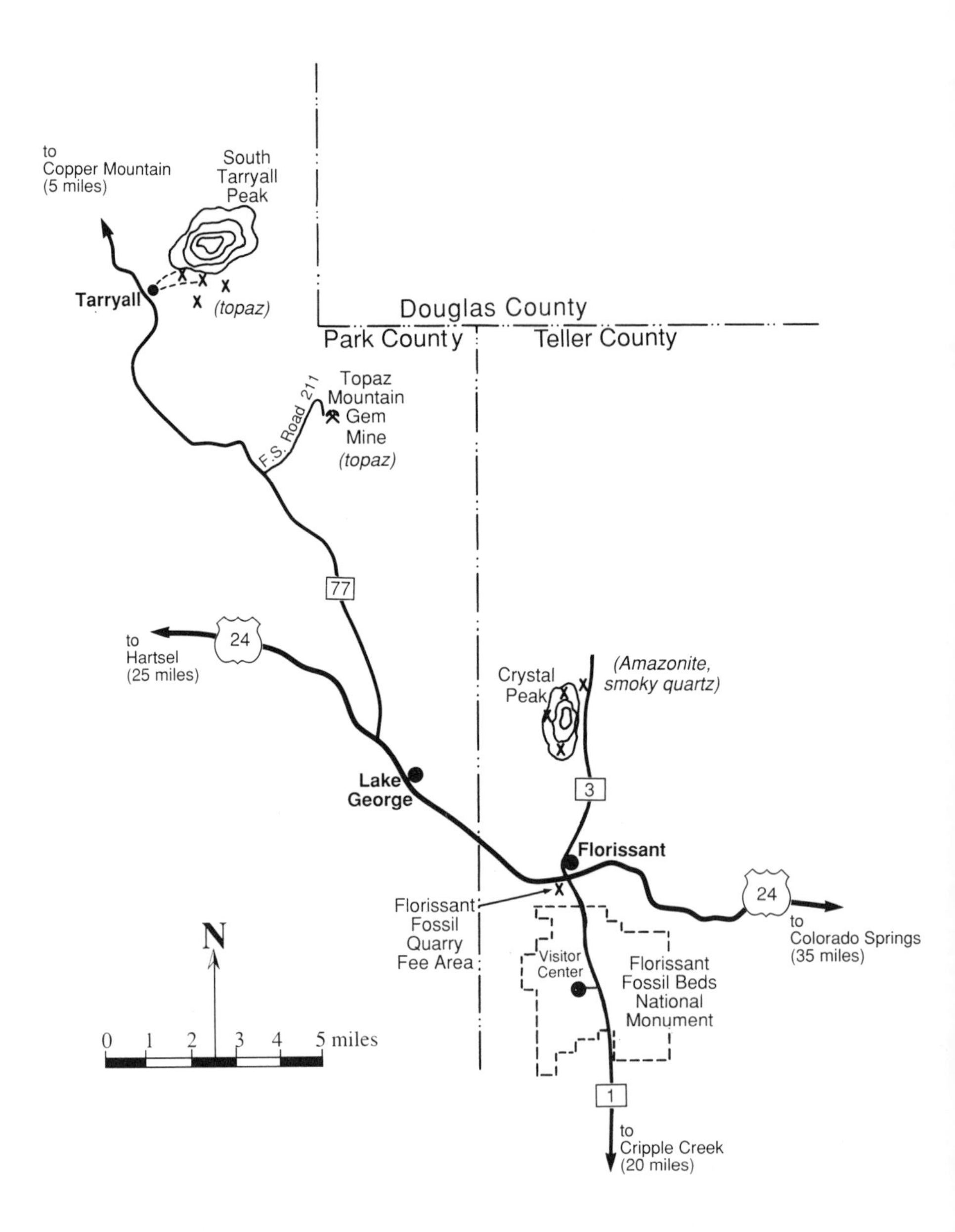

to
Copper Mountain
(5 miles)
South
Tarryall
Peak
Tarryall
(topaz)
Douglas County
Park County
Teller County
Topaz
Mountain
Gem
Mine
(topaz)
F.S. Road 211
77
to
Hartsel
(25 miles)
24
Crystal
Peak
(Amazonite,
smoky quartz)
3
Lake
George
Florissant
Florissant
Fossil
Quarry
Fee Area
Visitor
Center
Florissant
Fossil Beds
National
Monument
24
to
Colorado Springs
(35 miles)
N
0 1 2 3 4 5 miles
1
to
Cripple Creek
(20 miles)

Caterpillar fossil from Florissant.

Five-inch fish fossil from Florissant.

layers of a coarse, crumbly, yellow "mudstone," while the final mudflow became a durable, protective layer of caprock.

Ten million years later, the region uplifted to its current 8,000-foot elevation. The remains of the ancient lake became a sedimentary "island" resting atop granite country rock. The climate became cooler and drier, and surface erosion accelerated. Erosion finally penetrated the caprock in relatively recent times, revealing the fossil treasure in the softer shales below.

Fossil impression of Oligocene plant from Florissant.

In the 1860s, Florissant pioneers found abundant fossils and huge petrified tree trunks still in their original upright positions. Thomas Mead, a New York entomologist, visited Florissant on an 1871 butterfly-collecting expedition. Mead sketched the great stone stumps and collected insect fossils for Dr. Samuel H. Scudder, the nation's leading paleoentomologist. Dr. A. C. Peale, a geologist with the Hayden Survey, visited Florissant in 1874, further publicizing the fossil deposits.

Scudder supervised a major dig at Florissant in 1877, recovering in one summer more fossil insect species than had been collected in thirty years at Switzerland's Oeningen deposit, previously the world's leading insect-fossil source.

By the 1930s, Florissant had yielded 80,000 museum-grade specimens representing 1,100 insect species, 140 plant species, and dozens of species of birds, fish, and small mammals.

But the number of scientifically cataloged specimens paled before those removed by amateurs, who dug and sold tons of fossil-bearing shale. Souvenir hunters hammered whole petrified sequoia trunks to pieces, while commercial "collectors" used diamond-bonded circular and chain saw blades to cut petrified stumps into moveable pieces.

"The Trio," three huge petrified sequoia trunks still stand in their original upright positions at Florissant Fossil Beds National Monument.

During the post-World War II boom in western automobile tourism, entrepreneurs promoted the petrified trees as tourist attractions, as described by Richard Pearl in *The Mineralogist* in April 1953.

> Among the largest known petrified tree stumps, and a remarkable group of three standing stumps, called The Trio and unequalled anywhere else in the world, are the outstanding features of two petrified forests in the Pikes Peak region of Colorado.
>
> . . . you will reach the entrance to the Colorado Petrified Forest, the first of two commercial enterprises. Just beyond is the Pike Petrified Forest, which contains the more spectacular

> exhibits; Formerly called the Henderson Petrified Forest, and before that, the New Petrified Forest . . . The two "forests" are competitors, and the elderly owner of the Colorado forest has been known to use vigorous methods to try to "persuade" visitors to enter his property instead of the others.

In 1969, Colorado lawmakers urged the federal government to establish Florissant Fossil Beds National Monument to protect the fossil deposits. The museum at the visitor center has a fine collection of insect and plant fossils. Enlarged photographs show remarkable fossil detail, which in some cases has preserved the impression of individual hairs on the legs of bee-sized insects. Other exhibits trace the geologic events that created the fossil deposits 35 million years ago.

Visitors may view fossil-bearing shale exposures and fifteen large petrified sequoia stumps, the largest thirteen feet in diameter. The Florissant stumps are larger than those at Arizona's Petrified Forest National Park, but not as colorful or well silicified. The "Big Stump," eleven feet high and ten feet in diameter, is still embedded with rusted saw blades, left by "collectors" who tried to saw the great stump into pieces for shipment to the 1892 Chicago World's Fair.

Fossil collecting is prohibited at Florissant Fossil Beds National Monument.

REFERENCES: 12, 33, 62, 63, 68, 72, 95

NATURE'S WEALTH FOSSIL SHOP AND FLORISSANT FOSSIL QUARRY FEE COLLECTING AREA

Located on Teller County Route 1 just south of Florissant, this shop offers hundreds of Florissant fossil specimens for sale and displays a fine private collection.

A nearby outcrop of fossil-rich lake-bed shale is a popular fee collecting site. Collectors do not dig, but simply gather and split shale fragments. Florissant shale splits readily along its planes to cleanly expose fossil-bearing layers. The site is bladed regularly to uncover new sections of fossil-bearing shale.

Fossils range from the tiniest insects to five-inch specimens of plant stems, leaves, and fish. The most sought-after fossils are those of insects as long as two inches. Even small fossils reveal fascinating detail under 10X magnification.

The Florissant shales continue to provide fossil records of new insect species. Unusual fossil finds are sent to Dr. F. Martin Brown, a research associate with the University of Colorado Museum at Boulder and currently the leading authority on Florissant insect fossils.

Hundreds of collectors, including many regulars and club and school groups, visit the Florissant Fossil Quarry each year. For information contact: Nature's Wealth Fossil Shop, P.O. Box 5, Florissant, Colorado 80816 (719)748-3805. For information on fee collecting contact: Florissant Fossil Quarry, P.O. Box 126, Florissant, Colorado 80816 (719)748-3275.

TELLER COUNTY PEGMATITES

Teller County lies entirely within the Pikes Peak Batholith, a granitic intrusion covering 1,200 square miles. Teller County's granite pegmatites are a major source of amazonite, gem topaz, phenakite, and smoky quartz for both commercial and amateur collectors.

CRYSTAL PEAK–LAKE GEORGE AREA

Crystal Peak, three miles north of Florissant in northwestern Teller County, and the area north of Lake George in adjacent Park County have the most productive pegmatites.

Early settlers found float amazonite, smoky quartz, and topaz in alluvial gravels and traced them to in-situ pegmatite sources. The most popular collecting locale was Crystal Peak, also known as Crystal Butte, Topaz Butte, and the Florissant Crystal Beds.

In 1867, mineralogists formally recognized amazonite, the green-blue variety of microcline feldspar, as a Colorado mineral. Commercial collecting at Crystal Peak began in the 1870s, when Dr. A. E. Foote, of Philadelphia's Foote Mineral Company, hired miners to gather specimens. Experienced Crystal Peak collectors guided visitors on collecting expeditions in the 1880s. Commercial mining in the 1930s, usually by trenching into decomposed granite, was profitable and an important local tourist attraction. Crystal Gem Mines, a Crystal Peak fee collecting area, operated until the 1960s.

Collectors have recovered numerous superb museum-grade specimens of amazonite, smoky quartz, and topaz in the Crystal Peak-Lake George area. Crystal Peak pegmatites have yielded smoky quartz crystals weighing over 100 pounds, and one purported to weigh 400 pounds. The National Museum of Natural History (Smithsonian) has an elliptical, brilliant-cut smoky quartz gem from Crystal Peak weighing 785 carats.

In *Colorado Gem Trails and Mineral Guide,* Richard Pearl described typical pegmatite pockets as about six feet long and containing

Alluvial topaz crystal screened from gravel at South Tarryall Peak.

from 500 to 1,000 amazonite and smoky quartz crystals. A fifteen-foot-long pocket found in 1909 yielded $3,500 worth of smoky quartz, amazonite, and fluorite crystals.

Other pegmatite minerals present are phenakite, goethite, pseudomorphs of goethite after siderite, pale amethyst and rock crystal varieties of quartz, and muscovite and biotite mica.

Today, the Crystal Peak area is rural-residential. Known pegmatites are privately owned or under claim by individuals and mineral clubs, and permission to collect is required. Recent recoveries include well-formed, doubly terminated smoky quartz crystals as long as sixteen inches.

SPRUCE GROVE–SOUTH TARRYALL PEAK AREA

The area north of Lake George and east of Tarryall Creek (in eastern Park County) is noted for topaz. Early literature mentions topaz crystals weighing over one pound. The topaz is usually clear or light shades of blue and sherry.

Since 1909, collectors have recovered topaz and smoky quartz from the western and southern slopes of South Tarryall Peak, north of Lake George on Park County Road 77. Most crystals are now recovered from alluvial gravels and talus rock at the base of granite cliffs and slopes. Digging and screening decomposed granite gravel is the most productive collecting method.

From Lake George, follow Park County Road 77 thirteen miles

Cut topaz gems from Topaz Mountain Gem Mine fee collecting area.

north to the Spruce Grove Campground. The general collecting areas are within 1.5 miles east and southeast of the campground, and are reached by foot trails and rough jeep roads. Some sites are under claim. Adjacent land in Pike National Forest provides excellent pegmatite-prospecting opportunities.

TOPAZ MOUNTAIN GEM MINE FEE COLLECTING AREA

The Topaz Mountain Gem Mine (Park County) is an alluvial source of topaz crystals that have weathered from nearby granite pegmatite dikes.

Miners have trenched through ten feet of overburden to alluvial topaz concentrations in a red clay matrix on decomposed granite bedrock. Collectors complete the work on a fee basis, washing presorted bedrock gravel in water to loosen the clay, then inspecting it on screens. Any topaz crystals present, because of vitreous luster and brilliant transparency, are immediately recognizable.

Like all types of prospecting, finding topaz crystals in the gravel is a matter of luck. Some buckets contain none, others have yielded as many as fifteen pieces collectively weighing 125 carats. An average five-gallon (fifty-pound) bucket of gravel contains one to four crystals with a total weight of about twenty carats. Not all crystals are of gem quality. Because of hardness (Mohs 8.0) and short alluvial travel, crystal faces are only lightly abraded.

The topaz is clear, light blue, and light sherry. Larger crystals of more than fifty carats often display an interesting bicolor effect due partly to optical properties. Doubly refractive orthorhombic crystals, such as topaz, are usually trichroic, but the third color shift is rarely apparent. As viewing angle changes with rotation of the crystal, the color shifts between light blue and sherry. Cutting services are available for gem-quality rough topaz.

The Topaz Mountain Gem Mine is located seven miles north of Lake George on Park County Road 77. Turn east on Forest Service Road 211 (Matukat Road) and proceed two miles to the mine.

For information contact: Topaz Mountain Gem Mine, 2010 Wold Avenue, Colorado Springs, Colorado 80909 (719)596-5492.

REFERENCES: 1, 5, 19, 32, 41, 51, 58, 64, 66, 67, 73, 74, 84, 96

WELD COUNTY

STONEHAM BARITE

The Chalk Cliffs, near Stoneham, thirty miles north of Fort Morgan, are a well-known source of large, beautifully formed, blue barite crystals.

During the Oligocene epoch 25 million years ago, intense volcanic activity in the Rockies hundreds of miles to the west produced dense clouds of ash. The ash settled on the plains in layers, becoming interbedded with shale and eventually altering into a montmorillonite clay. Groundwater circulation later deposited barite within seams of the shale layers.

Erosion has since generally reduced the surface of the plains to the level of upper Cretaceous sediments. The Chalk Cliffs, however, remained as an isolated section of the once-overlying Oligocene shales and clays. Unlike the surrounding flat terrain, the Chalk Cliffs are badlands of steep, rapidly eroding bluffs and deep gullies exposing shale layers containing seams of both calcite and barite.

The Stoneham barite locality is in the Chalk Cliffs, an area of deep, rapidly eroding bluffs and gullies.

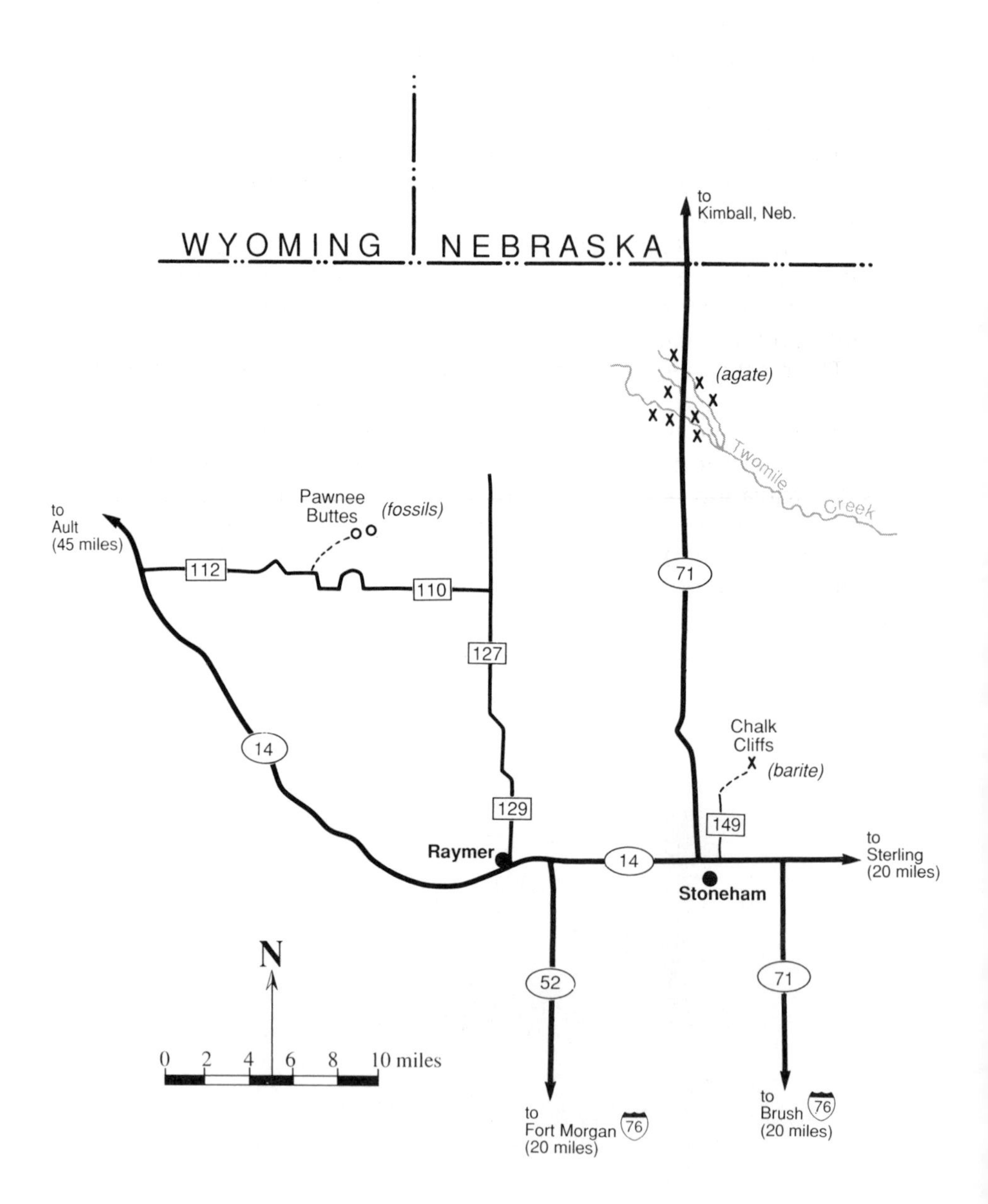
WYOMING
NEBRASKA
to
Kimball, Neb.
(agate)
Twomile
Creek
to
Ault
(45 miles)
Pawnee
Buttes
(fossils)
112
110
127
71
14
Chalk
Cliffs
(barite)
129
149
Raymer
14
Stoneham
to
Sterling
(20 miles)
N
52
71
0 2 4 6 8 10 miles
to
Fort Morgan 76
(20 miles)
to
Brush 76
(20 miles)

Stoneham barite crystals are transparent, usually very well developed, terminated, and sometimes doubly terminated. Their delicate blue color, comparable to that of fine aquamarine, is caused by the trace presence of radioactive minerals. Thin tabular crystals appear colorless.

The crystals average an inch or two in length, although exceptional specimens may measure six inches. The barite, found in fractures, seams, and cavities in the shale layers, is often associated with white to yellow drusy calcite. In the most attractive specimens, clean blue barite crystals project upward from a drusy calcite base. Bladed barite crystals, usually white or very pale blue, are present in roselike clusters.

The most productive section of the twenty-acre collecting area is along the western edge of the Chalk Cliffs. Crystal-filled seams and pockets in the weathered shale range in size from only a few inches to ten feet. Finding mineralized seams can require considerable excavation. Although hand digging can be productive, professionals employing mechanical excavation equipment have collected most of the Stoneham barite specimens on the market today. A system of deep gullies extending northward from the bluffs contains much fragmented barite.

Sand-calcite crystals occur in sediments above the barite-bearing

Blue barite on white drusy calcite from Stoneham.

Blue barite crystal in white calcite from Stoneham.

Three-inch blue barite crystals from Stoneham.

shales. As a cementing material, calcite forms euhedral crystals, incorporating clastic material (sand) in the process. The sand-calcite crystals appear as light gray, weather-rounded rhombohedrons.

Collectors in the Chalk Cliffs also find fossils, mostly of teeth, and colorful jasper and agate.

The Chalk Cliffs include both privately owned and public land. Most recent barite recoveries have come from private land. Although policies have varied, permission to collect is usually granted. Collectors should check with landowners who reside locally.

Stoneham is located twenty-five miles north of Fort Morgan and west of Sterling on Colorado Route 14. From the junction of Colorado Routes 71 and 14, drive one mile east on Colorado Route 14, then turn north on Weld County Road 149 and proceed three miles toward the low, buff-colored Chalk Cliffs. Take the right fork at the cattle guard and continue another mile. A faint track to the right leads to the rim of the Chalk Cliff gullies and the collecting sites.

The Stoneham locality is representative of other barite sites in the region. Any section of "badlands," locally called "blowouts," with buff-colored low bluffs and gullies, may have seams of barite and calcite in exposed Tertiary sediments.

REFERENCES: 5, 8, 11, 46, 64, 67, 87

TWOMILE CREEK AGATE

Banded agate and jasper are present along upper Twomile Creek. From Stoneham, follow Colorado Route 71 twenty miles north. North of the intersection of Weld County Road 124, and five miles south of the Wyoming line, Colorado Route 71 crosses three broad gullies of the Twomile Creek drainage. Collectors find agate and jasper in both the gully bottoms and on adjacent terraces.

REFERENCES: 66

PAWNEE BUTTES

The Pawnee Buttes, forty miles north of Fort Morgan in northeastern Weld County, are prominent regional landmarks. Exposed Tertiary sandstones are a rich source of Oligocene and Miocene mammal fossils.

Paleontologists discovered the fossils in 1907. By 1920, university and museum paleontologists had recovered the fossilized skeletons of over a hundred species of Oligocene land vertebrates, including camels, horses, and rhinoceroses. Agate, petrified wood, and a locally common brown jasper are found in the gullies.

Pawnee Buttes, landmarks of the high plains, have yielded numerous Tertiary mammal fossils.

Pawnee Buttes, a part of the Pawnee National Grasslands, is north and east of Colorado Route 14; access routes along a network of graded county gravel roads are marked by U.S. Forest Service signs. From the intersection of County Roads 110½ and 113, Forest Service Road 807 leads directly to the base of the buttes.

REFERENCES: 64

COLORADO MINERAL GUIDE

ACTINOLITE** see **TREMOLITE-ACTINOLITE

AGATE** see **QUARTZ, cryptocrystalline

ALABASTER** see **GYPSUM

AMAZONITE** see **ORTHOCLASE, var. microcline

AMETHYST** see **QUARTZ, crystalline

ANALCIME** see **ZEOLITE GROUP

APATITE — *Calcium fluorine-chlorine-hydroxyl phosphate*

Color/Luster: Green, red, yellow, violet, pink, colorless, white; vitreous, greasy.
Hardness: 5 **S.G.:** 3.2 **Streak:** White
Crystals: Hexagonal, prismatic, or tabular. Also massive, granular, compact, earthy. Cleavage poor in one direction crosswise.
General: Brittle; transparent to translucent; fracture uneven, conchoidal.
Notable County Occurrences: Chaffee, Larimer.

AQUAMARINE** see **BERYL

ARAGONITE — *Calcium carbonate*

Color/Luster: White, gray, colorless or pale shades of pink, yellow, violet, or brown; vitreous, resinous.
Hardness: 3.5-4 **S.G.:** 2.9-3 **Streak:** White
Crystals: Orthorhombic, usually short to long prisms; also stalactic aggregates and columnar. Twinning common, often showing pseudohexagonal cross section. Cleavage good in one direction, poor in two others.
General: Brittle; fracture subconchoidal; transparent to translucent; usually fluorescent. Aragonite is chemically identical to calcite but crystallizes in the orthorhombic system. It often occurs in evaporite beds with calcite and gypsum.
Notable County Occurrences: Larimer.

ARGENTITE (ACANTHITE) — *Silver sulfide*

Color/Luster: Lead gray to black; metallic.
Hardness: 2-2.5 **S.G.:** 7.2-7.4 **Streak:** Shiny black
Crystals: Isometric, (rare) as cubes or octahedrons. Usually as granular masses, coatings, or crusts. Cleavage indistinct.
General: Opaque; fracture indistinct. Acanthite, chemically identical to argentite, is the orthorhombic form existing at lower temperatures.
Notable County Occurrences: Boulder, Clear Creek, Custer, Dolores, Hinsdale, Lake, Mineral, Ouray, Park, San Juan, Summit.

ARSENOPYRITE ***Iron arsenide sulfide***

Color/Luster: Silver-white to steel gray; metallic.
Hardness: 5.5-6 **S.G.:** 6-6.2 **Streak:** Gray-black
Crystals: Monoclinic, usually as prisms with striated faces. Also granular and compact. Cleavage distinct in one direction.
General: Brittle; fracture uneven.
Notable County Occurrences: Gunnison, Ouray.

AZURITE ***Basic copper carbonate***

Color/Luster: Azure blue to dark blue; vitreous, dull.
Hardness: 3.5-4 **S.G.:** 3.8 **Streak:** Blue
Crystals: Monoclinic, usually equidimensional or tabular, small, well-formed, brilliant, deep blue, transparent crystals. Cleavage good in two directions.
General: Brittle; fracture conchoidal. Occurs in radiating, earthy, and botryoidal form. Often associated with malachite. Azurite, an alteration product of copper sulfides and formerly an ore of copper, has ornamental use.
Notable County Occurrences: Custer, Hinsdale, Lake, Montezuma, Ouray.

BARITE ***Barium sulfate***

Color/Luster: White, gray, colorless, blue, and light shades of green, brown, and red; vitreous, pearly.
Hardness: 3-3.5 **S.G.:** 4.5 **Streak:** White
Crystals: Orthorhombic, usually tabular, sometimes prismatic. Also fibrous, compact, and granular. Cleavage perfect in one direction, good in a second, distinct in a third.
General: Fracture uneven; transparent to translucent. Density is unusually high for a light-colored mineral. Used in glass and paint manufacturing and high-density drilling muds. Barite is a common gangue mineral in some metal sulfide mines.
Notable County Occurrences: Boulder, Chaffee, Eagle, Gunnison, Hinsdale, Lake, La Plata, Logan, Mesa, Mineral, Otero, Ouray, Park, Pueblo, Rio Grande, Saguache, San Miguel, Summit, Weld.

BERYL ***Beryllium aluminum silicate***

Color/Luster: Blue, green, pink, yellow, brown, white, colorless; vitreous.
Hardness: 7.5-8 **S.G.:** 2.66-2.9 **Streak:** Colorless
Crystals: Hexagonal, as six-sided prisms striated lengthwise. Cleavage indistinct in one direction. Also massive and as impure crystals of "common" beryl often several feet in length.
General: Fracture uneven to conchoidal; transparent to translucent. Beryl crystals are valuable gemstones named according to color: aquamarine, blue to blue-green; emerald, green; morganite, pink; goshenite, colorless; heliodor and golden beryl, brown to golden yellow. Aquamarine is the Colorado State Gem and the birthstone for October. Beryl is the primary ore of beryllium. Colorado beryl occurs almost exclusively in granite pegmatites.
Notable County Occurrences: Chaffee, Fremont, Grand, Gunnison, Larimer.

BIOTITE
Basic potassium magnesium iron aluminum silicate (mica group)

Color/Luster: Black, brownish black, greenish black; pearly, submetallic.
Hardness: 2.5-3 **S.G.:** 2.8-3.4 **Streak:** Colorless
Crystals: Monoclinic, usually tabular, also plates and scales. Cleavage perfect in one direction.
General: Opaque to translucent; forms tough, elastic, thin plates. Biotite is a common accessory mineral in many intrusive and volcanic rocks; large, well-formed crystals occur in granite pegmatites.
Notable County Occurrences: Chaffee, Park, Teller.

BISMUTHINITE
Bismuth trisulfide

Color/Luster: Lead gray to light gray; metallic.
Hardness: 2 **S.G.:** 6.8 **Streak:** Lead gray
Crystals: Orthorhombic, as long, striated prismatic crystals (rare); usually massive with foliated or fibrous texture. Cleavage perfect in one direction lengthwise.
General: Brittle; fracture uneven; crystals flexible.
Notable County Occurrences: Chaffee, Larimer.

BORNITE
Copper iron sulfide

Color/Luster: Copper red to bronze brown, tarnishes to deep blue and purple; metallic.
Hardness: 3 **S.G.:** 4.9-5.1 **Streak:** Gray-black
Crystals: Isometric (rare).
General: Brittle; fracture conchoidal, uneven. Usually compact or granular. Known as "peacock copper" for its bright tarnish. Bornite is an ore of copper.
Notable County Occurrences: Hinsdale, Saguache.

CALAVERITE
Gold ditelluride

Color/Luster: Brass yellow to silver-white; metallic.
Hardness: 2.5-3 **S.G.:** 9.1-9.4 **Streak:** Greenish or yellowish gray
Crystals: Monoclinic, usually as small blades or short striated prisms (unlike pyrite). No cleavage.
General: Brittle; fracture conchoidal. Calaverite is an ore of gold and often contains some silver.
Notable County Occurrences: La Plata, Teller.

CALCITE
Calcium carbonate

Color/Luster: White, colorless, and pale shades of red, brown, yellow, and blue; vitreous, dull.
Hardness: 3 **S.G.:** 2.7 **Streak:** White
Crystals: Hexagonal, usually as rhombohedrons, scalohedrons, and prisms; also oolitic, stalactitic, and tabular. Cleavage perfect in three directions to form rhombohedrons.
General: Transparent to translucent; fracture conchoidal. Effervesces vigorously in acid. Calcite, the most abundant carbonate mineral, is common within nodules and seams of sedimentary rock, in fossiliferous marine sediments, and is a gangue mineral throughout the Colorado Mineral Belt.

Notable County Occurrences: Chaffee, Cheyenne, Fremont, Gunnison, Hinsdale, Kit Carson, Lake, La Plata, Las Animas, Lincoln, Mesa, Mineral, Moffat, Otero, Ouray, Pueblo, Rio Grande, Routt, Saguache, San Juan, San Miguel, Weld.

CARNOTITE — *Hydrous potassium uranium vanadate*

Color/Luster: Bright yellow to greenish yellow; earthy, silky when coarsely crystalline, usually powdery.
Hardness: 2 **S.G.:** 4-5 **Streak:** Yellow
Crystals: Monoclinic, as very small crystals in loose aggregates.
General: Radioactive. A secondary uranium-vanadium mineral found primarily within sandstone. Important Colorado occurrences are in the Uravan Mineral Belt. Carnotite is also a former source of radium.
Notable County Occurrences: Mesa, Montrose, San Miguel.

CERARGYRITE see CHLORARGYRITE

CERUSSITE — *Lead carbonate*

Color/Luster: White, gray, yellow, brown; adamantine, greasy.
Hardness: 3-3.5 **S.G.:** 6.5-6.6 **Streak:** White
Crystals: Orthorhombic, as small, flat, striated, tabular plates, often intergrown in groups and lattices. Cleavage good in one direction. Also compact, massive, and granular.
General: Brittle; transparent to translucent; fracture conchoidal. Usually fluorescent. Cerussite, an oxidation product of galena, is an ore of lead. Cerussite is common in metal sulfide districts throughout the Colorado Mineral Belt.
Notable County Occurrences: Hinsdale, Lake, Mineral.

CHABAZITE see ZEOLITE GROUP

CHALCEDONY see QUARTZ, Cryptocrystalline

CHALCOCITE — *Copper sulfide*

Color/Luster: Dark lead gray, tarnishes to dull black: metallic.
Hardness: 2.5-3 **S.G.:** 5.5-5.8 **Streak:** Shiny dark gray to near-black.
Crystals: Orthorhombic, in short prismatic or tabular crystals. Cleavage poor in one direction. Also compact and granular.
General: Brittle; fracture conchoidal. Chalcocite is an ore of copper.
Notable County Occurrences: La Plata, Montezuma.

CHALCOPYRITE — *Copper iron sulfide*

Color/Luster: Brass yellow to golden yellow, often iridescent, tarnishes to blue, purple, or black; metallic.
Hardness: 3.5-4 **S.G.:** 4.1-4.3 **Streak:** Greenish black
Crystals: Tetragonal; usually as tetrahedons, faces uneven, striated in different directions. Cleavage poor in one direction.
General: Brittle; fracture uneven. Chalcopyrite, an ore of copper, is common throughout the Colorado Mineral Belt.
Notable County Occurrences: Boulder, Chaffee, Clear Creek, Dolores, Gilpin, Gunnison, Hinsdale, La Plata, Lake, Ouray, Park, San Juan, San Miguel, Summit.

CHLORARGYRITE (CERARGYRITE) — ***Silver chloride***

Color/Luster: Pearl gray to brown, may be colorless when first exposed, but photosensitivity soon produces brown or violet color; resinous, waxy.
Hardness: 2.5 **S.G.:** 5.5 **Streak:** White
Crystals: Isometric, usually cubic. No cleavage. Also massive.
General: Transparent to translucent; fracture uneven. Chlorargyrite is an ore of silver.
Notable County Occurrences: Boulder, Custer, Lake.

CHRYSOBERYL — ***Beryllium aluminum oxide***

Color/Luster: Green to greenish yellow; vitreous.
Hardness: 8.5 **S.G.:** 3.5-3.8 **Streak:** White
Crystals: Orthorhombic, usually tabular or prismatic. Cleavage good in one direction and poor in two others.
General: Brittle; fracture conchoidal, uneven; transparent to translucent. Chrysoberyl, a valuable gemstone, occurs in granite pegmatites.
Notable County Occurrences: Larimer.

CHRYSOCOLLA — ***Basic copper silicate***

Color/Luster: Green to blue, brown to black when impure; vitreous, greasy.
Hardness: 2-4 **S.G.:** 2-2.4 **Streak:** White to pale blue-green.
Crystals: Monoclinic, microcrystalline. No cleavage. Occurs in compact and botryoidal masses, often banded and with some opalescence.
General: Translucent; brittle; no cleavage. Chrysocolla, an alteration product of copper sulfides, is an ornamental stone.
Notable County Occurrences: Chaffee, Lake.

CHRYSOPRASE ***see QUARTZ, cryptocrystalline***

COLUMBITE–TANTALITE — ***A two-member, graded series of niobium and tantalum oxide with iron and manganese***

Color/Luster: Brownish black to iron black; submetallic, dull.
Hardness: 6 **S.G.:** 5.3-7.9 **Streak:** Dark red to brown-black
Crystals: Orthorhombic, usually as short prisms; also granular and disseminated. Cleavage good in one direction.
General: Fracture subconchoidal, uneven. Usually opaque. Magnetic. Occurs in granite pegmatites. Columbite-tantalite is an ore of niobium and tantalum.
Notable County Occurrences: Chaffee, Gunnison, Larimer, Park.

COPPER — ***Native metal***

Color/Luster: Copper red, but tarnishes to black, blue, and green; metallic.
Hardness: 2.5-3 **S.G.:** 8.9 **Streak:** Copper red, shiny
Crystals: Isometric, usually cubic and dodecahedral. Also as plates, scales, grains, and branching aggregates.
General: Ductile and malleable. Fracture hackly. Native copper usually contains small amounts of iron and silver.
Notable County Occurrences: La Plata.

CORUNDUM — ***Aluminum oxide***

Color/Luster: Colorless, white, red, pink, blue, yellow, green, gray; vitreous, adamantine.
Hardness: 9 **S.G.:** 3.9-4.1 **Streak:** White
Crystals: Hexagonal, usually tapering prisms, also tabular, often striated. No cleavage.
General: Brittle; fracture conchoidal, uneven. Usually fluorescent. Transparent to translucent. Transparent crystalline red corundum is ruby; all other colors are sapphire. Ruby is the birthstone for July; sapphire is the birthstone for September.
Notable County Occurrences: Chaffee.

COVELLITE — ***Copper sulfide (often with iron)***

Color/Luster: Indigo blue to near-black, iridescent brass yellow to red; submetallic.
Hardness: 1.5-2 **S.G.:** 4.6-4.8 **Streak:** Dark gray to black
Crystals: Hexagonal, well-formed crystals rare. Cleavage perfect in one direction. Often massive.
General: Brittle; fracture uneven. Often appears platy, thin plates may be flexible. Covellite is a minor ore of copper.
Notable County Occurrences: Rio Grande.

CRYOLITE — ***Sodium aluminum fluoride***

Color/Luster: White, colorless; vitreous.
Hardness: 2.5 **S.G.:** 2.9-3 **Streak:** White
Crystals: Monoclinic, as prisms. Usually massive or granular. No cleavage, but parts in three directions at near-right angles (pseudocubic).
General: Brittle; transparent to translucent. Fracture uneven. Because of similar indices of refraction, ground cryolite is nearly invisible in water. Occurs in pegmatites.
Notable County Occurrences: El Paso.

CUPRITE — ***Copper oxide***

Color/Luster: Red to reddish black; submetallic.
Hardness: 3.5-4 **S.G.:** 6.1 **Streak:** Brownish red
Crystals: Isometric, usually octahedral. Also compact, earthy, granular. Cleavage poor.
General: Brittle; fracture uneven. Translucent. Formerly an ore of copper.
Notable County Occurrences: Chaffee.

DIAMOND — ***Carbon***

Color/Luster: White, colorless, black and pale shades of pink, yellow, blue, green, and brown; adamantine and greasy.
Hardness: 10 **S.G.:** 3.5
Crystals: Isometric, usually octahedral with slightly curved faces. Cleavage perfect in four directions.
General: Brittle; fracture conchoidal; transparent to translucent. Sometimes fluorescent. Diamond is a precious gemstone and the birthstone for April.
Notable County Occurrences: Larimer.

DOLOMITE ***Calcium magnesium carbonate***

Color/Luster: White, gray, colorless, black, green, brown; vitreous, pearly.
Hardness: 3.5-4 **S.G.:** 2.8-2.9 **Streak:** White
Crystals: Hexagonal, usually rhombohedral with slightly curved faces. Also compact, granular, and massive. Cleavage perfect in three directions.
General: Brittle; transparent to translucent; fracture conchoidal.
Notable County Occurrences: Lake.

ENARGITE ***Copper arsenic sulfide***

Color/Luster: Gray-black to iron black; metallic when fresh but tarnishes to dull; metallic.
Hardness: 3 **S.G.:** 4.4 **Streak:** Gray-black
Crystals: Orthorhombic, usually tabular, also prismatic; striated on several faces. Cleavage perfect in one direction, distinct in two. Also granular, compact, and columnar.
General: Brittle; fracture uneven. Enargite is an ore of copper.
Notable County Occurrences: Rio Grande, Saguache.

EPIDOTE ***Basic calcium aluminum iron silicate***

Color/Luster: Yellow-green to greenish black; vitreous.
Hardness: 6-7 **S.G.:** 3.3-3.6 **Streak:** Gray
Crystals: Monoclinic, usually in long, grooved prisms and thick tabular crystals; termination often two sloping faces. Cleavage perfect in one direction. Also massive, granular, and columnar.
General: Brittle; fracture uneven; transparent to translucent. Often occurs in contact and regional metamorphic rocks.
Notable County Occurrences: Custer, Chaffee, Gunnison, Jefferson, Ouray.

FERBERITE ***Iron tungstate (the iron-rich end member of the wolframite series)***

Color/Luster: Reddish brown to black; submetallic, resinous. Twinning common. Also massive and granular.
Hardness: 4-4.5 **S.G.:** 7.5 **Streak:** Black
Crystals: Monoclinic, usually elongated and striated. Cleavage perfect in one direction lengthwise.
General: Fracture uneven; opaque; weakly magnetic.
Notable County Occurrences: Boulder.

FLUORITE ***Calcium fluoride***

Color/Luster: Violet, green, white, colorless, yellow, pink, brown; vitreous.
Hardness: 4 **S.G.:** 3-3.2 **Streak:** White
Crystals: Isometric; usually cubic. Cleavage perfect in four directions to form octahedrons.
General: Brittle, fracture uneven. Fluorspar, the impure granular form, is valuable as a flux in reduction smelting and steel manufacture. Crystals often occur in seams and pockets of fluorspar deposits and in pegmatites. Fluorite is a gangue mineral in many metal sulfide mining districts.
Notable County Occurrences: Boulder, Chaffee, Douglas, El Paso, Hinsdale, Jackson, Jefferson, Lake, Larimer, Mineral, Ouray, Park, San Juan, San Miguel, Summit, Teller.

GALENA ***Lead sulfide***

Color/Luster: Dark lead gray, bright silver on fresh cleavage surfaces; metallic.
Hardness: 2.5 **S.G.:** 7.4-7.5 **Streak:** Dark gray
Crystals: Isometric, cubic, or combination of cubic and octahedral. Cleavage perfect in three directions. Also compact and granular.
General: Brittle. Often contains significant amounts of silver (argentiferous galena). Galena, the primary ore of lead, is abundant in metal sulfide mining districts.
Notable County Occurrences: Boulder, Chaffee, Clear Creek, Custer, Dolores, Gilpin, Gunnison, Hinsdale, La Plata, Mineral, Ouray, Park, Saguache, San Juan, San Miguel.

GARNET ***A group of six related calcium aluminum silicates, with iron, magnesium, manganese, and chromium***

Color/Luster: Pyrope, deep red to reddish black; almandine, deep red to brown; spessartine, brownish red to hyacinth red; grossular, colorless to white, pink, brown, yellow, and green; andradite, dark red, green; uvarovite, green; vitreous.
Hardness: 6.5-7.5 **S.G.:** 3.56-4.32 **Streak:** Colorless
Crystals: Isometric, often dodecahedral. No distinct cleavage. Also compact, disseminated, and granular.
General: Brittle; fracture conchoidal, uneven; transparent to opaque. Garnet is an industrial abrasive. Red, crystalline garnet is a gemstone and the birthstone for January.
Notable County Occurrences: Chaffee, Clear Creek, Custer, Grand, Gunnison, Jefferson, Larimer, Ouray.

GOETHITE ***Hydrogen iron oxide***

Color/Luster: Yellow, yellowish brown, and dark brown to black; dull metallic.
Hardness: 5-5.5 **S.G.:** 3.3-4.3 **Streak:** Yellow-brown
Crystals: Orthorhombic, usually as flattened plates, tablets, scales, needles, and radial aggregates. Also in radial and botryoidal masses. Cleavage perfect in one direction lengthwise.
General: Brittle; opaque.
Notable County Occurrences: Chaffee, Douglas, El Paso, Teller.

GOLD ***Native metal***

Color/Luster: Gold yellow to pale yellow; bright metallic luster.
Hardness: 2.5-3 **S.G.:** 19.3 (pure)
Crystals: Isometric, usually octahedral, dedecahedral, or cubic. Often distorted into dendritic or leaf forms.
General: Gold is the only malleable yellow mineral. Native gold is always alloyed with other metals, usually silver, copper, and iron, thus specific gravity may vary from about 14.0 to 19.0. Gold forms a limited number of mineral compounds, primarily with the element tellurium. Placer gold occurs in fine particles (dust), grains, flakes, and nuggets.

Notable County Occurrences: Alamosa, Boulder, Chaffee, Clear Creek, Costilla, Dolores, Douglas, Eagle, Gilpin, Grand, Gunnison, Jefferson, Lake, La Plata, Moffat, Montezuma, Montrose, Ouray, Park, Rio Grande, Routt, San Juan, San Miguel, Summit, Teller.

GYPSUM — *Hydrous calcium sulfate*

Color/Luster: White, colorless, gray, and light shades of red, yellow, and brown; vitreous, pearly on cleavage surfaces.
Hardness: 1.5-2 **S.G.:** 2.3-2.4 **Streak:** White
Crystals: Monoclinic, usually rhombic, sometimes as twins with beveled edges in "arrowhead" shape. Also fibrous, massive, and granular.
General: Brittle; transparent to translucent; fracture conchoidal and splintery. Mostly of chemical origin within sedimentary beds from accumulations of marine salts. Gypsum varieties include selenite, colorless and transparent; satin spar, fibrous and lustrous; gypsum rock, an impure granular form mined extensively for cement and plaster manufacture; and alabaster, a compact, pure, granular form used for carving and ornamental purposes.
Notable County Occurrences: Cheyenne, Douglas, Eagle, El Paso, Larimer, Las Animas, Lincoln, Mesa, Moffat, Park.

HEMATITE — *Iron oxide*

Color/Luster: Black and steel gray to red and reddish brown; metallic.
Hardness: 5-6 **S.G.:** 4.9-5.3 **Streak:** Dark red to brownish red
Crystals: Hexagonal, often tabular as rosettes. Also columnar, as radiating masses and earthy. No cleavage.
General: Brittle; fracture uneven, splintery. Hematite is abundant. In massive deposits hematite is an important ore of iron.
Notable County Occurrences: Douglas, El Paso.

HUEBNERITE — *Manganese tungstate (the manganese-rich end member of the wolframite series)*

Color/Luster: Reddish brown; submetallic.
Hardness: 4-4.5 **S.G.:** 7.4 **Streak:** Reddish brown
Crystals: Monoclinic, usually in long, striated prisms, twinning common. Also massive and granular. Cleavage perfect in one direction lengthwise.
General: Fracture uneven; transparent. Huebnerite is an ore of tungsten.
Notable County Occurrences: San Miguel.

LAPIS LAZULI (LAZURITE) — *Sodium calcium aluminum silicate, with some sulfur*

Color/Luster: Azure blue to violet blue; dull to greasy.
Hardness: 5-5.5 **S.G.:** 2.4-2.5 **Streak:** Blue
Crystals: Isometric (rare). Usually granular or massive.
General: Brittle; fracture uneven. Often contains bits of pyrite. Lapis lazuli, a mixture of lazurite, calcite, pyrite, and diopside, is an ornamental stone and gemstone and is a birthstone for December.
Notable County Occurrences: Gunnison.

LEPIDOLITE ***Basic potassium lithium aluminum fluorosilicate (mica group)***

Color/Luster: Pink, lilac, yellowish; pearly.
Hardness: 2.5-3 **S.G.:** 2.8-2.9 **Streak:** Colorless
Crystals: Monoclinic, often as scaly aggregates; tough and elastic; cleavage perfect in one direction. Also foliated, granular, and compact.
General: Translucent to transparent. Laminae tough and elastic. Occurs in granite pegmatites.
Notable County Occurrences: Clear Creek, Gunnison, Fremont, Larimer.

LIMONITE ***A mixture of hydrous iron oxides***

Color/Luster: Yellow to brown; glassy to dull.
Hardness: 4-4.5 **S.G.:** 2.7-4.3 **Streak:** Yellowish brown
Crystals: Amorphous, as earthy, fibrous, botryoidal, stalactic, masses, and crusts. No cleavage.
General: Fracture varies. Abundant oxidation product of iron sulfides. Limonite is a common rust-colored staining agent in rocks and soils and often forms pseudomorphs after other iron minerals. Found statewide.

MAGNETITE ***Iron oxide***

Color/Luster: Iron black; metallic.
Hardness: 5.5-6.5 **S.G.:** 4.9-5-2 **Streak:** Black
Crystals: Isometric, usually as octahedrons. No cleavage.
General: Brittle; fracture subconchoidal. Strongly magnetic. Also massive and granular.
Notable County Occurrences: Alamosa, Chaffee.

MALACHITE ***Basic copper carbonate***

Color/Luster: Emerald green to dark green; adamantine, silky.
Hardness: 3.5-4 **S.G.:** 4 **Streak:** Light green
Crystals: Monoclinic (rare), as prisms. Cleavage perfect in one direction crosswise. Usually botryoidal and massive, often with characteristic color banding of light and dark greens.
General: Brittle; fracture conchoidal, splintery; translucent. Malachite, an alteration product of copper sulfides, is valuable as a gemstone and was formerly important an ore of copper.
Notable County Occurrences: Custer, Gunnison, Hinsdale, Lake, Mineral, Montezuma, Ouray.

MARCASITE ***Iron disulfide***

Color/Luster: Pale brass yellow to near-white; metallic.
Hardness: 6-6.5 **S.G.:** 4.8 **Streak:** Dark greenish to brownish.
Crystals: Orthorhombic, often as dipyramidal prisms or tabular. Also in stalactitic, radiating aggregates and botryoidal crusts. Cleavage distinct in two directions.
General: Brittle; fracture uneven. Marcasite, chemically identical to pyrite, crystallizes in a different system and lacks the cubic form and brass yellow color of pyrite. Commonly forms in high-acid, low-temperature sedimentary environments and is often associated with fossilized organic matter. Marcasite oxidizes readily in moist air.
Notable County Occurrences: El Paso, Jackson, Otero, Pueblo.

MESOLITE see ZEOLITE GROUP

MOLYBDENITE — *Molybdenum disulfide*

Color/Luster: Bluish lead gray; metallic
Hardness: 1-1.5 **S.G.:** 4.6-5.1 **Streak:** Greenish
Crystals: Hexagonal; good crystals rare, often distorted. Cleavage perfect in one direction. Also as scales, foliated masses, and granular vein fillings.
General: Flakes slightly flexible with greasy feel. Molybdenite is the primary ore of molybdenum. The oxidation product of molybdenite is molybdite, or molybdenum oxide, a bright yellow powdery mineral often present in small quantities on weathered specimens of molybdenite.
Notable County Occurrences: Lake.

MONAZITE — *Cerium lanthanum phosphate, usually with thorium and uranium*

Color/Luster: Yellowish brown to reddish brown; vitreous, resinous.
Hardness: 5-5.5 **S.G.:** 4.6-5.4 **Streak:** White
Crystals: Monoclinic, usually equant. Occurs as small grains in igneous rock; large crystals found in certain granite pegmatites.
General: Fracture conchoidal, uneven. Radioactive; brittle. Transparent to translucent. Monazite is an ore of thorium and other rare earth elements.
Notable County Occurrences: Chaffee.

MUSCOVITE — *Basic potassium aluminum silicate (mica group)*

Color/Luster: White or colorless, also slightly yellowish or brownish; vitreous, pearly.
Hardness: 2-2.5 **S.G.:** 2.7-3.0 **Streak:** Colorless
Crystals: Monoclinic, thin and tabular, often hexagonal in outline. Cleavage perfect in one direction.
General: Translucent to transparent; flakes are elastic, flexible, and tough. Also occurs in scaly, foliated, and fibrous form. Formerly mined for electrical insulation and Christmas tree "snow"; now mined as a decorative rock. Muscovite is a widely distributed rock-forming mineral. It is common in granite and forms large crystal "books" of crystals in granite pegmatites.
Notable County Occurrences: Chaffee, Fremont, Jefferson, Lake, Park, Teller, Larimer.

NATROLITE see ZEOLITE GROUP

OPAL — *Hydrous silica*

Color/Luster: White, colorless, pink, gray, green, blue, brown; vitreous, pearly.
Hardness: 5.5-6.5 **S.G.:** 2-2.2 **Streak:** White
Crystals: None. Massive, botryoidal, and compact.
General: Brittle; fracture conchoidal. Opalescent. Transparent to translucent. Occurs as amygdules in certain volcanic rocks and as cell replacement material in fossils, such as opalized wood. Common opal is usually white or pale and lacks the internal play of light, called "fire," characteristic of fire opal and precious opal. Opal is a valuable gemstone and the birthstone for October.
Notable County Occurrences: Elbert, Mesa, Park.

ORTHOCLASE ***Potassium aluminum silicate (potash feldspar)***
Color/Luster: White, gray, colorless, pink, brown, green; vitreous.
Hardness: 6-6.5 **S.G.:** 2.5-2.6 **Streak:** White
Crystals: Triclinic (microcline), monoclinic (orthoclase, sanidine), usually as single crystals with square or rectangular cross section. Orthoclase Carlsbad twins common. Perfect cleavage in two directions at right angles.
General: Fracture uneven, transparent to translucent. Potash feldspar varieties include andularia (moonstone), transparent or slightly opalescent; sanidine, usually white and glassy; and microcline. Microcline may be white, pale yellow, or green-blue. Deeply colored green or green-blue microcline is amazonite or amazonstone. Orthoclase is a widely distributed rock-forming mineral present in many intrusive, volcanic, and metamorphic rocks. It is a common component of granite; large crystals are abundant in granite pegmatites. Orthoclase is valuable as a "gentle" abrasive and in ceramic manufacturing.
Notable County Occurrences: Chaffee, Clear Creek, Douglas, El Paso, Fremont, Grand, Jefferson, Lake, Park, Teller.

PETZITE ***Silver gold telluride***
Color/Luster: Steel gray to iron black; metallic.
Hardness: 2.5-3 **S.G.:** 8.7-9 **Streak:** Black
Crystals: Isometric (?). Usually as granular and compact masses. Cleavage indistinct.
General: Fracture subconchoidal. Petzite is an ore of silver and gold.
Notable County Occurrences: Boulder, Hinsdale, La Plata, San Miguel.

PHENAKITE ***Beryllium silicate***
Color/Luster: Colorless, white, and light shades of pink and yellow; vitreous.
Hardness: 7.5-8 **S.G.:** 3.0 **Streak:** Colorless
Crystals: Hexagonal, in well-developed long prisms or flat rhombohedrons; lengthwise striations. Cleavage poor in one direction.
General: Transparent to translucent; fracture conchoidal and uneven. Colorado phenakite occurs in certain granite pegmatites. Phenakite is a gemstone.
Notable County Occurrences: Chaffee, El Paso, Teller.

PROUSTITE ***Silver arsenic sulfide***
Color/Luster: Scarlet to vermillion, darkens on exposure to light; adamantine to submetallic.
Hardness: 2-2.5 **S.G.:** 5.6 **Streak:** Bright red
Crystals: Hexagonal, as poorly formed prisms. Fracture conchoidal, uneven; translucent. Cleavage distinct in one direction. Also massive and compact.
General: Brittle. Proustite is a minor ore of silver.
Notable County Occurrences: Dolores, Hinsdale, Lake, La Plata.

PYRITE ***Iron disulfide***
Color/Luster: Rich to pale brass yellow; metallic.
Hardness: 6-6.5 **S.G.:** 4.9-5.2 **Streak:** Greenish black
Crystals: Isometric, usually cubic with parallel striations.
General: Brittle; fracture uneven; may be slightly iridescent; will spark with steel. Formerly valuable in sulfuric acid manufacturing; used as coloring agent in glass manufacturing. Also known as "fool's gold." Pyrite is the most abundant metal sulfide.

Notable County Occurrences: Boulder, Chaffee, Clear Creek, Custer, Eagle, Gilpin, Gunnison, Hinsdale, Jackson, Lake, La Plata, Mineral, Moffat, Otero, Ouray, Pueblo, Rio Grande, Saguache, San Juan, San Miguel, Summit.

QUARTZ, crystalline — *Silicon dioxide*

Color/Luster: Colorless, white, brown, black, yellow, purple, pink; vitreous, greasy.
Hardness: 7 **S.G.:** 2.65 **Streak:** White
Crystals: Hexagonal, usually prisms striated crosswise; typically terminated with double rhombohedrons appearing as hexagonal pyramids. No cleavage. Quartz gemstones are named by color: amethyst, purple; rose quartz, pink; rock crystal, colorless and transparent; citrine, yellow; and smoky quartz, light brown to near-black. Other crystalline quartz gemstones are named for inclusion effects: cat's eye, opalescent from asbestos inclusions; and tigereye, with lustrous yellow-brown parallel fibers.
General: Fracture conchoidal; transparent to subtranslucent. Quartz is an abundant rock-forming mineral. Large well-formed crystals occur in granite pegmatites. Also granular, massive, and disseminated. Colorado is a rich source of fine crystals of smoky quartz, rock crystal, and amethyst.
Notable County Occurrences: Chaffee, Custer, Dolores, Douglas, Eagle, El Paso, Fremont, Gilpin, Grand, Gunnison, Hinsdale, Jefferson, Lake, La Plata, Larimer, Mesa, Mineral, Ouray, Park, Rio Grande, Routt, Saguache, San Juan, San Miguel, Summit, Teller.

QUARTZ, cryptocrystalline (CHALCEDONY) — *Silicon dioxide, usually with aluminum and iron*

Color/Luster: Wide range of colors; waxy, vitreous, dull.
Hardness: 7 **S.G.:** 2.6-2.64 **Streak:** White
General: Transparent to translucent; fracture conchoidal; occurs as crusts, nodules, and vein fillings in botryoidal and mammillary form. Often occurs with fossils as cell replacement material. Gemstone and ornamental varieties of chalcedony are classified by color or pattern: agate, banded, often concentric or wavelike, with varying colors and degrees of translucency; jasper, opaque in deep mottled reds, yellows, and browns; chrysoprase, apple green; bloodstone, spots of red in green; moss agate, with dendritic mosslike inclusions; carnelian, dark red to brownish red; sard, brown; onyx, layered with straight, parallel bands.
Notable County Occurrences: Bent, Elbert, El Paso, Fremont, Garfield, Grand, Jackson, La Plata, Larimer, Las Animas, Mesa, Mineral, Moffat, Montezuma, Montrose, Prowers, Rio Blanco, Rio Grande, Saguache, Weld.

RHODOCHROSITE — *Manganese carbonate*

Color/Luster: Pink to dark red; vitreous, pearly.
Hardness: 3.5-4 **S.G.:** 3.4-3.6 **Streak:** White
Crystals: Hexagonal, usually as rhombohedrons; cleavage perfect in three directions, forming a rhombohedron. Also granular, botryoidal, and as cleavage masses.
General: Brittle; fracture uneven, subtransparent to translucent. Rhodochrosite is an ore of manganese and has lapidary and jewelry uses.

Colorado rhodochrosite most often occurs with metal sulfide minerals and quartz in vein and replacement deposits.
Notable County Occurrences: Chaffee, Clear Creek, Dolores, Eagle, Hinsdale, Lake, Ouray, Saguache, San Juan.

RHODONITE — *Manganese silicate, usually with calcium*

Color/Luster: Brownish red to pink, may tarnish to black or brown; vitreous.
Hardness: 5.5-6 **S.G.:** 3.5-3.7 **Streak:** Colorless
Crystals: Triclinic, rare in tabular or prismatic form; usually fine-grained granular, compact, or cleavable masses. Cleavage good in two directions.
General: Fracture conchoidal, uneven; transparent to translucent. Often attractively "laced" with thin veinlets of black manganese oxide. Rhodonite is an ornamental stone and gemstone.
Notable County Occurrences: Chaffee, Ouray, San Juan.

SAPPHIRE see CORUNDUM
SATIN SPAR see GYPSUM

SCHEELITE — *Calcium tungstate*

Color/Luster: White, colorless, also gray, yellowish, greenish, brownish; vitreous, adamantine.
Hardness: 4.5-5 **S.G.:** 6 **Streak:** Yellowish white
Crystals: Tetragonal, usually dipyramidal or tabular. Also as massive and granular aggregates. Cleavage distinct in one direction and poor in two others.
General: Fracture conchoidal, uneven. Transparent to translucent. Fluorescent. Scheelite is an ore of tungsten.
Notable County Occurrences: Boulder, Jefferson, Larimer.

SELENITE see GYPSUM

SIDERITE — *Iron carbonate*

Color/Luster: Brown, reddish brown, white; vitreous, dull, pearly.
Hardness: 3.5-4 **S.G.:** 3.8 **Streak:** White to pale yellow
Crystals: Hexagonal, often as rhombohedrons with curved faces. Cleavage perfect in three directions.
General: Brittle; translucent; fracture uneven. Becomes magnetic when heated. Also granular, compact, massive, fibrous, and botryoidal. Occurs in vein and replacement metal sulfide deposits.
Notable County Occurrences: Clear Creek, Eagle, Mineral, San Miguel, Teller.

SILVER — *Native metal*

Color/Luster: Silver-white to gray-black; metallic luster.
Hardness: 2.5-3 **S.G.:** 10-11
Crystals: Isometric (rare), usually octahedral, dodecahedral, or cubic. Occurs in grains, plates, wires, and arborescent forms.
General: Malleable. Native silver occurs much less frequently than native gold. Silver forms many sulfide minerals and also combines with tellurium.
Notable County Occurrences: Custer, Dolores, Lake, Mineral, Ouray.

SMITHSONITE *Zinc carbonate*

Color/Luster: Usually white, also brown, green, gray, or blue; pearly, adamantine.
Hardness: 4-4.5 **S.G.:** 4.3-4.5 **Streak:** White
Crystals: Hexagonal (rare). Usually botryoidal, stalactitic, granular, or massive.
General: Brittle; fracture uneven, splintery; translucent. Smithsonite, an alteration product of sphalerite, is an ore of zinc.
Notable County Occurrences: Lake.

SPHALERITE *Zinc sulfide*

Color/Luster: Brown, yellow, reddish, greenish; submetallic.
Hardness: 3.5-4.0 **S.G.:** 3.9-4.0 **Streak:** Light brown
Crystals: Isometric, usually tetrahedral or dodecahedral; faces may be slightly rounded. Cleavage perfect in six directions. Also granular, massive, and botryoidal.
General: Brittle; fracture conchoidal; transparent to nearly opaque (with high iron content). Sphalerite, the primary ore of zinc, is common in metal sulfide mining districts.
Notable County Occurrences: Boulder, Chaffee, Clear Creek, Custer, Dolores, Eagle, Gunnison, Hinsdale, Lake, La Plata, Mineral, Ouray, Park, Rio Grande, Saguache, San Juan, San Miguel, Summit.

SPHENE see TITANITE

SPODUMENE *Lithium aluminum silicate*

Color/Luster: White or gray, also yellowish, green, pink to purple; vitreous, pearly.
Hardness: 6.5-7 **S.G.:** 3.1 **Streak:** White
Crystals: Monoclinic, as stubby prisms, often lathlike and flattened; deeply striated parallel to elongation. Also occurs as cleavage masses. Cleavage good in two directions at nearly right angles.
General: Brittle; fracture uneven and splintery; transparent to translucent. Spodumene occurs only in lithium-rich granite pegmatites. Spodumene is a valuable gemstone; Kunzite is pink, lilac, or amethystine; hiddenite is green.
Notable County Occurrences: Larimer.

STIBNITE *Antimony trisulfide*

Color/Luster: Lead gray, tarnishes to black with occasional iridescence; metallic.
Hardness: 2 **S.G.:** 4.6 **Streak:** Dark gray
Crystals: Orthorhombic, as aggregates or radiating masses of prisms. Striated lengthwise. Also granular. Cleavage perfect in one direction lengthwise.
General: Brittle; fracture uneven. Crystals may be flexible with slight iridescent tarnish.
Notable County Occurrences: Gunnison, Hinsdale.

SYLVANITE — *Gold silver telluride*

Color/Luster: Silver-white to steel gray, sometimes with hint of yellow; metallic.
Hardness: 1.5-2 **S.G.:** 8.2 **Streak:** Black
Crystals: Monoclinic, as small prisms; also bladed, columnar, and granular. No striations on faces (unlike calaverite).
General: Brittle; fracture uneven. Sylvanite is an ore of gold and silver.
Notable County Occurrences: Boulder, Gilpin, Hinsdale, La Plata, San Miguel, Teller.

TELLURIDE** see **CALAVERITE, PETZITE, and SYLVANITE

TENNANTITE** see **TETRAHEDRITE SERIES

TETRAHEDRITE SERIES (TETRAHEDRITE-TENNANTITE) — *Copper antimony sulfide–copper iron arsenic sulfide*

Color/Luster: Steel gray to black; metallic.
Hardness: 3-4 **S.G.:** 4.6-5.1 **Streak:** Dark gray and black to reddish brown
Crystals: Isometric, usually as tetrahedrons. No cleavage.
General: Brittle; fracture uneven to subconchoidal. Also compact and granular. Tetrahedrite and tennantite are ores of copper.
Notable County Occurrences: Dolores, Hinsdale, Lake, La Plata, Ouray, Park, Saguache, San Juan, San Miguel.

THOMSONITE** see **ZEOLITE GROUP

TITANITE (SPHENE) — *Calcium titanium silicate*

Color/Luster: Brown, black, or gray, also with yellowish or greenish shades; resinous, adamantine.
Hardness: 5-5.5 **S.G.:** 3.4-3.6 **Streak:** White
Crystals: Monoclinic, usually tabular; also compact. Cleavage distinct in two directions.
General: Brittle; fracture conchoidal; transparent to translucent.
Notable County Occurrences: Jefferson.

TOPAZ — *Aluminum fluorosilicate*

Color/Luster: Colorless, blue, yellow, sherry; vitreous.
Hardness: 8 **S.G.:** 3.4-3.6 **Streak:** Colorless
Crystals: Orthorhombic, usually stubby to medium-long prisms, lengthwise striations. Cleavage perfect in one direction.
General: Transparent to translucent; fracture subconchoidal and uneven. Most Colorado topaz occurs within mariolitic cavities of granite pegmatite and rhyolite. Sufficiently high-density forms loose alluvial concentrations. Topaz is a precious gemstone and the birthstone for November.
Notable County Occurrences: Chaffee, Douglas, El Paso, Larimer, Park, Teller.

TOURMALINE
A complex boron aluminum silicate group

Color/Luster: Pink, red, blue, green, black, and multicolored; vitreous.
Hardness: 7-7.5 **S.G.:** 3-3.3 **Streak:** White
Crystals: Hexagonal, in short to long prisms, with typically rounded triangular cross sections. Lengthwise striations. Also radiating and columnar. No cleavage.
General: Brittle; fracture uneven, conchoidal; transparent to opaque. The three primary varieties are dravite, a brown, magnesium-rich tourmaline; elbaite, a blue, green, yellow, red, or colorless lithium-rich tourmaline; and schorl, a black or greenish black iron- and manganese-rich tourmaline. Elbaite tourmaline is a valuable gemstone. Schorl was popular in Victorian mourning jewelry.
Notable County Occurrences: Clear Creek, El Paso, Fremont, Gunnison, Lake.

TREMOLITE–ACTINOLITE
Basic calcium magnesium iron silicate series. Actinolite is rich in iron.

Color/Luster: Tremolite is white to dark gray, actinolite is bright to dark green; vitreous, silky.
Hardness: 5-6 **S.G.:** 2.9-3.5 **Streak:** Colorless
Crystals: Monoclinic, usually as long prisms in fibrous, radiating, and bladed aggregates. Cleavage perfect in two directions to form characteristic diamond shape.
General: Fracture splintery, uneven. Transparent to translucent.
Notable County Occurrences: Chaffee, Gunnison.

TURQUOISE
Hydrous basic copper aluminum phosphate

Color/Luster: Sky blue through pale green, often mottled and veined; dull, waxy.
Hardness: 5-6 **S.G.:** 2.6-2.8 **Streak:** White to pale green
Crystals: Triclinic (rare).
General: Subtransparent to opaque; fracture conchoidal. Occurs as nodules and thin vein fillings. Turquoise is a gemstone and a birthstone for December.
Notable County Occurrences: Conejos, Lake, Mineral, Saguache.

URANINITE
Uranium oxide

Color/Luster: Brownish or greenish black; submetallic, with greasy, dull, "tarlike" luster.
Hardness: 5-6 **S.G.:** 6.5-10 **Streak:** Brownish to grayish black to dark olive green
Crystals: Isometric, usually cubic and octahedral; also in massive and botryoidal form (pitchblende). No cleavage.
General: Brittle; opaque; fracture conchoidal. Radioactive. Uraninite is an important ore of uranium and was formerly valuable for its trace radium content.
Notable County Occurrences: Gilpin.

ZEOLITE GROUP: ***Analcime, Chabazite, Mesolite, Natrolite, and Thomsonite***

Zeolites are a group of twenty-two related hydrous tectosilicate minerals commonly occurring in cavities within basalt. Water can be driven off or replaced without altering the basic structure, and composition of sodium, calcium, and aluminum varies.

ANALCIME ***Hydrous sodium aluminum silicate***

Color/Luster: White, gray, colorless, sometimes with slight reddish, greenish, or yellowish tint; vitreous.
Hardness: 5.5 **S.G.:** 2.2-2.3 **Streak:** White
Crystals: Isometric, usually as trapezohedrons, also massive and compact. No cleavage.
General: Brittle; fracture uneven to subconchoidal; transparent to translucent.
Notable County Occurrences: Jefferson.

CHABAZITE ***Hydrous calcium sodium aluminum silicate***

Color/Luster: White, yellow, pink; vitreous.
Hardness: 4-5 **S.G.:** 2-2.2 **Streak:** White
Crystals: Hexagonal, usually rhombohedral, also compact. Cleavage poor in three directions.
General: Brittle; fracture uneven; transparent to translucent.
Notable County Occurrences: Jefferson.

MESOLITE ***Hydrous sodium calcium aluminum silicate***

Color/Luster: White or colorless; vitreous.
Hardness: 5 **S.G.:** 2.3 **Streak:** White
Crystals: Monoclinic, usually capillary and acicular, also fibrous; vitreous, silky. Cleavage perfect in two directions.
General: Fracture uneven; transparent to translucent.
Notable County Occurrences: Jefferson.

NATROLITE ***Hydrous sodium aluminum silicate***

Color/Luster: White, colorless, sometimes yellowish or grayish; vitreous, silky.
Hardness: 5-5.5 **S.G.:** 2.2-2.3 **Streak:** Colorless
Crystals: Orthorhombic, usually slender; also radial, granular, and compact. Cleavage perfect in one direction.
General: Brittle; fracture uneven; transparent to translucent.
Notable County Occurrences: Jefferson.

THOMSONITE ***Hydrous calcium sodium aluminum silicate***

Color/Luster: White; vitreous, pearly.
Hardness: 5-5.5 **S.G.:** 2.3-2.4 **Streak:** Colorless
Crystals: Orthorhombic, as prisms with rectangular outline, distinct crystals rare. Usually columnar or radiating. Cleavage perfect in one direction.
General: Brittle; fracture uneven to subconchoidal; transparent to translucent.
Notable County Occurrences: Jefferson.

ZIRCON — ***Zirconium silicate***

Color/Luster: Colorless and pale shades of gray, brown (smoky), yellow, blue, green, and red; adamantine, vitreous.
Hardness: 7.5 **S.G.:** 4.7 **Streak:** Colorless
Crystals: Tetragonal, typically in stubby prisms showing square cross section and pyramidal termination. Fracture uneven.
General: Brittle; transparent to translucent. Zircon is a gemstone and is a birthstone for December.
Notable County Occurrences: El Paso.

GLOSSARY

Accessory mineral Any mineral component of a rock, usually present in small quantities, which does not affect the rock's classification or name.

Alluvial Pertaining to sedimentary deposits of soil, sand, gravel, or similar detrital material deposited by running water or gravitational movement, as in a riverbed or alluvial fan.

Alteration Physical or chemical changes to minerals or rocks that result in formation of new minerals or textural changes in the rocks.

Ammonite Any fossilized portion of an ammonoid cephalopod

Amygdule A gas-formed, mineral-filled cavity, usually in volcanic rock.

Argentiferous Containing silver.

Auriferous Containing gold.

Batholith A very large body of igneous rock intruded deep within the earth's crust and subsequently exposed by erosion.

Book A term describing groups of muscovite or biotite mica crystals from which individual thin sheets can be stripped.

Botryoidal A surface shaped with spherical bulges, resembling bunches of grapes.

Breccia A rock composed of angular fragments cemented together.

Cabbing The fashioning of a cabochon.

Cabochon A gem cut in convex form and highly polished but not faceted.

Calcareous Limy, containing calcium carbonate.

Caldera A collapsed volcanic system.

Carbonization A process of fossilization in which organic remains are reduced to carbon.

Cleavage The tendency of certain minerals, when struck with a sharp blow, to break along one or more definite planes to form regular, smooth surfaces.

Conchoidal Shell-shaped; used to describe the fracture pattern of such materials as agate and obsidian.

Concretion A rounded, often spherical, cemented accumulation of mineral matter, often pyrite, calcite, silica, or gypsum.

Contact The geologic term for the surface marking the junction of two bodies of rock.

Contact metamorphism Metamorphism caused directly by intrusion of magma and taking place at or near the contact with the magma.

Coprolite Fossilized excrement.

Country rock The general mass of adjacent rock, as distinguished from that of intrusions, veins, etc.

Cryptocrystalline rock A rock consisting of microscopic crystals (also microcrystalline).

Crystal A solid mass of mineral with a regular geometric shape determined by an orderly internal arrangement of atoms and bounded by smooth, flat surfaces, or faces.

Dendrite A branching pattern within a rock or mineral, usually formed by manganese oxide filling a fissure.

Dendritic Branching, resembling a tree.

Dendritic agate Agate with dendritic inclusions resembling the general shapes of moss, flowers, and trees; moss agate, plume agate.

Density Weight per unit volume, usually expressed as specific gravity.

Dike A wall-like body of igneous rock cutting through surrounding rock.

Dodecahedral Having twelve faces, as in certain garnet crystals.

Essential minerals The major mineral constituents of a rock that are used to name or classify the rock.

Exposure A body of mineral matter exposed to view; an outcrop.

Extrusive rock An igneous rock that solidifies on the surface of the earth; volcanic rock.

Fault In geology, a crack in the earth's crust accompanied by relative displacement of the opposite sides.

Float Loose pieces of mineral or fossils that have been transported from their place of in-situ origin by water, erosion, or gravity.

Fluorescence The ability of certain minerals to emit light when exposed to ultraviolet radiation.

Fortification agate Banded agates exhibiting parallel banding with acute cornering, resembling the angular structure of a fort.

Fracture The irregular breakage of a mineral.

Gangue The economically worthless rock in which ore minerals occur.

Gem Any precious or semiprecious mineral or fossil material that has been cut, polished, or otherwise fashioned into a form suitable for jewelry use.

Gemstone Any natural precious or semiprecious mineral or fossil material that may be cut, polished, or otherwise fashioned into a form suitable for jewelry use.

Geode A hollow nodule or concretion, the interior of which is lined with crystals or filled entirely with mineral matter.

Hardness The degree of cohesion of particles on the surface of a mineral; resistance of a mineral to abrasion or scratching.

Heap leaching In gold mining, the use of a solvent, such as sodium cyanide, to inexpensively extract gold from crushed, low-grade ores.

Hydraulicking A type of placer mining utilizing high-pressure water jets to erode away large volumes of sediments.

Hydrocarbon A natural organic compound containing only hydrogen and carbon and occurring in such materials as petroleum coal, and kerogen.

Igneous Formed by direct crystallization or solidification of magma. One of the three primary classifications of rock.

Intrusive rock An igneous rock that solidified underground from magma emplaced into cracks or crevices of existing rocks to form an intrusive body.

Laccolith A lens-shaped intrusion with a dome-shaped upper surface and flat bottom surface parallel to the bedding or foliation of the surrounding rock.

Lapidary A cutter, polisher, or engraver of gemstones; the art of working gemstone materials.

Lava Molten rock, or magma, extruded onto the surface that will solidify into volcanic rock.

Lava flow A mass of lava; the body of rock formed by its solidification.

Lens An ore or mineral body generally elliptical in outline or lenticular in shape.

Lenticular Lens-shaped.

Luster The manner in which the surface of a mineral reflects light.

Magma Molten rock material originating beneath the earth's crust that solidifies to form igneous rocks at or below the surface.

Massive (form) Any mineral occurring without definite external crystal form or as poorly defined masses of small crystals.

Matrix The fine-grained material that surrounds the larger crystals or particles in a porphyritic or sedimentary rock; any material in which a fossil or crystal is embedded.

Metamorphic Formed from a pre-existing rock through heat, pressure, chemical action, or the effect of superheated fluids.

Micromount A mineral specimen small enough to require magnification for observation and study.

Mineralization The formation or introduction of ore minerals into pre-existing rock masses.

Mohs scale A semilogarithmic scale of mineral hardness ranging from 1 (softest) to 10 (hardest). The representative minerals on the Mohs scale are 1, talc; 2, gypsum; 3, calcite; 4, fluorite; 5, apatite; 6, orthoclase feldspar; 7, quartz; 8, topaz; 9, corundum; 10, diamond. Devised by the German mineralogist Frederick Mohs (1773-1839).

Ore Any mineral or group of minerals containing sufficient quantities of metals to make them profitable to mine.

Outcrop Any mass of exposed bedrock.

Oxidized zone The upper part of an ore body that has been altered by oxygen and the downward percolation of groundwater.

Pegmatite An igneous rock with very coarse grain size, or a body of such rock. Pegmatites, usually found as smaller bodies within large igneous or metamorphic rock masses, are noted sources of gemstones and well-developed crystals.

Pelecypod Any bivalved aquatic invertebrate, including clams and oysters.

Phenocryst Any large crystal in a porphyritic rock.

Pipe A vertically emplaced, cylindrical mass of igneous rock.

Placer Any alluvial deposit of heavy mineral particles, especially of gold and certain gemstones, that has been concentrated gravitationally, usually by the action of water.

Plug The solidified core of an extinct volcano.

Porphyritic Characterized by distinct crystals (phenocrysts) enclosed in a finer-grained crystalline or glassy matrix.

Precipitation The process by which a dissolved or suspended solid is separated from a liquid.

Pseudomorph A crystal, or apparent crystal, having the outward appearance of another mineral that has been replaced by substitution or chemical alteration.

Quarry A surface excavation from which nonmetallic minerals are mined.

Sedimentary Formed by the accumulation of sediments.

Silica Silicon dioxide.

Silicified Any mineral or organic compound that has been altered or replaced by silica.

Specific gravity A relative expression of the density of a mineral based upon comparison with an equal volume of water.

Streak The color of the powder of a mineral created by rubbing the mineral over the surface of unglazed, white porcelain.

Striations Fine parallel grooves or lines on crystal faces.

Tailings Discarded refuse material produced from the milling or other treatment of ores; tailings rarely have collectible minerals.

Talus Rock debris at the base of a cliff or steep slope.

Telluride A binary compound of tellurium with gold or silver, such as calaverite and sylvanite.

Trilobite An extinct marine arthropod with a flattened segmented body covered by a dorsal exoskeleton marked into three lobes.

Tuff A stratified rock formed from deposition of fine volcanic detritus, such as cinder and ash.

Twin Two or more single crystals of the same mineral intergrown in a systematic arrangement.

Vein A tabular or sheetlike body of mineral matter cutting across or through surrounding rock, and with a different composition from the surrounding rock.

Volcanic rock Any extrusive igneous rock.

Volcanism The movement of molten rock through volcanoes and fissures to the earth's surface, where it cools into extrusive igneous rock.

Volcano A vent in the earth's crust through which lava and associated gases and ash escape onto the surface.

Vug A small, hollow cavity in rock, usually lined with a crystalline incrustation.

Waste In mining, the worthless gangue minerals and country rock removed and discarded in mining operations. Mine waste dumps can provide many collectible mineral specimens.

REFERENCES

1. Adams, J. W. 1968. Pegmatite minerals. *Mineral and water resources of Colorado.* Denver: USGS in collaboration with Colorado Mineral Industrial Development Board.

2. Anon. 1881. *Biennial Report of the State Geologist of the State of Colorado.* Denver: Tribune Publishing Company.

3. ———. 1987. *Faults, fossils, and canyons: significant geological features on public lands,* D. W. Kuntz, H. J. Armstrong, and F. J. Athearn, editors. Colorado State Office. Bureau of Land Management.

4. ———. 1991. *Rockhound guide.* Buena Vista (Colorado) Chamber of Commerce.

5. Baldwin, C. 1979. *Colorado gem and mineral collecting locales,* Vol. 1. Boulder: Johnson Publishing Company.

6. Barb, C. F. 1958. Some Colorado gem trails. *The Mineralogist,* Jun-Jul-Aug, Sep.

7. Barnes, R. 1985. The mines and minerals of Rico. *The Mineralogical Record,* May-Jun.

8. Brobst, D. A. 1968. Barite. *Mineral and water resources of Colorado.* Denver: USGS in collaboration with Colorado Mineral Industrial Development Board.

9. Burbank, W. S. 1932. *Geology and ore deposits of the Bonanza Mining District, Colorado.* USGS professional paper 169.

10. Butler, A. P. 1968. Uranium. *Mineral and water resources of Colorado.* Denver: USGS in collaboration with Colorado Mining Industrial Development Board.

11. Campbell, F. H., and Mitchell, R. S. 1961. Sand-calcite crystals from Stoneham, Colorado. *Rocks & Minerals,* Jan-Feb.

12. Chronic, H. 1988. *Roadside geology of Colorado.* Missoula, Montana: Mountain Press Publishing Company.

13. Chronic, J. 1960. Geology of south-central Colorado. *Guide to the geology of Colorado,* R. J. Weimer and J. D. Haun, editors. Denver: Geological Society of America, Rocky Mountain Association of Geologists, Colorado Scientific Society.

14. Cobban, W. A. 1973. Significant ammonite finds in uppermost Mancos shale and overlying formations between Barker Dome, New Mexico, and Grand Junction, Colorado. *Cretaceous and Tertiary rocks of the southern Colorado plateau.* Durango, Colorado: Four Corners Geological Society.

15. Collins, D. S., and Heyl, A. V. 1984. History of the Colorado-Wyoming state line diatremes. *Rocks & Minerals,* Jan-Feb.

16. Dayvault, R. D., and Goodknight, C. S. 1987. Supplemental road log Douglas Pass to FAA radar station. *Paleontology and geology of the dinosaur triangle.* Grand Junction: Museum of Western Colorado.

17. Dayvault, R. D., and Gorski, A. 1989. Fossils from the Green River Formation, Douglas Pass area, Colorado. *Rocks & Minerals,* Mar-Apr.

18. Del Rio, S. M. 1960. *Mineral resources of Colorado, first sequel.* Denver: State of Colorado Mineral Resources Board.

19. Eckel, E. 1961. *Minerals of Colorado: A 100-year record.* USGS bulletin 1114.

20. Fahl, R. 1948. Colorado plume agate. *The Mineralogist,* Jan.

21. Finley, G. I. 1916. *Colorado Springs folio, Colorado.* USGS geologic atlas of the United States, no. 203.

22. Gilbert, G. K. 1897. *Pueblo folio, Colorado.* USGS geologic atlas of the United States, no. 36.

23. Henderson, C. W. 1926. *Mining in Colorado.* USGS professional paper 138.

24. Hollister, O. J. 1867. *The mines of Colorado.* Springfield, Massachusetts: Samuel Bowles and Company.

25. Holmes, R. W., and Kennedy, M. B. 1983. *Mines and minerals of the great American rift.* New York: Van Nostrand Reinhold.

26. Hurst, C. T. 1935. The lost mine. *Southwestern Lore,* Dec.

27. Ives, R. L. 1941. The Green Ridge pegmatite, Grand County, Colorado. *Rocks & Minerals,* Jan.

28. Jacobson, M. I. 1985. Kings Canyon lithium pegmatite, Crystal Mountain District, Larimer County, Colorado. *Rocks & Minerals,* Sep-Oct.

29. ———. 1987. Minerals from the pegmatites of the Crystal Mountain District, Larimer County, Colorado. *Rocks & Minerals,* Jul-Aug.

30. Jacobson, M. I., and Tilander, N. G. 1982. A lithium-bearing pegmatite in the Clear Creek District, Clear Creek County, Colorado. *Rocks & Minerals,* Nov-Dec.

31. Kile, D. E., and Modreski, P. J. 1988. Zeolites and related minerals from the Table Mountain lava flows near Golden, Colorado. *The Mineralogical Record,* May-Jun.

32. Kile, D. E., Modreski, P. J., and Kile, D. L. 1991. Colorado quartz, occurrence and discovery. *Rocks & Minerals,* Sep-Oct.

33. Kleeman, T. H. 1941. Some Colorado fossil wood localities. *The Mineralogist,* Sep.

34. Knowlton, F. H. *Fossil floras of the Vermejo and Raton Formations of Colorado and New Mexico.* USGS professional paper 101.

35. ———. 1934. *Fossil plants of the Tertiary lake beds of south-central Colorado.* USGS professional paper 131-G.

36. Kunz, G. F. 1883. Precious stones. *Mineral Resources of the United States for 1882.* USGS.

37. ———. 1885. Precious stones. *Mineral Resources of the United States for 1883 and 1884.* USGS.

38. Kushner, E. F. 1972. *A guide to mineral collecting at Ouray, Colorado.* Privately published.

39. Lee, W. T., and Knowlton, F. H. 1917. *Geology and paleontology of the Raton Mesa area and other regions in Colorado and New Mexico.* USGS professional paper 101.

40. Longyear, B. O. 1939. Collecting Amethyst in Colorado. *The Mineralogist,* Jun.

41. McKinney, L., and McKinney, T. 1987. *Colorado Gems and Minerals.* Colorado Travel Guidebook Series. Frederick, Colorado: Renaissance House.

42. Michalski, T. C. 1984. Colorado amethyst. *Rocks & Minerals.* Jan-Feb.

43. Minch, R. S. 1983. Epidote from the Calumet Iron Mine in the Turret District, Salida, Colorado. *Rocks & Minerals,* Jul-Aug.

44. Minor, W. C. 1945. Pinon Mesa trails. *Rocks & Minerals,* Nov.

45. Mitchell, R. S. 1960. Small barite nodules from Ovid, Colorado. *Rocks & Minerals.* Jan-Feb.

46. Modreski, P., Lees, B., and Wilson, D. 1990. New explorations at Stoneham, Colorado. *Rocks & Minerals,* May-Jun.

47. Montgomery, A. 1938. Storm over Antero. *Rocks & Minerals,* Dec.

48. Muntyan, B., and Muntyan, J. 1985. Minerals of the Pikes Peak granite. *The Mineralogical Record,* May-Jun.

49. ———. 1988. Recent collecting activity in San Juan County, Colorado. *Rocks & Minerals,* Jul-Aug.

50. Nelson, C., and Riemyer, W. D. 1983. Geology of the Anaconda-Gunnison mine area, Gunnison County, Colorado. *Gunnison gold belt and Powderhorn carbonatite field trip guidebook,* R. C. Handfield, editor. Denver: Denver Region Exploration Geologists Society.

51. Odiorne, H. H. 1978. *Colorado amazonstone, the treasure of Crystal Peak.* Denver: Forum Publishing Company.

52. Over, E., Jr. 1929. Mineral localities of Colorado. *Rocks & Minerals,* Dec.

53. Parker, B. H. 1974. Gold placers of Colorado. *Colorado School of Mines Quarterly.* Vol. 69, No. 3.

54. ———. 1974. Gold placers of Colorado. *Colorado School of Mines Quarterly.* Vol. 69, No. 4.

55. Pearl, R. M. 1939. Gem collecting at Nathrop, Colorado. *The Mineralogist,* Oct.

56. ———. 1941. Colorado turquoise localities. *The Mineralogist.* Jan.

57. ———. 1941. Rare minerals—St. Peters Dome, Colorado. *The Mineralogist,* Jun.

58. ———. 1941. Florissant, Colorado gem locality. *The Mineralogist,* Aug.

59. ———. 1941. Turquoise deposits of Colorado. *Economic Geology,* Vol. 36, No. 3.

60. ———. 1942. Minerals of the San Luis Valley, Colorado. *The Mineralogist.* Aug.

61. ———. 1947. Largest turquoise nugget: a Colorado find. *The Mineralogist,* Jun.

62. ———. 1953. A Colorado petrified forest. *The Mineralogist,* Apr.

63. ———. 1969. *Exploring rocks, minerals, and fossils in Colorado.* Chicago: Sage Books/The Swallow Press.

64. ———. 1972. *Colorado gem trails and mineral guide.* Athens, Ohio: Sage Books/The Swallow Press.

65. Quick, L. 1963. *The book of agates.* New York: The Chilton Book Company.

66. Ransome, J. G. 1962. *The rock-hunter's range guide.* New York: Harper and Brothers.

67. ———. 1964. *A range guide to mines and minerals.* New York: Harper and Row.

68. ———. 1964. *Fossils in America.* New York: Harper and Row.

69. Read, C. R. 1934. *A flora of Pottsville age from the Mosquito Range, Colorado.* USGS professional paper 185-D.

70. Reitsch, C. W. 1939. Smoky quartz and amazonstone at Pine Creek, Colorado. *Rocks & Minerals,* Nov.

71. Rosemyer, T. 1988. The Sunnyside Mine, Eureka Mining District, San Juan County, Colorado. *Rocks & Minerals,* Sep-Oct.

72. Sanger, W. 1982. *Florissant Fossil Beds National Monument: Window to the past.* Estes Park, Colorado: Rocky Mountain Nature Association.

73. Schlegel, D. M. 1957. *Gem stones of the United States.* USGS bulletin 1042-G.

74. Scott, G. R. 1968. Gem stones. *Mineral and water resources of Colorado.* Denver: USGS in collaboration with Colorado Mineral Industrial Development Board.

75. Seaman, M. S. 1933. Creede District, Colorado, yields fine mineral specimens. *The Mineralogist,* Aug.

76. ———. 1935. Fluorite deposits of Wagon Wheel Gap, Colorado. *The Mineralogist,* May.

77. Shannon, J. M., and Shannon, G. C. 1985. The mines and minerals of Leadville. *The Mineralogical Record,* May-Jun.

78. Stose, G. W. 1916. *Apishapa folio, Colorado.* USGS geologic atlas of the United States, No. 227.

79. Truebe, H. A. 1984. Minerals of the Italian Mountain area, Colorado. *The Mineralogical Record,* Mar-Apr.

80. Van Alstine, R. E. 1968. Fluorspar. *Mineral and water resources of Colorado.* Denver: USGS in collaboration with Colorado Mineral Industrial Board.

81. Vanderwilt, J.W. 1947. *Mineral resources of Colorado.* Denver: State of Colorado Mineral Resources Board.

82. Voynick, S. 1986. Hartsel barites and pegmatites. *Rock & Gem,* Feb.

83. ———. 1986. A Golden museum. *Rock & Gem,* Mar.

84. ———. 1986. Colorado amazonite. *Rock & Gem,* May.

85. ———. 1987. Colorado's Blue Wrinkle lapis mine. *Rock & Gem,* Apr.

86. ———. 1988. Summitville gold. *Rock & Gem,* Jan.

87. ———. 1988. High plains barite. *Rock & Gem,* Mar.

88. ———. 1988. The great diamond hoax. *Rock & Gem,* May.

89. ———. 1988. There will never be another Leadville. *Rock & Gem,* Jun.

90. ———. 1988. Dinosaurs in Utah. *Rock & Gem,* Oct.

91. ———. 1989. The gold of Cripple Creek. *Rock & Gem,* Jan.

92. ———. 1989. Colorado oil shale. *Rock & Gem,* Mar.

93. ———. 1989. Clamming the high plains. *Rock & Gem,* Nov.

94. ———. 1989. Colorado carnotite. *Rock & Gem,* Dec.

95. ———. 1990. The fossils of Florissant. *Rock & Gem,* May.

96. ———. 1990. Finds on Topaz Mountain. *Rock & Gem,* Nov.

97. ———. 1991. Gold in Cache Creek Park. *Rock & Gem,* Jan.

98. ———. 1991. Central City and Idaho Springs. *Rock & Gem,* Feb.

99. ———. 1991. Royal Gorge pegmatites. *Rock & Gem,* Mar.

100. ———. 1991. Colorado amethyst. *Rock & Gem,* Apr.

101. ———. 1991. Tracking dinosaurs. *Rock & Gem,* Sep.

102. ———. 1991. The crystallized gold of Breckenridge. *Rock & Gem,* Oct.

103. ———. 1991. Owl Canyon alabaster. *Rock & Gem,* Nov.

104. ———. 1991. Creede, Colorado. The Summitville gold boulder. *Rock & Gem,* Dec.

105. ———. 1992. Keep all you find! *Rock & Gem,* Jan.

106. ———. 1992. Alma mining district. *Rock & Gem,* Mar.

107. ———. 1992. Colorado's dinosaur country. *Rock & Gem,* May.

108. ———. 1992. *Colorado Gold.* Missoula, Montana: Mountain Press Publishing Company.

INDEX

ABOUT THE AUTHOR

Lynda La Rocca photo

Stephen M. Voynick has worked as an advertising, industrial, and technical phototgrapher; marine salvage diver in the Caribbean and South America; and hardrock miner in Colorado, Wyoming, and Arizona. Since 1982, he has worked as a full-time freelance writer from his home in Leadville, Colorado. Voynick writes regularly for a number of national magazines on topics of natural resources, western history, mining, and applied technology. With a lifelong interest in minerals, gemstones, and fossils, he has written more than 100 articles for Rock & Gem magazine and is currently a contributing editor.

Colorado Rockhounding is Voynick's fourth book from Mountain Press Publishing Company, following *Leadville: A Miner's Epic* (1984); *Yogo: The Great American Sapphire* (1985); and *Colorado Gold* (1992). His earlier works include *The Making of a Hardrock Miner* (Howell-North Books, 1978); *In Search of Gold* (Paladin Press, 1982); and *The Mid-Atlantic Treasure Coast* (Middle Atlantic Press, 1984). Voynick has also researched and written a book-length manuscript detailing the history of Colorado's Climax Mine on commission for the Climax Molybdenum Company.

Dedicated to the memory of

Gus Seppi
1912–1988

Miner, mineral collector, gentleman, and friend

Life is worth living for such moments alone, for it is in such a high mountain world as Antero's, . . . timeless and separate from outside things, that one can put in each proper place bad weather, bad luck, hard work, and even mineral collecting. If we had not collected a single mineral, still it all would have been very much worth-while.

Arthur Montgomery, on collecting
aquamarine on Mt. Antero in 1937

CONTENTS

PREFACE

In 1958 Richard M. Pearl, a geology professor at Colorado College, wrote and published *Colorado Gem Trails and Mineral Guide,* the first statewide field guide to Colorado's mineral- and gemstone-collecting locales. Thousands of mineral and gemstone collectors have since used Pearl's guidebook. In testimony to its usefulness and popularity, after thirty-five years, *Colorado Gem Trails and Mineral Guide,* in its third revised edition, is still in print.

Thanks in part to Richard Pearl's work, field mineral collecting, fossil collecting, and recreational gold panning and mining have never been more popular. Colorado remains one of the best states in the nation for amateur collection of gold, gemstones, minerals, and fossils.

When Pearl's guidebook first appeared in the 1950s, Colorado's population was 1.3 million, and automobile tourism was in its infancy. Rural land and wilderness was just that—rural and wild. Recreational use demands on private rural land were minimal and "No Trespassing" signs were the exception rather than the rule. Many mining districts were active, and hundreds of mines added collectible ore and gangue mineral specimens to already voluminous dumps. The few mineral collectors who visited active mines were often welcomed, permitted to search mine dumps, and sometimes even advised on where to look for the "good stuff."

But Colorado's population has since tripled. Both tourism and recreational use of public and private rural land are at record high levels. Mountain towns struggle in the difficult transition from mining economies to tourism. Residential and commercial development of rural lands and posting of private property increases each year.

Colorado's mining industry has declined radically. Although annual mineral production is still in the hundreds of millions of dollars, few metal mines, traditionally the best source for many mineral specimens, remain active. Due to increasing concerns about security and liability, both active mines and mines on standby or caretaker status now generally prohibit trespassing and mineral collecting, unless by special arrangement.

Collectors have scoured Colorado's old mine dumps for decades, recovering many of the best and most readily accessible specimens.

Some mineral-collecting sites popular in the 1950s no longer exist because of environmental restrictions, land development, and mined land reclamation.

Nevertheless, public interest in amateur mineral and fossil collecting has boomed. The popularity of gold panning and recreational gold mining has soared in our current era of high-priced, "free market" gold.

Exciting new dinosaur theories formulated in the 1970s renewed both scientific and public interest in the "terrible lizards," as well as in general fossil collecting. Simultaneously, however, new federal regulations restricted the unauthorized collecting of vertebrate fossils, including fossilized dinosaur bone, on public lands. Federal agencies have withdrawn many former collecting sites on public land from amateur collecting to preserve irreplaceable paleontological resources for scientific investigation.

Accordingly, this book is intended as a guide to selected Colorado gold-, gemstone-, mineral-, and fossil-collecting localities and related sites of interest for the 1990s.

INTRODUCTION

When the term "rockhound" originated about 1915, it referred to both traditional prospectors and to the formally trained field geologists who were rapidly replacing them. Today, the term generally refers to the growing ranks of amateur field collectors of gold, gemstones, minerals, and fossils.

Amateur mineral and fossil collecting in Colorado began in the 1860s soon after the Pikes Peak gold rush. For nearly a century, rockhounding had a relatively limited following, for serious rockhounds needed not only a basic interest in minerals, but the time and resources to pursue their interests.

Rockhounding activity increased sharply in the 1950s. A postwar economic boom provided many Americans with higher disposable incomes, more time for recreation, affordable automobiles, and better highway access to once-remote mining districts and collecting sites. With higher general educational levels, more people could understand and appreciate the basics of geology, gemology, mineralogy, and paleontology—the sciences that make rockhounding a most fascinating hobby.

Rockhounds are served by hundreds of local and regional club newsletters and four nationally circulated magazines: *Lapidary Journal, Rock & Gem, Rocks & Minerals,* and *The Mineralogical Record.* W. R. C. Shedenhelm, graduate geologist, former twenty-year senior editor of *Rock and Gem,* estimates that the United States has more than a thousand active rockhounding and lapidary clubs. "The total number of people who collect in the field or buy specimens at shows," Shedenhelm estimates, "runs into the millions."

Rockhounding's unusually broad appeal begins with a hands-on association with mineral variety and beauty. But rockhounding can also mean fresh air and outdoor exercise; field trips may be a pleasant afternoon visit to a familiar site, or a challenging expedition into rugged country. "Prospector's luck" adds the enticing element of intrigue, for no one knows what may turn up in the next pan, over the next ridge, or in the next mine dump. Collections and knowledge grow with each field trip and specimen. Lapidaries take special pride in fashioning gems from rough gemstones they personally collected. History buffs collect ores and minerals that underwrote great prospecting and mining adventures, while fossil collectors seek specimens that are a record of the evolution of life itself. Rockhounding may be pursued on any level of expertise, from that of the casual once-a-year field collector to the serious and knowledgeable amateur mineralogist and paleontologist.

Rockhounds have few geographical limitations, for minerals and fossils occur virtually everywhere. Gold panners still wash nuggets from California gravels once worked by the forty-niners, while Eastern collectors recover zeolite crystals from basalt formations in the shadow of New York City's skyscrapers.

Rockhounding opportunities, of course, vary widely from state to state. Because of remarkable geological, topographical, mineralogical, and paleoenvironmental diversity, together with an epic history of intensive mining and fossil quarrying, the richness and variety of Colorado's mineral and fossil sources are unexcelled. Even Colorado's place names whisper promises of collecting opportunities—Crystal Peak, Topaz Point, Fossil Ridge, Baculite Mesa, Gold Run, Ruby Mountain, Agate Creek, Opal Hill, and so on. Colorado has more than 20,000 inactive or abandoned mine sites, a thousand miles of gold-bearing streams, and a wonderful array of colorful mining towns, historic mining districts, fossil quarries and dinosaur trackways, research natural areas, interesting rock shops and mineral, mining, and paleontological museums, and collecting localities where no one comes away empty-handed.

In Colorado, rockhounds may collect agate and petrified wood on the vastness of the Great Plains, follow dinosaur trackways on hogback ridges, observe paleontologists excavating fossilized bones of the dinosaurs, pan gold in mountain streams, search for gemstones on towering peaks, or collect along mining-district roads that are literally paved with glittering metal sulfide minerals. This book will help you in all these adventures.

Part One provides general background information that will be valuable in planning field collecting trips. Chapter One is a thumbnail sketch of the geologic origins that account for the richness and diversity of Colorado's gold, gemstone, mineral, and fossil resources.

Chapter Two traces the history of Colorado mining, digging, and collecting efforts, which have yielded countless mineral and fossil specimens. Special sections cover Colorado's gold fields; the Colorado Mineral Belt; the gemstone-rich pegmatite belt; the Uravan Mineral Belt, where the great 1950s uranium rush was played out; the oil shale country of the Piceance Creek Basin; and the "boneyards," where paleontologists still excavate superb fossils of dinosaurs and other forms of early life.

Chapter Three addresses the issues of legality and safety in field mineral and fossil collecting. It is important to realize that mention of any specific locality in this book does not necessarily mean that site is open to collecting. The responsibility of collecting legally and safely rests ultimately with collectors themselves.

Part Two lists and describes more than 350 collecting localities and related sites of mineralogical and paleontological interest in fifty-two of Colorado's sixty-three counties. Counties, or geographical groups of counties, appear alphabetically. Included are:

- More than 120 specific or general rock-, mineral-, or gemstone-collecting localities
- Seventy gold-bearing creeks and rivers where gold concentrations are high enough to have warranted at least some commercial placer mining
- Fifty-five mining districts with thousands of old mines and mine dumps with collectible mineral specimens
- Forty-four fossil-collecting localities
- Sixty-five related sites of mineralogical, mining, or paleontological interest, including national monuments; rock shops and museums with notable collections of Colorado minerals, gemstones, and fossils; dinosaur-fossil quarries and trackways; Bureau of Land Management paleontological research natural areas; public underground mine tours; mines of unusual historic, economic, or mineralogical interest; and geological points of interest

Seventy-one maps show locations of all mentioned localities and sites in relation to U.S. and interstate highways, Colorado highways, county and Forest Service roads, and major topographical features. Field collectors should use these maps in conjunction with detailed topographical maps.

The Colorado Mineral Guide provides brief physical descriptions useful in field identification of seventy-six Colorado minerals, with notation of mentioned county occurrences.

The glossary provides definitions of terms that may be unfamiliar to the reader. The listed references describe collecting localities and other sites and provide the reader with additional background and information.

PART ONE: COLLECTING IN COLORADO

ERA	PERIOD	EPOCHS	AGE (millions of years ago)
Cenozoic	Quaternary (Pleistocene Ice Age)		
			3
	Tertiary	Pliocene	
			12
		Miocene	
			26
		Oligocene	
			38
		Eocene	
			54
		Paleocene	
			65
Mesozoic	Cretaceous		
			135
	Jurassic		
			200
	Triassic		
			240
Paleozoic	Permian		
			280
	Pennsylvanian		
			325
	Mississippian		
			370
	Devonian		
			415
	Silurian		
			445
	Ordovician		
			515
	Cambrian		
			600
Pre-cambrian	Lipalian Interval		
			? 1,000
			? 5,000

1 GEOLOGY OF COLORADO

The extraordinary diversity and richness of Colorado's mineral and fossil deposits are not due to any one geological event or simple sequence of events, but to lengthy, complex, and often repetitive processes of emplacement, deposition, uplifting, faulting, erosion, and alteration.

Colorado's oldest rocks are about two billion years old. When they were formed, North America was simply a part of the earth's crust, rather than an identifiable continent. The crust is composed of a dozen large, rigid plates, each about sixty miles thick. These tectonic plates, floating atop a mass of viscous magma, are bordered by midocean ridges and deep ocean trenches. The midocean ridges are points of crustal origin, where magma surges slowly upward to solidify into new sections of crust. The opposite occurs at the trenches; edges of plates are drawn under those of adjacent plates, forced downward, and remelted into magma.

This continuous circulation of crustal and magmatic material imparts a slow, powerful motion to the plates. Although movement is only inches per year, when measured against hundreds of millions of years of geologic time, tectonic plate travel, or continental drift, becomes global in scope and accounts for the relative positions of the continents today.

During the Precambrian era, when the continents were clustered in the general area of Europe-Africa-Asia, at least two lengthy periods of mountain building occurred near the center of the North American Plate. Surface erosion associated with long periods of stability completely leveled those Precambrian ranges. By the dawn of the Paleozoic era, 600 million years ago, seas spread eastward across the North American Plate, depositing marine sediments that became massive formations of shale, limestone, and sandstone.

The North American Plate adjoined the Arctic Ocean on the north, the Caribbean Sea on the south, the Mid-Atlantic Ridge on the east, and the East Pacific Rise on the west. The eastern section of the plate was a stable craton, or continental shield, while the west developed a less stable basin-and-range foundation. Colorado was located on a structurally weak, narrow north-south belt that separated the two sections.

Late in the Paleozoic era, perhaps 300 million years ago, Colorado emerged from receding seas as a highland. At the same time, the Mid-Atlantic Ridge began spreading, creating an embryonic Atlantic Ocean and sending the North American Plate on a slow westerly drift. The separation of North America from Europe and Africa generated tectonic stresses that buckled the structurally weakened central belt upward to form the ancestral Rocky Mountains.

Colorado's ancestral Rockies formed two large ranges: one paralleled the present Front Range, and the other stretched northwest from the San Luis Valley toward Utah. Within 50 million years, erosion reduced both ranges to coarse gravels that formed floodplains, deltas, and the floors of broad valleys.

Midway through the Mesozoic era, 200 million years ago, a semitropical climate fostered an explosion in diversity and richness of plant and animal life. During the Jurassic and Cretaceous periods, dinosaurs thrived in lush forests and along the warm lakes and rivers that covered much of central and western Colorado. Eastern Colorado lay beneath a warm, shallow interior seaway that teemed with marine life and deposited thick layers of shale, sandstone, and limestone. As sea levels rose and fell, a broad shoreline of sandy beaches, barrier bars, and lagoons moved slowly across much of Colorado.

When the dinosaurs neared extinction at the close of the Cretaceous period 65 million years ago, the North American Plate had drifted 1,500 miles westward, colliding with the Pacific Plate and generating new tectonic stresses and another mountain-building episode. Again, the weak central belt buckled, lifting great blocks of Precambrian basement rock and Paleozoic and Mesozoic sediments upward to form the new Rocky Mountains, the Rockies we know today.

After the age of the dinosaurs, mammals evolved rapidly to become the predominant life-form of the new mountain, forest, and plain environment of the Tertiary period. Forty million years ago, stresses fractured and faulted the weakened central belt, permitting magma to surge upward from the depths. Enormous surges of magma extruded onto the surface in southwestern Colorado to form the San Juan Mountains. Volcanism was less frequent in central Colorado; nevertheless, great masses of magma forced into crustal faults and crevices solidified into such porphyritic intrusions as batholiths, sills, dikes, and stockworks.

Both the magmatic intrusions and extrusions further fractured the surrounding country rock, permitting the passage of superheated, mineral-rich solutions. These hydrothermal events physically and chemically altered the country rock, while depositing minerals in bodies of widely varying composition, concentration, and configuration. Hydrothermal mineralization was concentrated in a narrow 150-mile-long zone stretching diagonally across the Colorado Rockies. This highly mineralized zone—now known as the Colorado Mineral Belt—contained a wealth of precious, base, and alloying metals.

Continued erosion of the Rockies deposited huge amounts of silt, sand, and gravel on the eastern plains, the intermontane valleys, and the western canyon-and-plateau country. Twenty million years ago, as the North American Plate continued grinding over the edge of the Pacific Plate, crustal stresses caused a massive regional uplift, raising much of the southwestern United States, including Colorado, about 5,000 feet to its present elevation.

The final sculpting of Colorado's mountains and valleys occurred when global cooling spawned the Pleistocene ice ages. Colorado was not swept by the great northern continental ice sheet, but by regional alpine glaciation over most of its higher elevations. Deep snowfall accumulated and compressed into perpetual snowfields, then into mobile glacial masses that scoured away enormous volumes of rock. When the glaciers finally retreated under moderating temperatures, they released torrents of water that cut deep canyons and sorted the gravels and glacial debris. Most of Colorado's gold placers, destined to play such a big role in its later history, were created during this period. At the end of the ice ages, a mere 15,000 years ago, Colorado's topography appeared much as it does today.

Colorado is divided into three general topographical zones: the eastern plains, the central mountains, and the western plateau-and-canyon country. Colorado's elevations range from a low point of 3,300 feet near Lamar, where the Arkansas River flows into Kansas, to 14,433 feet at the summit of Mt. Elbert near Leadville. Although the Colorado climate is generally semiarid, the elevation range gives rise to widely varying ecological life zones, from plains grasslands, piñon-juniper woodlands, and deep mountain forests of pine, spruce, fir, and aspen to the largest expanse of alpine tundra south of Alaska.

The forces of erosion that sculpted Colorado's diverse modern topography have also exposed enormous formations of sedimentary, metamorphic, and igneous rock with their contained treasures of gold, gemstones, minerals, and fossils. Mining, digging, and collecting those treasures is an ongoing adventure.

2 MINING, DIGGING & COLLECTING: A HISTORY

Human beings began utilizing Colorado's minerals over 10,000 years ago, flaking pieces of chalcedony into tools and projectile points. Native Americans later mined turquoise from the San Luis Valley and collected amazonite from the Pikes Peak region for ornamental use.

The Pikes Peak gold rush of 1858-59 quickly turned Colorado from a wilderness into a territory, and prospectors embarked on an odyssey of mineral discovery that continues today. By the turn of the century, Colorado ranked among the nation's great mineral sources, with hundreds of mines turning out millions of dollars in gold, silver, lead, zinc, and copper each year.

Prospectors also discovered commercial deposits of nonmetallic rocks and minerals, including coal, oil shale, gypsum, alabaster, marble, and beautiful gemstones such as aquamarine, amazonite, topaz, smoky quartz, rhodochrosite, and phenakite. Bone hunters and paleontologists discovered fossils including those of dinosaurs and insects that helped revolutionize the science of paleontology.

One of Colorado's more than 20,000 old mine sites.

The light yellow color of most metal mine dumps is caused by iron hydroxide, a product of the oxidation of metal sulfide minerals.

In the twentieth century, Colorado's miners produced radium, vanadium, tungsten, mica, feldspar, rare earth minerals, and the greatest treasure of all—molybdenum. And in 1948 Colorado hosted the last great frontier-style mineral rush for the indispensable element of the atomic age—uranium.

The discovery and recovery of Colorado's mineral and fossil treasures have left behind a colorful history, hundreds of fading mining districts, and more than 20,000 diggings ranging from gold mines to dinosaur quarries. Mineral collectors, gemstone aficionados, lapidaries, mining-history buffs, and amateur paleontologists will be particularly interested in six specific areas of exploration and development: the gold fields, the Colorado Mineral Belt, the pegmatite belt, the Piceance Creek Basin oil shale country, the Uravan Mineral Belt, and the fossil "boneyards."

THE GOLD FIELDS

Legends tell that early Spanish and French explorers first discovered gold in what is now Colorado, but American mountain men made the first substantiated finds in the early 1800s. Their discoveries proved inconsequential until a modest strike near the present site of Denver in June 1858 triggered the Pikes Peak rush.

The Gold Fields

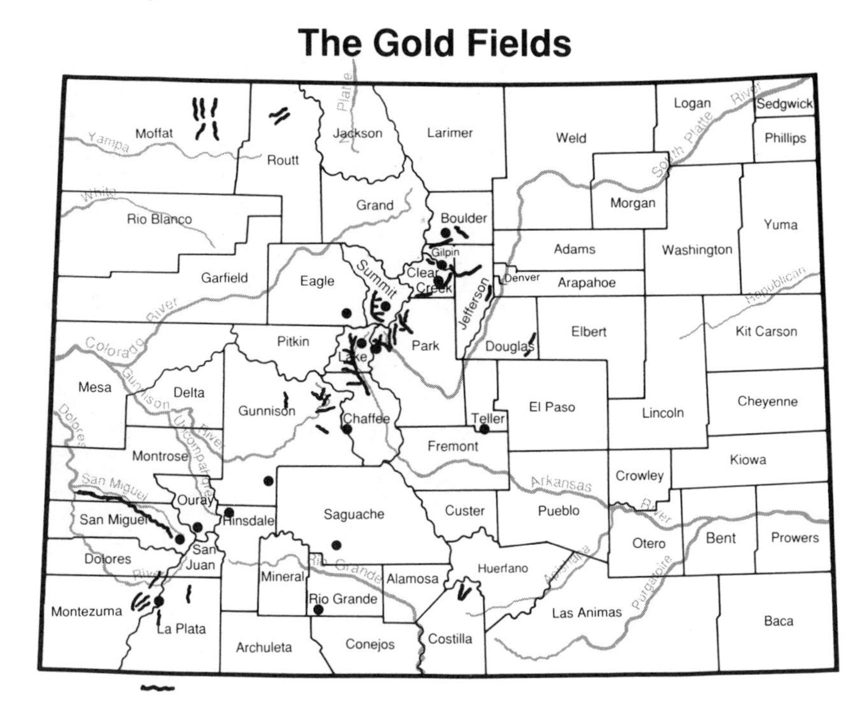

Thousands of gold seekers rushed to the overpromoted "Cherry Creek diggings," their wagons emblazoned with "Pikes Peak or Bust!" Many did bust, for only 100 ounces of gold were actually mined in 1858, and the Pikes Peak rush seemed destined for disaster.

But in 1859, prospectors redeemed the rush at its darkest moment with big strikes near the sites of Idaho Springs and Central City. By 1860, prospectors had struck gold and founded dozens of rough mining camps along the upper reaches of Clear Creek and the Blue, South Platte, and Arkansas rivers. In just ten years, Colorado placer miners recovered 1.2 million ounces—forty standard tons—of gold.

Colorado gold mining followed a recurring pattern of boom and bust. The gold rush itself quickly turned to bust in the 1860s, when miners depleted the rich discovery placers and high-grade outcrops. Although prospectors also discovered many gold-bearing metal sulfide deposits, their development awaited the improved underground mining and smelting technology of the 1880s.

Gold production soared in the 1890s with development of Cripple Creek, one of North America's greatest gold deposits. By 1900, Colorado led the nation in gold production, mining over a million troy ounces per year.

Lone recreational gold miner at work on a Colorado river.

Even placer mining recovered strongly when a fleet of floating-bucketline dredges worked the deep gravels of many old placers.

But after World War I, gold production declined rapidly. Since 1837, Congress had fixed the price of gold at $20.67 per troy ounce, and decades of inflation had finally eroded its real value. With gold literally "not worth mining," Colorado's annual production plummeted to only 200,000 troy ounces.

Mining boomed again following the revaluation of gold to $35 per ounce during the Depression years. Hundreds of gold mines reopened and thousands of otherwise jobless men and women panned just enough gold from old placers to survive. By 1941, Colorado's annual gold production had doubled to 400,000 ounces.

The recovery was cut short by World War II. After the War Production Board's limitation order L-208 closed most gold mines to divert mining manpower and equipment to emergency production of base and alloying metals, Colorado's gold production fell to its lowest level since 1858.

L-208 was rescinded in 1945, but gold mining never recovered, for inflation had already overtaken "$35 gold." By the 1960s, Colorado gold mining had all but ceased.

After governments made gold a free-market metal commodity in the early 1970s, the metal reached a historic high of $800 per troy ounce in 1980, stimulating both commercial and recreational gold mining. With the price of gold now stabilized at $300 to $400 per troy ounce, Colorado's current annual gold production has risen to about 100,000 ounces per year, and may increase further through the 1990s.

Since 1858, Colorado has produced 41 million troy ounces, or 1,400 standard tons, of gold. Because of the metal's great density, those 1,400 tons could fit into an eleven-foot cube. Although prospectors have found the bonanza placers and high-grade vein outcrops, geologists believe more far more gold remains in the ground than has ever been mined.

Colorado offers many opportunities for gold panners, recreational gold miners, and mineral collectors seeking specimens of native gold. More than half of Colorado's metal mines produced gold either as a primary product or a by-product, and with a little luck and know-how collectors can find mineral specimens with bits of visible gold in mine dumps.

Panners and recreational placer miners have even better chances for success. Since 1858, placer miners have recovered two million ounces of gold from 1,000 miles of Colorado streams and rivers. Thousands of panners and recreational gold miners find a little more each year.

The basic appeal of panning and recreational placer mining rests in its simplicity. Gold's great density permits easy recovery from stream gravels by hydraulic gravitational separation, or "washing." The specific gravity of pure gold is 19.3, meaning it is 19.3 times more dense than an equal volume of water. Native gold never occurs in pure form, but always in combination with small quantities of other metals (usually silver, copper, and iron), and therefore has a somewhat lower specific gravity. When a pan of gravel is mixed thoroughly with water and agitated, any gold present settles quickly to the bottom. Repetitive agitation, with continuous removal of the barren top gravels, produces a concentrate of gold and other dense minerals, such as iron-based "black sands." Sluice-type recovery devices employ the same simple principle of hydraulic gravitational separation.

Colorado's gold nuggets are considerably smaller than those of California or Alaska. Colorado's gold fields lacked both massive veins of native gold and a suitable "trap and hold" topography. Undocumented early reports tell of Colorado nuggets weighing thirty and forty troy ounces. The largest Colorado gold nugget known to exist, the "Penn Hill nugget," mined from Pennsylvania Mountain in Park County and now displayed at the Denver Museum of Natural History, weighs just twelve troy ounces.

Although half of Colorado's sixty-three counties have yielded at least some placer gold, six counties account for 95 percent of Colorado's total placer gold production. Not surprisingly, those six counties—Summit, Lake, Park, Clear Creek, Chaffee, and Gilpin—have the best opportunities for gold panners and recreational gold miners today.

THE COLORADO MINERAL BELT

Gold-rush prospectors found placer gold in alluvial gravels and lode gold in oxidized vein outcrops. The oxidized portions of the vein outcrops were shallow and represented only the surface tip of far larger unoxidized multimetal sulfide deposits that were also rich in gold, silver, and base metals.

Almost all of Colorado's metal sulfide deposits fell within a belt extending from Boulder County 150 miles southwest into the San Juan Mountains. Although enormous in extent and richness, the discoveries were initially more frustrating than exciting. Existing smelting technology was unsuited to the so-called refractory (literally, "stubborn") ores. Not even gold could always be extracted economically. Owners sometimes abandoned bonanzas, failing to recognize their potential.

By the 1880s, after the necessary advances in assaying, smelting, and underground mining methods, hundreds of mining districts

These picturesque mine ruins are among thousands of others in the Colorado Mineral Belt.

Colorado Mineral Belt

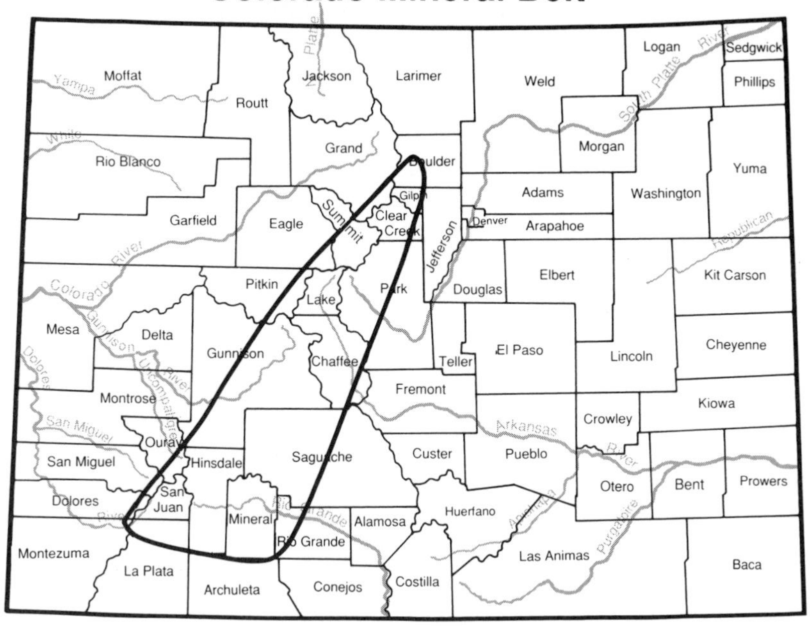

stretched from Caribou and Central City through Leadville, Alma, and Aspen all the way to Lake City, Silverton, Ouray, Telluride, and Rico, a vast area known as the Colorado Mineral Belt. Soaring silver production attracted railroads and investment capital, but also contributed to national overproduction, a big factor in the 1893 silver-market crash, which closed many mines.

By 1900, when the rich, "direct smelting" ores were exhausted, the new flotation separation process, which concentrated low-grade ores, brought renewed prosperity to the Colorado Mineral Belt. Colorado's lead, zinc, and copper deposits helped satisfy the growing industrial demand for base metals. Mining of tungsten and molybdenum began during World War I.

By the 1920s, when mining wallowed in a deep postwar depression, the thousands of Colorado Mineral Belt mines had produced 10 million troy ounces of gold worth $200 million; 630 million troy ounces (22,000 tons) of silver worth $500 million; 2.1 million tons of lead worth $250 million; 870,000 tons of zinc worth $1.5 billion; and substantial quantities of copper, tungsten, and molybdenum.

Many Mineral Belt mines worked around the clock to help satisfy emergency metal demands of World War II and the Korean

War. Molybdenum surpassed gold as Colorado's most valuable metal in the 1960s, when cumulative production at Lake County's Climax molybdenum mine reached $2 billion.

The Colorado Mineral Belt covers 3,000 square miles over seventeen counties: Boulder, Chaffee, Clear Creek, Eagle, Gilpin, Gunnison, Hinsdale, Lake, Mineral, Ouray, Park, Pitkin, Rio Grande, Saguache, San Juan, San Miguel, and Summit. With the exceptions of Teller County's Cripple Creek gold deposit and Custer County's Silver Cliff silver-mining district, the Mineral Belt contains all of Colorado's important metal deposits.

Although only a few dozen mines remain active, the Colorado Mineral Belt has an estimated 20,000 inactive or abandoned mines. Some earned millions, others never shipped paying ore; some were large operations with miles of workings, others were simple tunnels and shafts. Most metal mines are clustered in districts, but others are isolated. Across the entire 3,000-square-mile Colorado Mineral Belt, there are nearly 400 mining districts and an average of five mines per square mile.

Most of Colorado's 750 minerals occur in the Mineral Belt. In variety of minerals and sheer number of specimens, the Colorado Mineral Belt's thousands of mine dumps are among the most productive collecting areas in the United States.

Mineral collectors should understand the process of oxidation, a natural chemical process affecting most metal mine dumps within the Colorado Mineral Belt. Upon exposure to water and atmosphere, metal sulfide minerals react with oxygen to form sulfuric acid and free metal ions. The mineral that oxidizes most rapidly is the most common sulfide in the Mineral Belt—pyrite, or iron disulfide, also known as "fool's gold." Oxidation of pyrite produces sulfuric acid and free iron ions. Dilution eventually decreases the acidity, precipitating the dissolved iron as yellow iron hydroxide. Accordingly, most Mineral Belt metal mine dumps are stained a characteristic brown-yellow to yellow-orange color.

Sulfide mineral oxidation creates a serious mine-drainage pollution problem in the Mineral Belt. The acidic mine drainage water dissolves heavy metals such as silver, lead, copper, cadmium, and zinc, which in sufficient concentration are detrimental or toxic to aquatic life. The rocks and banks of streams polluted by mine drainage water are stained a telltale brown-yellow color.

Oxidation occurs most rapidly on the surface of mine dumps and in flooded underground workings. Because of the combined effects of physical and chemical deterioration, mine dumps may appear to be heaps of mud or sand with little collecting potential.

But digging beyond the heavily oxidized surface layer can reveal unoxidized sulfide minerals. Large pieces of mine dump rock may also have heavy surface oxidation. Breaking large specimens will often reveal unoxidized interiors with collectible crystals of metal sulfide and gangue minerals.

Colorado Mineral Belt mines have provided countless fine mineral specimens for museums and top private collections. Typically, Mineral Belt specimens are composites containing several minerals. Typical combinations include rhodochrosite on tetrahedrite; argentiferous galena in association with golden barite; sphalerite and galena on white dolomite; chalcopyrite in white quartz; galena and native silver in amethyst; and ores with more than half their overall weight consisting of pyrite, galena, and sphalerite.

Also present on the mine dumps are the oxide and carbonate alteration products of sulfide minerals. Common examples are smithsonite, the green carbonate of zinc; siderite, the brownish carbonate of iron; cerussite, the pinkish carbonate of lead; and malachite and azurite, the two bright green and blue basic carbonates of copper.

Collectors have searched the Mineral Belt mine dumps for decades. Commercial collectors have gathered carload lots of gangue and ore minerals for rock shops and scientific supply houses. During the 1950s and early 1960s, some dumps were "remined" for their pyrite content, then valuable for the manufacture of sulfuric acid.

Today, the best collecting potential is for micromount and miniature ("thumbnail") specimens. Collectors familiar with Mineral Belt mining districts tailor their field trips for specific minerals. Boulder County is a good source of tungsten minerals, including ferberite, huebnerite, and scheelite. The Central City and Idaho Springs districts have a variety of base metal sulfides and pyrite often associated with clear quartz. The Park County mines near Alma have provided silver minerals and superb red rhodochrosite. Lake County has two major districts: Climax, the world's largest molybdenite deposit, and the historic Leadville district, with twenty square miles of dumps rich in pyrite, galena, and sphalerite. Creede is famous for its "amethyst silver" specimens—lead-silver sulfides and native silver in a matrix of banded amethyst quartz. The San Juan Mountain mining districts are prime sources of pink rhodonite and metal sulfides in white quartz. Finally, the Colorado Mineral Belt is interesting not only for its many mineral-collecting opportunities, but for its scenic beauty and mining history.

THE PEGMATITE BELT

Pegmatites, coarse varieties of granite, provide many of Colorado's most beautiful and valuable mineral crystals and gemstones, including aquamarine, topaz, phenakite, amazonite, and smoky quartz.

United States Army explorers first reported mineral crystals in Colorado in 1847. Gold-rush prospectors found many crystal occurrences, and commercial collecting began about 1865. In the early 1870s, Dr. A. E. Foote, of Philadelphia's Foote Mineral Company, the nation's largest distributor of mineral-crystal specimens, employed twenty men to dig and collect amazonite and smoky quartz from granite pegmatites northwest of Pikes Peak.

By 1880, entrepreneurs greeted arriving passengers at the Colorado Springs railroad station with enticing displays of "Pikes Peak Crystals." In what may have been Colorado's first fee mineral-collecting venture, $20 bought a three-day collecting expedition to Florissant's Crystal Peak, including horses, tent, meals, and "knowledgeable guidance in the gem fields."

Prospectors found gem topaz at Crystal Peak in 1880, then discovered gem aquamarine in pegmatites below the 14,256-foot summit of Mt. Antero near Buena Vista. Colorado's growing suite of pegmatite crystals and gemstones won top awards at prestigious state, national, and international trade fairs and expositions.

George Frederick Kunz, America's first recognized gemologist, reported extensively on Colorado pegmatite crystals in the early annual reports of the United States Geological Survey. In the 1882-83 *Mineral Resources*, Kunz wrote:

> The Pikes Peak topaz and phenakite locality has been searched to some extent, and the topaz taken from it thus far would be valued fully at $1,500, one crystal of topaz being held at $100, and one phenakite at fully this amount.

In 1885 Kunz reported:

> A number of beryls of fine blue color . . . have been found near Mt. Antero, in the Arkansas Valley, Chaffee County, Colorado. One of these was four inches long and three-eighths of an inch across. . . . The large beryl mentioned in "Mineral Resources" for 1883 and 1884 has afforded the finest aquamarine of American origin known. It weighs 133 3/4 carats and measures 35x35x20 millimeters. It is a brilliant cut gem and with the exception of a few internal hair-like striations it is absolutely perfect.

Pegmatite Belt

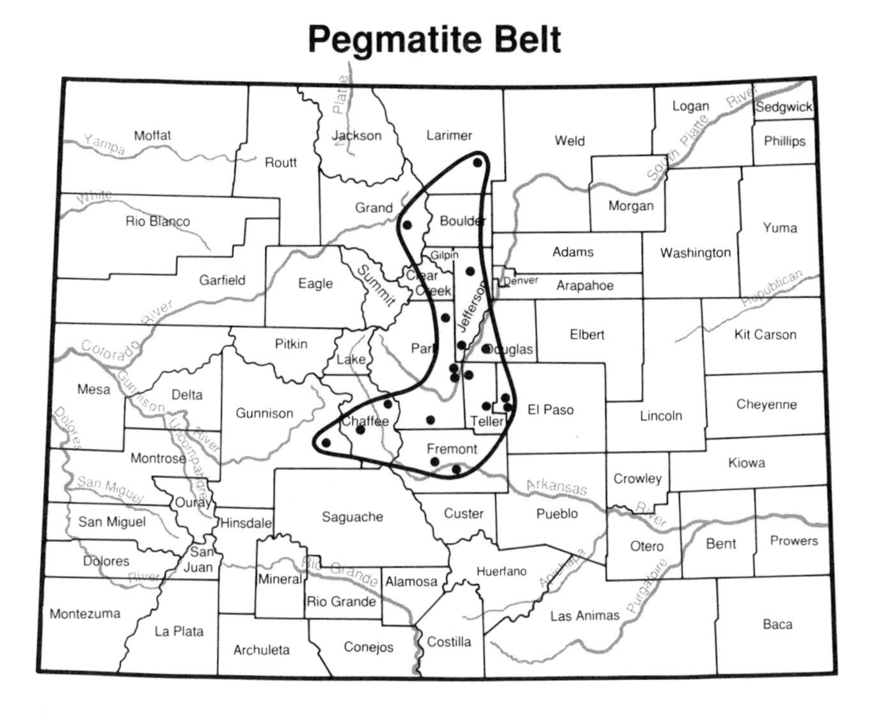

Kunz also noted:

> The Platte Mountain topaz locality, near Pikes Peak . . . has been prospected very extensively during the last fourteen months, and many fine crystals of topaz have been found, some of them yielding cut stones of 10 to 193 carats each in weight. . . . One of the larger ones is as fine a gem as America has produced of any kind.

Collectors recovered all of those superb gemstone crystals from granite pegmatites. Granite, an intrusive rock that solidified from magma, is composed primarily of feldspar, quartz, and smaller amounts of muscovite and biotite mica, hornblende, and other accessory minerals. Most granitic intrusions solidified relatively quickly into fine-grained granites with feldspar, quartz, and mica present in evenly dispersed crystals (phenocrysts) not larger than one-third of an inch.

Granite pegmatite, however, refers to coarse-grained granite, or an emplaced body of coarse-grained granite. Pegmatites formed from pockets of magma that retained heat to cool and solidify very slowly. Instead of quickly "freezing" into a uniform granite with well-dispersed crystalline mineral components, the residual magma crystallized on a fractional, or mineral-by-mineral, basis, forming horizontal pods and lenses, as well as irregular veins and dikes.

Residual magma was often enriched with concentrations of accessory and rare minerals. Gas sometimes created mariolitic cavities, or vugs, providing space for the growth of large and extraordinarily well-developed crystals. Many pegmatites are zoned, with concentric sections of mineral groups that crystallized at different intervals during the cooling process. Outer zones may consist of increasingly coarse granite followed by graphic granite, then massive feldspar, quartz, mica, and, finally, interior mariolitic cavities lined with spectacular mineral crystals.

Pegmatite size and composition varies greatly. Colorado pegmatites range from only a few feet to more than a mile long and are categorized by the mineral of greatest economic or specimen value present, such as feldspar, mica, quartz, beryl, lithium, and rare earth. Most Colorado pegmatites occur within a large belt extending from Crystal Mountain in Larimer County south to the Royal Gorge in Fremont County, then west into Gunnison County. The heart of the Pegmatite Belt is the Pikes Peak Batholith, a sixty-five-mile-long intrusion of coarse, pink Pikes Peak granite covering 1,200 square miles in Douglas, Teller, Park, Fremont, and El Paso counties. The Pikes Peak Batholith has yielded most of Colorado's pegmatite minerals and is world-famous for its amazonite, topaz, and smoky quartz.

Although Colorado's pegmatites are best known for crystal specimens and gemstones, some were mined commercially for feldspar, mica, and rare earth and radioactive minerals such as lepidolite, beryl, columbite-tantalite, and monazite. Specimens of many unusual minerals are still found in pegmatite quarries and on adjacent dumps.

Eroded or otherwise exposed pegmatites are easily recognized, for their primary minerals—feldspar, quartz, and mica—are highly reflective and considerably lighter in color than the surrounding rock. Collectors recover the finest crystals by carefully excavating pegmatite cavities. Any ground subsidence or cavelike pocket in a cliff may be a partially collapsed pegmatite cavity. An exhibit at the Denver Museum of Natural History depicts a partially exposed Pikes Peak granite pegmatite cavity containing amazonite. In the field, extracting crystals intact from such cavities requires proper tools, experience, patience, and effort.

Casual collectors find loose crystals in talus slopes and drainages below known pegmatite occurrences. Although usually abraded or fragmented, loose pegmatite crystals such as smoky quartz, amazonite, or topaz are still prized specimens.

THE PICEANCE CREEK BASIN: OIL SHALE COUNTRY

In the 1970s and early 1980s, Colorado's most celebrated rock was oil shale. Rock shops and highway tourist shops from Denver to Salt Lake City sold polished oil shale paperweights, bookends, brooches, and belt buckles. Sales were especially brisk in the bars, grocery stores, and gas stations in the towns between Glenwood Springs and Grand Junction along the southern edge of the Piceance Creek Basin. At the time, oil shale was not just an attractive sedimentary rock, but a symbol of imminent economic prosperity for western Colorado and "energy independence" for the United States.

The Piceance (PEE-ants) Creek Basin is the world's richest known hydrocarbon energy resource, but anticipated economic booms never materialized. Today, oil shale, with its delicate banding and soft pastel colors, is a reminder of one of the stranger stories in American mineral-resource development.

Oil shale, neither oil nor shale, is an organic marlstone, a fine-grained sedimentary rock containing hydrocarbon energy in the form of kerogen, a black, rubbery solid that is an undeveloped

Piceance Creek Basin (Oil Shale)

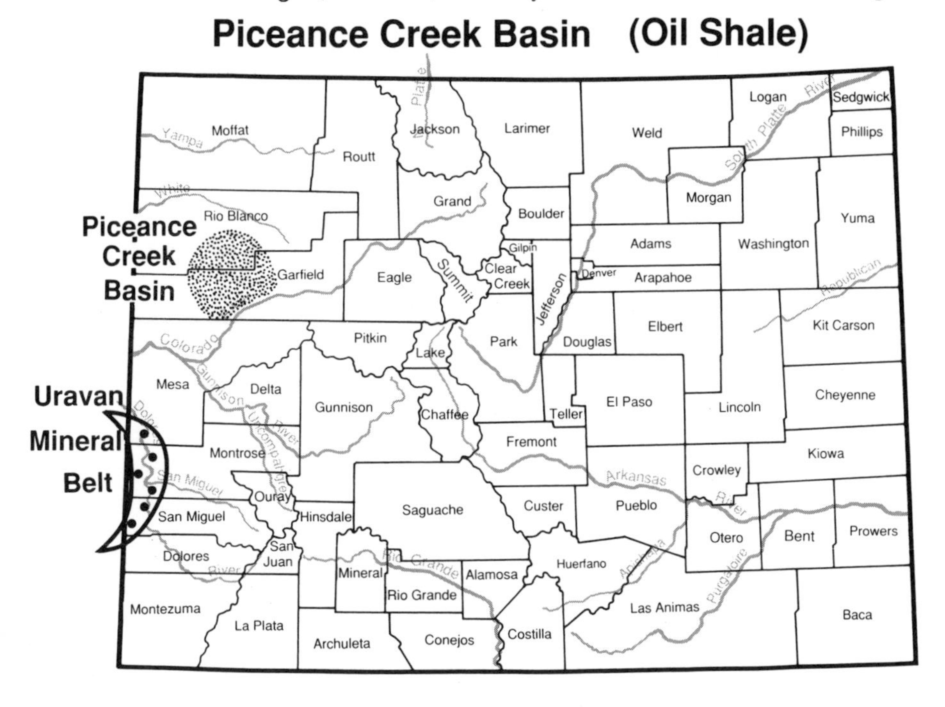

Green River Formation oil shale from the Piceance Creek Basin.

form of petroleum. To recover the hydrocarbon values, oil shale is simply crushed and heated to 900° F, the temperature necessary to achieve pyrolysis, or destructive distillation. Heat breaks the long molecular chains, converting the kerogen to a liquid, then to a gas. The condensed gas is shale oil, a crude petroleumlike liquid easily refined into lubricants and liquid fuels. Higher grades of oil shale contain twenty to fifty gallons of shale oil per ton and have great economic potential.

Oil shale occurs worldwide, but the largest and richest deposits ever discovered are within the Green River Formation that underlies much of northwest Colorado and adjacent Utah and Wyoming. The 70-million-year-old Green River shales were deposited by a regional inland sea that evaporated to form two large, shallow, algae-rich lakes. When the uplifting of the modern Rockies cut off water supply and outlets, the lake stagnated into vast mud flats where plant and animal life thrived. Over 10 million years, deposition of clay, sand, silt, volcanic ash, and great quantities of organic matter formed a deep lake bed. Heat and pressure eventually converted some organic matter into coal, petroleum, and natural gas. Most, however, never altered beyond the kerogen stage.

The Book Cliffs, near Parachute and I-70, are exposures of Piceance Creek Basin oil shale, the world's richest known hydrocarbon energy resource.

The Green River Formation oil shales are up to 3,000 feet thick. The energy "heart" of the formation is the Piceance Creek Basin in Garfield and Rio Blanco counties, where the United States Geological Survey estimates the currently recoverable shale oil could satisfy the United States' liquid fuel needs for the next century.

The amount of kerogen present determines the physical properties of oil shale. Lower grades are laminated and fissile, and outcrops weather into crumbling heaps of "paper" shale. Richer grades form durable, massive blocks. High-grade shales containing more than twenty gallons of shale oil per ton take on the properties of kerogen rather than the host marlstone. They are tough, resilient, and deform plastically under load. Pioneers called rich oil shale "rubber rock."

Kerogen also determines oil shale coloration. Lower-grade layers range from off-white to light tans and grays; richer layers are deep brown, mahogany, or near-black. Dramatic color variations in adjacent layers create an eye-catching appearance.

The energy potential of oil shale has been known since the English recovered "oyle" from stone in the 1600s. When whale oil became scarce, Scotland established the first important shale oil industry in 1850. In 1857, American "shalers" built fifty-three small shale oil plants near the low-grade oil shale deposits of Ohio and Kentucky. But Pennsylvania oil wells soon flooded the market with cheap petroleum, and the oil shale industry collapsed.

The Piceance Creek Basin oil shales are far richer than those of the East. Ute tribesmen told how summer lightning strikes actually ignited rich outcrops. Pioneers "cooked" rich oil shale in stills to recover a "tar" to lubricate wagon wheels and preserve leather. The Utes called oil shale "the rock that burns," and unfortunate white settlers soon learned why. In the 1880s, pioneer Mike Callahan built a hearth in his new cabin with rocks gathered from the Roan Cliffs near Parachute. According to legend, Mike invited his neighbors to a housewarming, built a roaring fire in his hearth, then watched helplessly as the rock ignited, leveling the entire cabin. Local fruit growers burned oil shale in their orchards to generate smudge for frost protection. Later, engineers studied "rubber rock" as a possible replacement for wooden railroad ties.

In 1912, the rapidly industrializing United States began importing petroleum. Oil prices soared and the Department of the Interior ordered investigation of Colorado's oil shales. After field surveys in 1915-16, the United States Geological Survey announced that the estimated 40 billion barrels of shale oil in the Piceance Creek Basin could supply the nation's liquid fuel needs "forever."

A 1918 *National Geographic* article titled "Billions of Barrels of Oil Locked Up in the Rock" reported: "These are mountains, indeed, ranges of mountains which for miles carry thick beds of rock that yield thirty to fifty gallons of oil to the ton." So immense was the resource, Congress established three vast U.S. Naval Oil Shale Reserves.

Promoters quickly replaced the name "rubber rock" with "oil shale." Hordes of prospectors, investors, speculators, and adventurers rushed in. Within two years, prospectors, under authority of the General Mining Law of 1872, staked 30,000 oil shale claims covering an area the size of New Jersey. When an uproar ensued over the federal land "giveaway," Congress passed the Mineral Leasing Act of 1920, making oil shale lands leasable only.

Two hundred oil shale companies raised millions of dollars selling stock. The twenty that actually took to the field produced a mere 2,500 barrels of shale oil. Drugstore chemists used some in sheep dip, hair tonic, and veterinary salves; most went to pave a few streets in Glenwood Springs. By the late 1920s, cheap petroleum from Texas, Oklahoma, and California turned Colorado's first oil shale boom to bust.

In 1943, the U.S. War Department learned that shale oil from a captured Chinese plant was the primary fuel for the Japanese Imperial Navy. Congress passed the Synthetic Fuels Act of 1944, authorizing construction of a federal shale oil plant at Anvil Points in Garfield County. Anvil Points achieved great success; in 1953, the government predicted a million-barrel-per-day Colorado shale oil industry within a decade.

After World War II, the United States again imported petroleum. Rising oil prices lured speculators back to the Piceance Creek Basin as thousands of grandfathered, pre-1920 oil shale claims appreciated rapidly. A major oil company built a successful shale oil plant in 1958; in the 1960s engineers proposed underground atomic detonations to recover shale oil at a fraction of the cost of petroleum.

Another land rush followed. By 1968, speculators had claimed or leased 7,000 square miles of prime Colorado oil shale land. Controversy raged again as huge tracts of land passed into private ownership, mostly to oil companies that had yet to produce any shale oil.

After the OPEC oil embargo of 1973-74, the government announced "Project Independence," an urgent plan to achieve national energy independence by 1980, largely through rapid development of oil shale. Oil companies, embarrassed by windfall profits derived from skyrocketing oil prices, grudgingly cooperated. Western Colorado boomed and *Time* magazine predicted that within twenty years oil shale would supply all of the nation's liquid fuel needs. But companies suspended development in 1976 when petroleum prices flattened.

During the 1979 energy crisis, massive federal subsidies triggered the biggest oil shale boom ever. Eight major oil companies attracted 10,000 workers to the Piceance Creek Basin's huge underground mines and surface retorting and support facilities. But by May 1982, plummeting oil prices turned that boom, too, into bust.

Oil shale, while not an economic bonanza, does make interesting specimens. Collecting in the canyons and along the cliffs of the Piceance Creek Basin is simply a matter of selecting desired colors, patterns, and grades.

And, just as pioneer Mike Callahan learned, high-grade oil shale will indeed burn. Simply preheat a small specimen to partially liquefy the kerogen, then ignite it with a match. Be sure to do this in a safe outdoor area, as burning oil shale sputters and smokes.

Cutting oil shale requires low blade-rotation speed to minimize friction temperature and prevent alteration of the kerogen. Cut oil shale takes a good polish and resembles exotic hardwood in appearance.

THE URAVAN MINERAL BELT AND RADIOACTIVE MINERALS

The story of Colorado uranium is interesting mineralogically, technologically, and historically, and the 1950s uranium rush has left behind hundreds of mine dumps where radioactive minerals may be collected.

Radiation is a general term referring to energy in the form of waves or particles, and includes heat, light, microwaves, and nuclear radiation. Nuclear radiation, the product of nuclear fusion, fission, or atomic decay, has the unique ability to ionize the atoms it strikes, splitting them into positively and negatively charged ions, or positively charged atoms and subatomic particles.

Materials that emit nuclear radiation are said to be radioactive. Their radiation may be of one or any combination of three types: alpha particles, beta particles, and gamma rays. Alpha and beta particles have high mass and low energy with little penetrating power. Alpha particles, for example, cannot penetrate a piece of paper or one inch of air. But gamma rays, similar to artificial X-rays with low mass and very high energy, can penetrate many metals and a foot of solid rock.

Natural nuclear radiation is cosmic or geophysical in origin. The most common source of geophysical radiation is uranium, a naturally radioactive element present in trace amounts in most igneous rock. Unstable uranium atoms disintegrate continuously into a "decay chain," a lengthy, orderly progression of naturally radioactive elements and isotopes that includes proactinium, thorium, radium, radon, polonium, bismuth, and lead. The chain ends with the common, nonradioactive, stable form of lead, Pb-206. Continuous atomic disintegration of the uranium decay chain emits varying amounts and intensities of alpha, beta, and gamma nuclear radiation.

Uranium, once believed to be rare, is actually more abundant in the earth's crust than gold and silver combined. It occurs in over one hundred minerals, but only a few vanadates, silicates, and complex oxides are valuable as ores. Because of very slow atomic decay rates, most uranium minerals are only mildly radioactive.

Uranium minerals had no use until 1898, when French scientist Marie Curie learned that some contained trace amounts of rare, highly radioactive radium. Pitchblende, the massive form of uraninite, or uranium oxide, was the first commercial source of radium. Central City miners hand cobbed high-grade pitchblende from local mines and shipped it to France.

In 1898, French chemist Charles Poulot obtained specimens of a

Uravan Mineral Belt (Uranium)

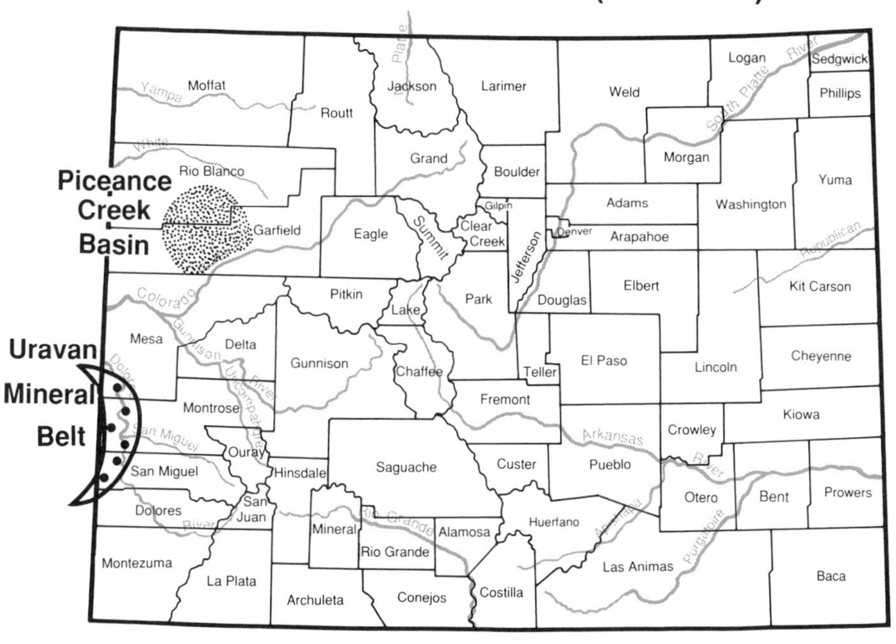

brightly stained sandstone from western Colorado. He identified the yellow mineral-staining agent as a vanadate containing uranium. Poulot paid $2,600 to custom mine and hand sort ten tons of the bright yellow sandstone. Researchers in Paris identified it as a hydrous potassium uranium vanadate and named it carnotite, after Marie-Adolphe Carnot, a French chemist.

Initially, Colorado carnotite was valuable only for its vanadium, a metal French metallurgists used to toughen steel. When the Ford Motor Company publicized vanadium steel as a feature of the Model A, U.S. demand grew. In 1905, the Vanadium Corporation of America began mining and milling carnotite ore in western Colorado.

In 1912, French physicists, desperate for radium, purchased 1,200 tons of high-grade, hand-sorted Colorado carnotite. From it they extracted eight grams (less than one-quarter ounce) of radium. Radium, worth $3 million per ounce, was the costliest material on earth. When vanadium mills began by-product radium recovery from carnotite, western Colorado was the world's only important commercial radium source. In 1930, 70,000 tons of Colorado carnotite yielded 202 grams (about half a pound) of radium, by far the largest amount ever isolated.

When foreign radium and vanadium suppliers took over the market, Colorado carnotite mining ceased. Thousands of miners lost their jobs and deserted towns like Uravan (URAnium-VANadium) and Vancorum (a loose acronym for VANadium CORporation of AMerica). Vanadium mining recovered modestly in the mid-1930s, but the uranium content of the carnotite ore was discarded.

The Uravan Mineral Belt is a seventy-mile-long, crescent-shaped zone of mineralized sandstone cutting across the canyon-and-plateau country of Mesa, Montrose, and San Miguel counties. The sandstone, in formations up to 6,000 feet thick, was deposited over 100 million years by a succession of shallow seas and deltas. When originally deposited, it contained only normal geochemical trace amounts of uranium. But groundwater carrying dissolved uranium salts percolated downward through the sandstone. Concentrations of organic matter in certain strata acted as a natural chemical reducing agent, precipitating the uranium and associated vanadium salts as the mineral carnotite. The Uravan Mineral Belt carnotite deposits are generally lens-shaped, horizontal, and about six feet thick and several hundred feet long. Carnotite occurs as a powdery coating or pore filling, staining the host sandstone a bright saffron yellow.

Mineralization of the Uravan Mineral Belt was complete by late Cretaceous times. Subsequent uplifting initiated surface erosion that eventually sculpted today's canyon-and-plateau topography, exposing large stratigraphic intervals of red and yellow sandstone. Remote, rugged, and colorful, the Uravan Mineral Belt was a perfect setting for a uranium rush.

In 1939, German physicists split the atom and Albert Einstein informed the U.S. president of the technological feasibility of an atomic bomb. The first self-sustaining nuclear reaction confirmed the nuclear potential in 1942, creating an emergency need for uranium. Across the Uravan Mineral Belt, vanadium mills reprocessed old tailings for their uranium content. The uranium in the atomic bombs detonated in 1945 at Alamogordo, Hiroshima, and Nagasaki came from three strategic sources: Canada, the Congo, and Colorado's Uravan Mineral Belt.

Strict postwar government controls, secrecy, and an uncertain market initially discouraged uranium prospecting and mining. But the Atomic Energy Commission (AEC), established in 1947, ordered priority development of secure domestic sources of uranium. The AEC announced an incentive program with guaranteed minimum prices for ores, substantial cash bonuses for discovery and production, mine-development allowances and low-interest loans, and even haulage allowances from remote mine sites. The thousands of fortune hunters that descended upon the Uravan Mineral Belt included everyone from professional geologists to out-of-work gold miners and, as one newspaper wrote, "farmers who couldn't tell carnotite from cornmeal." With them came teachers, ranchers, war veterans, laborers, accountants, speculators, con artists, and adventurers from across the country. Grand Junction, with its new AEC field office, was the headquarters for the great uranium rush.

Uranium prospecting, a mix of science, art, and luck, depended upon detection of natural radioactivity. Two instruments, miniaturized during the war years, did just that. The Geiger-Müller counter, the familiar "Geiger counter," detected ionization effects of gamma rays within a thin-walled glass vacuum tube containing an anode and a cathode in a rarified argon atmosphere. In operation, the tube was negatively charged; entering gamma rays collided with and ionized argon atoms, causing the tube to discharge with an electrical pulse. That tiny current was amplified to produce an audible click and move a meter needle. "Geigers," although inefficient and slow, were enormously popular because of their very affordable $20 price.

The more efficient and sensitive scintillation counters detected the collision of gamma rays with atoms of phosphors (such as the coatings on fluorescent light tubes). Collision caused activated phosphor atoms to fall to their "ground" energy levels by releasing energy as a burst of light. A photomultiplier tube converted the light energy to an amplified electrical current for measurement. Since radiation could be measured instantly, scintillators were suited for air- and ground-mobile radiometric surveying. Unlike the affordable Geigers, scintillators cost about $1,000 each—a big investment in 1948.

One newspaper described the "typical" uranium prospector as driving a $50 war-surplus jeep loaded with surplus C-rations, topographical maps, rock picks, sample bags, claim stakes, and, most importantly, a shoebox-sized Geiger counter. Grand Junction outfitters claimed that for only $300 they could "set a man up in the uranium sweepstakes."

Prospectors first "zeroed" their Geiger counters to background cosmic and geophysical radiation levels, then surveyed as many likely uranium-bearing formations as possible. The atomic-age version of "gold in the pan" was clicking earphones and swinging needles. Prospectors then staked claims and collected samples for

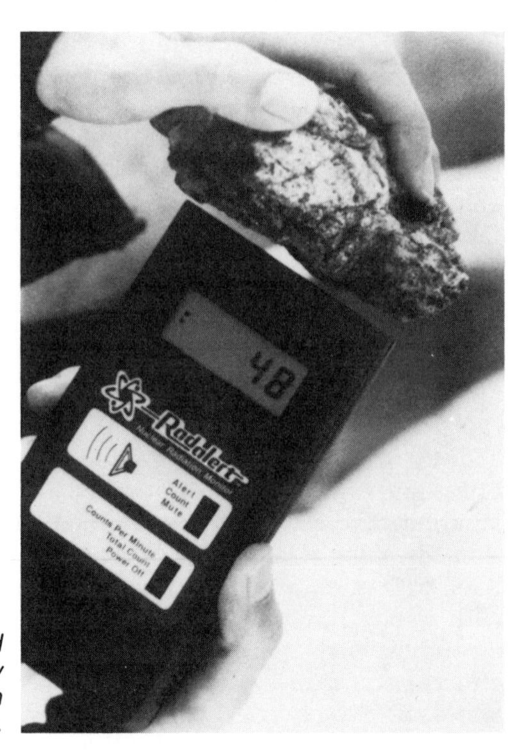

Modern miniaturized radiation monitors easily detect elevated radiation levels in radioactive minerals.

laboratory radiometric and chemical assay. High assay reports meant "hot" claims, which could be bought, sold, traded, borrowed against, or leased in any of the dozens of "uranium exchanges" throughout the Uravan Mineral Belt.

Some miners struck it rich on bonanza carnotite deposits—usually petrified tree trunks that had precipitated uranium salts from groundwater for millions of years. Some trunks, half silica and half pure carnotite, weighed twenty tons and brought $100,000 at the mill scales.

By 1953, 3,000 prospectors and 6,000 miners worked in the Uravan Mineral Belt. Nearly 900 small mines shipped carnotite ore to mills in Grand Junction, Uravan, Naturita, Vancorum, and Slick Rock, and in the nearby Utah towns of Blanding and Moab. Mills leached carnotite from crushed ore, precipitated the uranium salts as a mixture of insoluble oxides and uranates called "yellow cake" (a claylike paste containing about 75 percent uranium oxide), then shipped it in drums to the AEC.

Natural uranium is composed of three isotopes: 99.3 percent U-238, 0.7 percent U-235, and a trace of U-234. U-235, the only naturally fissionable material, is the most valuable isotope. Diffusion plants converted the refined yellow cake to uranium hexafluoride gas, then separated the isotopes by gas diffusion. The final product was enriched uranium; containing high levels of U-235, it was the basic material for weapons and fuel-rod fabrication, or for conversion into transuranic elements and isotopes.

By 1958, big open-pit mines in Wyoming and New Mexico dominated uranium production. Faced with large stockpiles of yellow cake, the AEC terminated its incentive program, officially ending the already fading uranium boom. Today, only a handful of uranium mines in the Uravan Mineral Belt produce sporadically.

The Colorado Geologic Survey lists 450 "major" uranium mines in the Uravan Mineral Belt in Mesa, Montrose, and San Miguel counties. The total number of mines actually exceeds a thousand. Carnotite specimens can be found in mine dumps and along mine haulage roads, giving rockhounds a fine opportunity to collect and study a radioactive mineral.

Modern, miniaturized, pocket-sized radiation monitors combining solid-state electronics and Geiger-Müller tubes provide useful relative measurements of natural radioactivity. Once zeroed to normal background cosmic and geophysical radiation levels, radiation monitors, which cost as little as $200, can easily detect radioactive specimens.

Interestingly, some petrified wood and dinosaur bone found in the Uravan Mineral Belt and adjacent areas of western Colorado emits radiation several times greater than normal background levels.

THE "BONEYARDS" AND FOSSIL FIELDS

Colorado pioneers found that Colorado was also rich in the fossilized remains of early life. By the late 1860s, Florissant, already famed for its petrified "forest," was a popular collecting site for fish, insect, and plant fossils.

Dinosaur-bone discoveries at Cañon City and Morrison in 1876 and 1877 established Colorado as a world-class fossil source. The discoveries attracted the attention of professors Othniel Marsh and Edward Drinker Cope, the leading U.S. paleontologists. Old World excavations had already proven the existence of dinosaurs, but sporadic recovery of incomplete skeletons excited neither scientists nor the public and created confusion about the true nature of the "dinosaurians," or "terrible lizards."

But the Morrison and Cañon City excavations yielded complete fossilized skeletons of the sauropod *Diplodocus*, the carnosaur *Allosaurus*, and the spined *Stegosaurus*. A. J. Desmond described the

Fossils and Petrified Wood

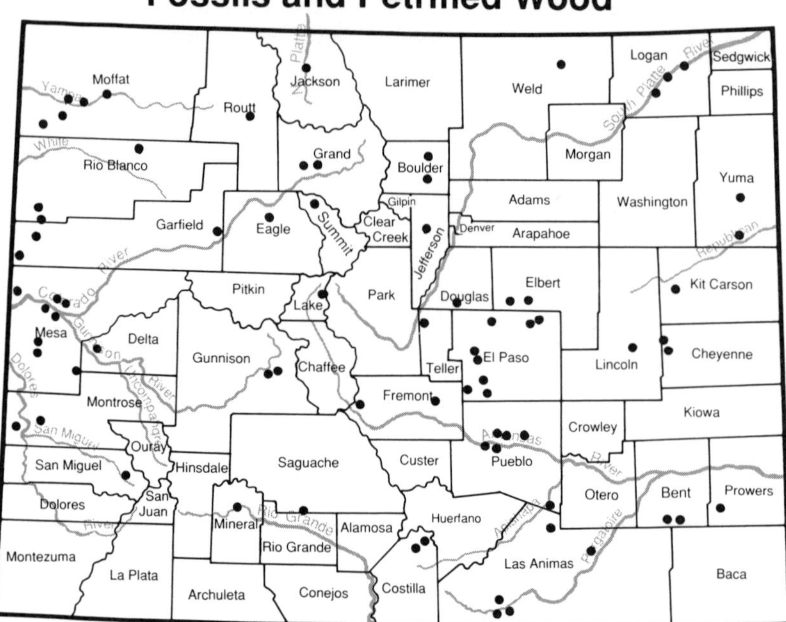

significance of the Morrison recoveries in his 1975 book, *The Hot-Blooded Dinosaurs*:

> . . . here the monsters sprang fully formed from the earth. It was as if a veil was dramatically lifted . . . no longer would paleontologists be forced to cope with crumbling fragmentary remains; from now on they were to unearth whole specimens perfectly preserved in a veritable giants' graveyard.

Geologists named the fossil-rich Jurassic sandstone the Morrison Formation, for the nearby town of Morrison. The spectacular recoveries started a paleontological "bone rush" wherever the Morrison Formation was exposed. Scientists were unable to assimilate and properly interpret the deluge of paleontological information that poured out of the Morrison sandstone. At Morrison, Marsh named a gigantic sauropod *Titanosaurus*, a name already assigned to an Old World fossil. He changed the name to *Atlantosaurus*, then quickly named another big Morrison skeleton *Apatosaurus*. Meanwhile, at Cañon City, Cope recovered a giant sauropod, naming it *Camarasaurus*. The race to publish professional papers continued, with little communication or cooperation between competing paleontologists.

Colorado is one of North America's richest sources of dinosaur fossils.

In 1877, Dr. Samuel H. Scudder's excavations at Florissant were equally remarkable. Florissant eventually yielded over 50,000 museum-grade specimens, including 1,100 different insect species, 140 plant species, and dozens of fossils of birds, small mammals, and fish.

In the late 1880s, paleontologists found fossilized horn cores in the Denver Formation just west of Denver. Excavation yielded a variety of herbivorous and carnivorous dinosaurs, including *Ornithominius velox*, a bird-footed dinosaur the size of a kangaroo. By the 1890s, most major North American and European museums displayed Colorado dinosaur and insect fossils.

In the twentieth century, paleontologists followed the leads of western Colorado's commercial "bone hunters." Elmer S. Riggs, of Chicago's Field Museum, discovered "boneyards" near Grand Junction, excavating the fossilized remains of *Brachiosaurus altithorax*, the largest dinosaur then known.

In 1909, paleontologist Earl Douglass discovered yet another remarkable dinosaur boneyard along the northwestern Colorado-Utah border in what is now Dinosaur National Monument. By 1925, excavations had produced 325 tons of fossilized dinosaur bones, including the most complete *Apatosaurus* skeleton ever found.

Leaf fossils are locally abundant in some of Colorado's sedimentary strata.

In the 1920s, paleontologists on the eastern plains recovered superb Tertiary mammal fossils from the Pawnee Buttes area in Weld County, including the horses *Eohippus* and *Hypohippus* and an early rhinoceros. And in the 1930s, paleontologists began studying well-preserved dinosaur trackways in hogback exposures of Dakota sandstone.

In 1982, *Stegosaurus*, found at Morrison, Cañon City, and at several sites near Grand Junction, was officially designated the state fossil of Colorado. *Stegosaurus* is familiar for its distinctive, flat, bony plates running dorsally from neck to tail. The *Stegosaurus* skeleton at the Denver Museum of Natural History was reconstructed from fossilized bones found in 1937 near Cañon City by a local high school teacher and his students.

Despite the many great historical discoveries, field paleontology in Colorado has not yet realized its full potential. Today, six major museum and university excavations are under way. Many historic sites, current digs, and dinosaur trackways have recently been opened to public tours for the first time.

Fossils of plants and marine life occur in many areas of Colorado, most notably in exposures of sedimentary formations on the plains, along the Dakota hogback ridge that divides Colorado's plains-mountain topography, and in the western canyon-plateau country. The mountains also have fossil-rich lake beds and isolated, uplifted sections of marine sediments. As an example, the Cretaceous ammonite locality at Kremmling is 9,000 feet above current sea level and other marine fossils have been found at elevations exceeding 13,000 feet.

The distinguished Harvard paleontologist Dr. George Simpson wrote that fossil hunting was "the most fascinating of all sports," and added:

> The hunter never knows what the bag may be—perhaps nothing, perhaps a creature never before seen by human eyes. Over the next hill may lie a great discovery. It requires knowledge, skill, and some degree of hardihood. And its results are so much more important, so worthwhile, and so enduring than those of any other sport. The fossil hunter does not kill; he resurrects. And the result of his sport is to add to the sum of human pleasure and to the treasure of human knowledge.

Geographically, the opportunities to "prospect" for fossils in Colorado are even greater than for minerals and gemstones, and erosion continues to expose new fossils each year. Collectors must be aware of important restrictions and responsibilities regarding fossil hunting (see Chapter Three).

Mineral collectors must respect private property.

Colorado's many mine dumps are popular mineral-collecting sites. Collectors have the responsibility to collect legally and safely.

3 COLLECTING LEGALITY, SAFETY & RESPONSIBILITY

LEGALITY

Collecting minerals, gold, gemstones, and fossils in Colorado, or anywhere in the United States, is an activity that may be controlled, restricted, or prohibited, depending upon personal policies of private landowners or, on public lands, upon municipal, county, state, or federal regulations.

During the Pikes Peak gold rush, most of the 104,091 square miles of land within the soon-to-be-territory of Colorado was public domain, that is, federally owned, but otherwise unassigned and unadministered. Public domain was loosely controlled by the General Land Office and "free for the taking" by any U.S. citizen, a policy designed to encourage settlement and development of the West. Much public domain passed into private ownership through the Homestead Act, the Railroad Act, and the General Mining Law of 1872. The federal government assigned additional tracts to Indian and military reservation use, and to statehood land grants.

From the huge tracts of public domain that still existed at the turn of the century, the government created the federal forest reserves in 1901 and assigned them to the newly created U.S. Forest Service in 1907. Public domain in Colorado disappeared entirely in 1934, when the Taylor Grazing Act assigned the remaining arid, unforested rangelands to administration of the U.S. Grazing Service. In 1947, the Department of the Interior combined the Grazing Service and the old General Land Office into the Bureau of Land Management.

The point is this: No matter where mineral and fossil collectors are in Colorado today, they are on "somebody's" land.

About 60 percent of Colorado's land is privately owned by individuals or corporations. This includes much of the agricultural land of the plains, the intermontane valleys and the western slope, most urban and suburban areas, and virtually all placer and lode mining districts.

The federal government still owns 36 percent of Colorado land, about 38,800 square miles, including most of the mountains, forests, and arid rangeland. The U.S. Forest Service (USFS) administers 23,000 square miles and the Bureau of Land Management (BLM) administers 13,000 square miles. National parks and monuments and national recreation areas take up another 850 square miles. The Southern Ute and Ute Mountains Ute reservations cover 1,700 square miles. Military and government testing installations include yet another 500 square miles.

The state owns over 2,000 square miles of Colorado's land, including 758 square miles of public wildlife areas and 233 square miles of state park and recreation areas. Colorado's remaining land is owned by municipalities and counties.

Collecting and prospecting is prohibited by federal law in all national parks and monuments. Colorado has eight national parks and monuments: Black Canyon of the Gunnison National Monument, Colorado National Monument, Florissant Fossil Beds National Monument, Great Sand Dunes National Monument, Hovenweep National Monument, Mesa Verde National Park, Rocky Mountain National Park, and the eastern two-thirds of Dinosaur National Monument.

Colorado's vast tracts of BLM and USFS lands, however, provide many opportunities to collect gold, gemstones, minerals, and fossils. Rockhounding regulations are set forth in the Code of Federal Regulations. Title 43 regulates commercial collecting, claiming, and mining; Title 36 regulates amateur collecting. The concept of amateur collecting, however, actually originates in CFR 43, in which requirements for locatable minerals specifically exempt ". . . individuals desiring to search for or occasionally collect small mineral specimens or samples." One need not locate (claim) land, obtain permits, or file plans of operation for ". . . prospecting and sampling which will not cause significant surface resource disturbance and will not involve removal of more than a reasonable amount of mineral deposit for analysis and study."

The government defines "rockhound" as "an amateur who hunts and collects rocks and minerals as a hobby." Both the USFS and BLM consider rockhounding a recreational activity in accordance with their multiuse land management concepts. Accordingly, removal of small quantities of rock and mineral samples from the surface of National Forests and BLM lands is generally permitted. No permit (written permission) or fee special permission (such as the Forest Service's firewood-collecting fees) is required, providing the specimens are "for personal, noncommercial use; no mechanical equipment is employed; no significant surface disturbance results; and collection does not interfere with existing mineral permits, leases, claims, sales, or private rights."

"Disturbance" becomes "significant" when (1) natural recovery would not be expected to take place in a reasonable period of time, (2) there is unacceptable air or water degradation, or (3) there is unnecessary loss or damage to resources. Some criteria are vague, but common sense is the best guide. Light manual excavation followed by filling holes is acceptable. But an unfilled four-foot-deep hole, for example, could obviously be construed as "significant disturbance."

An important difference between mining and rockhounding is end use of the collected material. If the material is sold, collecting may be interpreted as mining, which is illegal without the required permits.

Special restrictions apply to petrified wood. The BLM enforces a daily collection limit of 25 pounds, not to exceed 250 pounds annually. Noncommercial collecting of small amounts of petrified wood is permitted in unrestricted sections of the National Forests as well. Always check with USFS or BLM offices when in doubt about collecting regulations.

Both the USFS and BLM consider gold panning and small-scale recreational placer mining a form of rockhounding, providing there is no significant stream disturbance. Suction dredges with intake diameters measuring three inches or less are considered recreational equipment and within the purview of rockhounding. Larger dredges are classified as commercial mining equipment and require permits. Regulations may vary locally by specific stream and season.

The USFS and BLM have special restrictions on fossil collecting. Fossils were once generally protected under the Antiquities Act of 1906, with permission to collect on public lands reserved only for paleontologists and other approved professionals.

The Archaeological Resources Protection Act of 1979 gives amateur fossil collectors much more freedom. The act distinguished between paleontological resources (fossils) and archaeological resources (human artifacts). Plant and invertebrate fossils, unless specifically protected, may now be collected. The collection or disturbance of vertebrate (fish, mammal, bird, and reptile) fossils, because of their greater scientific value, is specifically prohibited.

Fossil collecting that disturbs archaeological resources, including any and all cultural remains, is prohibited. Furthermore, any fossils found within the context of an archaeological resource are also protected under the Antiquities Act.

Unauthorized collecting is specifically prohibited at a number of specially designated fossil occurrences on public lands, including research natural areas near Kremmling, Grand Junction, and Cañon City.

Nontribal members may not collect on the Southern Ute and Ute Mountain Ute reservations without special authorization, which is rarely granted, from the respective tribal councils at Ignacio and Towaoc.

Collecting is not permitted on military installations, including both Fort Carson near Colorado Springs and the larger Fort Carson maneuver area on the plains northeast of Trinidad.

Collecting on state wildlife area land is permitted, providing there is no significant disturbance of land and collected materials are intended for noncommercial use. Collecting on state Department of Highways land, which includes many interesting road cuts, is permitted providing there is no interference with traffic, public safety is not compromised, and surface disturbance is minimal.

The bottom line regarding legality of mineral and fossil collecting in Colorado is simply to check first with landowners or administrators.

SAFETY

Mineral collecting in Colorado, like other forms of outdoor recreation, involves a certain unavoidable element of physical risk that can be minimized by using common sense. Three areas, however, warrant special attention: abandoned mines, which are exceedingly dangerous; Colorado weather, which, especially in the high country, can change radically and rapidly; and the physiological effects of high altitude.

ABANDONED MINES

The greatest risks are associated with abandoned underground mines. Abandoned mines, perhaps because they have been so romanticized in literature and motion pictures, exude intrigue. They seem to whisper promises of adventure and mystery, of riches to be claimed, and, arguably, of mineral specimens that far surpass those on the surface dumps.

Colorado has more than 20,000 abandoned mines, many of which may still be entered. For all their diversity in age, size, configuration, and mineral product, they share one commonality: All are extremely dangerous. Unfortunately, the popular conception of that danger is limited largely to "cave-ins." While unstable ground is certainly dangerous, the actual hazards are far more extensive and include deteriorated ground support systems, bad air, bad water, high radiation, and the chance of fire or explosion. Only experienced miners or mine rescue specialists fully understand the extraordinary dangers of abandoned underground mines.

The hazards of old underground mines extend to the surface. Shafts, vertical workings opening to the surface, are the most dangerous part of any mine; 90 percent of accidents and entrapments are shaft-related. Collars, the tops of shafts, are particularly dangerous, for exposure to the elements may have rotted timbers and eroded adjacent ground, making sudden collapse possible.

Shafts may be hundreds of feet deep. A fall into a shaft, even if only ten or twenty feet down, can cause serious injury. Old shafts typically contain projecting timbers and rusted pipes; the bottoms, or sumps, are often flooded.

The portals, or surface entrances of tunnels, are also dangerous. Departing miners often dynamited tunnels closed just beyond the portal timbers to protect their work should they return. Although it may seem possible to climb over caved debris to enter a tunnel, dynamited roof sections are very unstable.

Never equate old mines with caves. Caves, natural subterranean chambers many thousands of years old, are chemically and physically stable. Mines, however, are not natural and are much more recent in origin. Blasting is a heavy-handed process that, by its nature, not only removes designated rock sections, but can fracture and loosen adjacent in-situ rock. Most mines are subject to ongoing chemical oxidation and physical rock stress. Roof collapse of caves, at least during historical times, has been rare; caving in abandoned and even working mines is quite common.

Miners must often employ ground support systems. Small workings and roof arching provide some degree of inherent ground support. In older mines, timbering was the preferred form of ground support. In many historic mining districts, demand for timbers to support "bad ground" consumed entire forests. Wooden timbers are susceptible to slow rotting processes: "Wet rot" refers to organic decomposition; "dry rot" is deterioration through desiccation. Both compromise the effectiveness of timber support. Ground near active faults or in certain sandstones may never stabilize, but continue to exert constricting forces making eventual collapse inevitable. When timbers "take weight" and become severely stressed, they may shatter suddenly and violently. In underground confines, shattering timbers can be as deadly as the rockfall that may follow.

Rock bolts, steel rods two to ten feet long, are a more recently developed ground support system. Miners insert bolts into overhead drill holes and tighten them against expanding heads to lock weakened rock sections together.

Tunnels with neither timbering nor rock bolts are said to be "bald-headed." This may indicate stable rock—or simply an attempt to cut costs. Also, rock that was stable a century ago is not necessarily stable today.

Underground cave-ins do not occur as portrayed in movies, where a bit of "warning" rock comes down first, providing just enough time for an exciting escape. In real life, massive cave-ins are over in the blink of an eye. Cave-ins are usually fatal, for rescue efforts are necessarily slow and may trigger more extensive caving. Cave-in entrapment can cause death by suffocation, thirst, or even sheer fright. Colorado rural law-enforcement officers know that answers to some missing-person cases can probably be found in collapsed abandoned mines.

The underground mine environment is hostile and foreign. Should lights fail, the absolute darkness may cause disorientation and even panic. Some underground "explorers" foolishly rely on matches, candles, makeshift torches, and flashlights for light. Open-flame light sources may be easily extinguished or may cause fire or an explosion. Conventional flashlights do not provide reliable, long-term light and are not designed for rugged, wet, underground use.

After abandonment, some mines flood. Shallow water covering a mine floor may conceal rusted rails and pipes or, worse, deep vertical workings.

One of the biggest problems in old mines is bad air, meaning air that will not sustain life. The danger is not usually from poisonous gases, but from oxygen depletion or displacement. Working metal mines are rarely classified as "gassy," that is, likely to accumulate potentially explosive levels of methane. But old mines may contain decaying organic matter, such as timbers or animal carcasses, which may generate methane gas under the right conditions. In unventilated mines methane can displace oxygen and continuous oxidation of sulfide minerals will consume oxygen. Oxygen deficiency can quickly cause light-headedness, shortness of breath, and general weakness. And an open-flame light source in a methane-enriched atmosphere can trigger an explosion.

The Colorado Bureau of Mines warns that bad air in shafts is a special hazard. Descending a shaft may require so little exertion that bad air may not be noticed. But the cumulative debilitating effects of oxygen deficiency at the bottom may make the strenuous climb out impossible.

Coal and uranium mines are often gassy, meaning methane may accumulate in unventilated workings. Many unventilated uranium mines have accumulated levels of highly radioactive radon isotopes that emit radiation thousands of times above "maximum safe" levels.

Natural chemical oxidation, by accelerating physical deterioration of rock, adversely affects rock stability, especially in metal sulfide mines. Metal sulfide-bearing rock, even if solid when first mined, will eventually crumble. Oxidation also produces sulfuric acid, which may weaken or completely dissolve rock bolts. Should you become trapped in an abandoned metal mine, your only water supply will be highly acidic, heavy-metal-laden mine drainage.

Underground fire can be deadly. Open-flame light sources may ignite highly flammable dry-rotted timbers as well as solvents, liquid fuels, or old explosives left behind by early miners. The greatest danger from underground fire is not the flames, but consumption of the limited oxygen supply and generation of deadly carbon monoxide. Carbon monoxide is so poisonous that levels of only one hundred parts per million may be fatal.

Old explosives, such as dynamite and detonating caps, are another hazard. Early caps used to initiate dynamite detonation contained fulminate of mercury, a compound so shock sensitive that merely dropping one may set it off. In unstable old mines, the concussion from a single detonated cap can bring the roof down.

Early dynamites were simply liquid nitroglycerine mixed with an inert, absorbent filler that reduced its inherent shock sensitivity to (barely) manageable limits. Time and temperature can separate the components, and old dynamites may drip with a clear-to-yellow syrupy liquid that is pure nitroglycerine with every bit of its notorious shock sensitivity.

Nitroglycerine fumes can quickly relax capillary walls, increasing blood flow through the brain to cause blinding headaches. A few moments of inhaling nitroglycerine fumes in an old mine can bring on a debilitating headache when you least need one.

If you still insist on exploring underground mines, never do it alone. Someone must remain outside to summon help should an accident occur. Regular search and rescue teams or local fire departments may not have the experience or the equipment to do the job, and a specialized mine rescue team may have to be summoned. Delays will be inevitable, for Colorado's only mobile underground mine rescue station is on inactive standby status. Even if you are rescued, legal difficulties may just be starting. Since most mines are on private or claimed property, trespassing charges could

The recently sealed shaft of a hundred-year-old Colorado mine. Three thousand such sites have been made safe by the Colorado Mined Land Reclamation Board.

be filed. You may also be liable for the considerable costs of your own rescue.

Since 1980, the Colorado Mined Land Reclamation Board has inventoried 21,000 old mines. Three thousand of those sites have been "made safe," that is, sealed or fenced to prevent entry and accidents.

ALTITUDE AND WEATHER

Altitude and weather can also affect collecting trips, especially in the high country. All of Colorado's major lode- and placer-mining districts are above 7,500 feet, and many are above 10,000 feet. The Mt. Antero aquamarine locality, at 14,000 feet, is the highest gemstone locality in North America.

Because of reduced partial pressure of oxygen, the human body, regardless of individual physical conditioning, must function with less oxygen than normal in the rarified atmosphere of high elevations. Pace of activity will decrease and fatigue will increase. At the worst, the symptoms of mountain, or altitude, sickness will develop, sometimes at elevations of only 8,000 feet. Flulike symptoms may include headache, dizziness, general weakness, nausea, and shortness of breath.

Fortunately, few individuals are afflicted by severe altitude sickness. Nevertheless, doctors recommend that anyone planning activity at elevations significantly higher than those to which they are accustomed should take a few days to acclimatize, drink plenty of fluids, avoid heavy foods and alcohol, get sufficient rest, plan a slower pace of activity, and allow more time for completion. Many high-elevation Colorado mineral-collecting trips have ended prematurely because collectors tried to do too much too quickly.

Automobile engines also lose efficiency and power from insufficient available oxygen at high elevations. Incomplete combustion may reduce power and foul spark-plug gaps and cylinder heads. Simple engine adjustments, such as "leaning" carburetor mixtures and advancing ignition timing, are recommended before setting out into higher elevations of the back country.

Colorado has thousands of miles of unmaintained four-wheel-drive roads at or above timberline (elevation approximately 11,500 feet). Highway maps deceptively classify some roads as "improved," thus implying they are suitable for highway cars. Always inquire locally about high-country road conditions before setting out. Many higher roads, regardless of map classifications, are suitable only for four-wheel-drive vehicles in top condition with an experienced rough-road driver at the wheel.

Weather also plays a big part in successful high-country collecting. Many higher mining districts and gemstone locales may be inaccessible because of winter snow, which can mean from October through May. Heavy snow may fall even in June and September, and snow squalls are common in July and August.

Violent electrical storms are the most serious summer weather concern above timberline. Storms may move at frontal speeds exceeding thirty miles per hour, and can turn a pleasant summer alpine day into a nightmare of snow, sleet, cold rain, lightning, high winds, and plummeting temperatures. The best rule when venturing above timberline is to be prepared for anything—including a rapid descent.

RESPONSIBILITY

Since the Pikes Peak gold rush, Colorado has yielded a wealth of superb mineral, gemstone, and fossil specimens. Whether Colorado will continue to be a mecca for collectors in the future depends largely upon how collectors act today. The legal collection of minerals, gemstones, and fossils on both public and private land is not a right, but a privilege.

Many current collecting restrictions stem directly from abuse of the collecting privilege. Landowners have been justified in closing property to collecting and refusing to grant permission upon request. Collectors have left behind trash and unfilled holes, open livestock gates, damaged fences, and polluted water. Others have vandalized cabins, razed old structures for firewood, and hauled away minerals and fossils in commercial quantities.

The BLM has imposed a twenty-five-pound daily collection limit on petrified wood specifically to counter "truckload" collecting. The BLM's Kremmling ammonite-fossil locality is now protected to prevent unlicensed commercial collecting that has previously damaged the site. Mineral and fossil collecting in Colorado can continue only if proper respect is given to private landowners and government regulations.

High-country collectors must also respect the fragility of the alpine environment. In Colorado, the land above timberline is alpine tundra, a treeless ecosystem typical of arctic, subarctic, and high mountain regions. Colorado has the largest area of alpine tundra in the contiguous forty-eight states. Tundra is composed of mosses, lichens, hardy grasses, sedges, and dwarfed shrubs, all low and compact enough to survive in the harsh alpine climate. With alpine growing seasons as short as eight weeks, annual plant growth may be only a fraction of an inch. Alpine tundra is extremely fragile and susceptible to permanent damage. When above timberline, stay on established tracks and trails, and limit collecting to mine dumps or already disturbed ground.

Fossil collectors have a special scientific responsibility. Amateurs have made many of Colorado's great dinosaur-fossil discoveries. Although federal regulations currently prohibit disturbing or collecting vertebrate fossils on public lands, collectors may discover previously unknown fossil occurrences of dinosaurs or other vertebrates. Every collector has a responsibility to report such discoveries.

"Enlightened amateurs can make worthwhile contributions to mineralogy and paleontology," wrote Ralph King, formerly with the Montana Bureau of Mines and Geology, "if they know enough to recognize significant finds and report them to public agencies. Mineralogists and paleontologists would rather take the time to look at hundreds of ordinary specimens than have one significant specimen escape them."

Some major discoveries certainly are unreported. Some amateurs fear looking foolish should their find turn out to be rather ordinary; others are reluctant to "waste the time" of a scientist.

Dr. Wade E. Miller, head of the Geology Department at Brigham Young University, has investigated many reports from amateurs, and hopes to investigate many more."I appreciate very much any report of a fossil by an amateur—as do my colleagues. I always appreciate the opportunity to discuss things with an amateur. Remember," Dr. Miller emphasizes, "those first bones at the Dry Mesa site [*Supersaurus*] were discovered and reported by amateurs."

Legal, safety, environmental, and scientific responsibility can assure that future generations will continue to enjoy the privilege of collecting minerals, gemstones, and fossils in Colorado.

PART TWO: COLLECTING LOCALITIES & RELATED SITES OF INTEREST

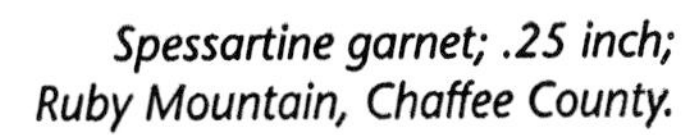

Spessartine garnet; .25 inch; Ruby Mountain, Chaffee County.

Beryl, var. aquamarine; 1.5 inches; Mt. Antero, Chaffee County.

Beryl, var. aquamarine; 1.5 inches; Mt. Antero, Chaffee County.

Placer gold; Cache Creek Park, Chaffee County.

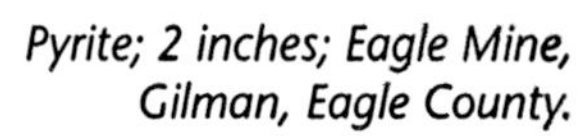

Pyrite; 2 inches; Eagle Mine, Gilman, Eagle County.

Galena and sphalerite; 3 inches; Eagle Mine, Gilman, Eagle County.

Agate geode; 4 inches; Curio Hill, Fremont County.

Eocene leaf fossil; 5 inches; Douglas Pass, Garfield County.

Lapis lazuli; 5 inches; North Italian Mountain, Gunnison County.

Lepidolite; 4 inches; Brown Derby Mine, Quartz Creek, Gunnison County.

Marble; 8 inches; Yule Quarry, Gunnison County.

Golden barite; 3 inches; Sherman Tunnel, Lake County.

Cerussite; 1.5 inches; Tucson Mine, Leadville, Lake County.

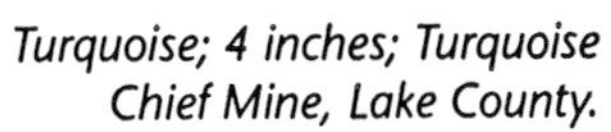

Turquoise; 4 inches; Turquoise Chief Mine, Lake County.

Selenite (gypsum), var. satin spar; 4 inches; Owl Canyon, Larimer County.

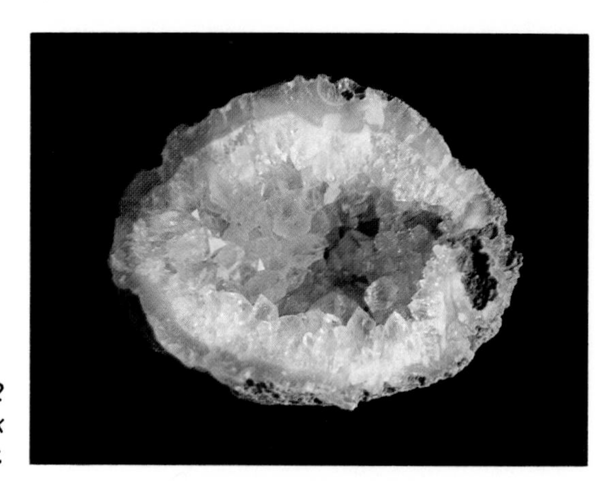

Quartz geode; 2 inches; Wolf Creek Pass, Mineral County.

Amazonite; 3 inches; Harris Park, Park County.

Rhodochrosite; 1 inch; Sweet Home Mine, Alma, Park County.

Rhodochrosite on tetrahedrite; 2 inches; Sweet Home Mine, Alma, Park County.

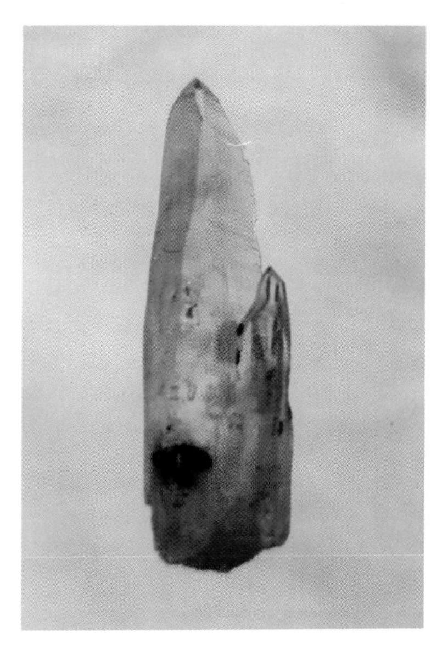

Quartz, var. amethyst; 2 inches; Crystal Hill, Saguache County.

Smoky Quartz; 3 inches; Crystal Peak, Teller County.

Topaz; 1 to 2 inches; Topaz Mountain, Teller County.

Topaz; 1 to 2 inches; Topaz Mountain; Teller County.

Barite; 3 inches; Stoneham, Weld County.

ALAMOSA COUNTY

GREAT SAND DUNES NATIONAL MONUMENT

The Great Sand Dunes have two distinctions: They are North America's highest dunes, and they once hosted Colorado's most bizarre gold rush.

Located twelve miles northeast of Alamosa along the eastern edge of the San Luis Valley, the 700-foot-high dunes cover 150 square miles and are visible from 50 miles away. The dunes were created by the accumulation of individual grains of sand.

Twenty-five million years ago, the earth's crust stretched and tilted sharply downward along a fault system at the base of the Sangre de Cristo Mountains, forming a deep rift. Over millions of years, erosion of the San Juan and Sangre de Cristo mountains filled the valley with sediments to its present elevation.

Two million years ago, the Rio Grande established itself as the valley's principal river. In channels that migrated laterally up to fifteen miles, the river sorted the alluvium, leaving behind dry lake beds and natural levees of silt and sand. Sparsely vegetated and arid, the loosely consolidated channel features were susceptible to wind erosion.

About 20,000 years ago, Pleistocene alpine glaciers carved several low passes in the crest of the Sangre de Cristo Range. Prevailing southwest winds swept the valley floor, funnelling sand and silt northeast toward the passes. The Sangre de Cristo Range compressed the winds, forcing them upward to burst violently through the passes. But as friction and turbulence decreased wind velocity, the heavier sands were deposited at the base of the mountains. The reversed winds of winter storms keep the dune fields stable, and several cubic miles of sand have now accumulated.

Because the creative geological and meteorological forces remain active, the dunes are a rare opportunity to study the dynamics of natural sand movement. High dunes "migrate" as far as 328 feet in a year. Winter storms can move dune crests backward twenty feet in a single day.

Half of the sand is composed of porphyritic and aphanitic volcanic fragments that originated in the San Juan Mountains. Thirty percent consists of bits of crystalline quartz derived from the Sangre de Cristo Mountains. Ten percent is fragments of plagioclase

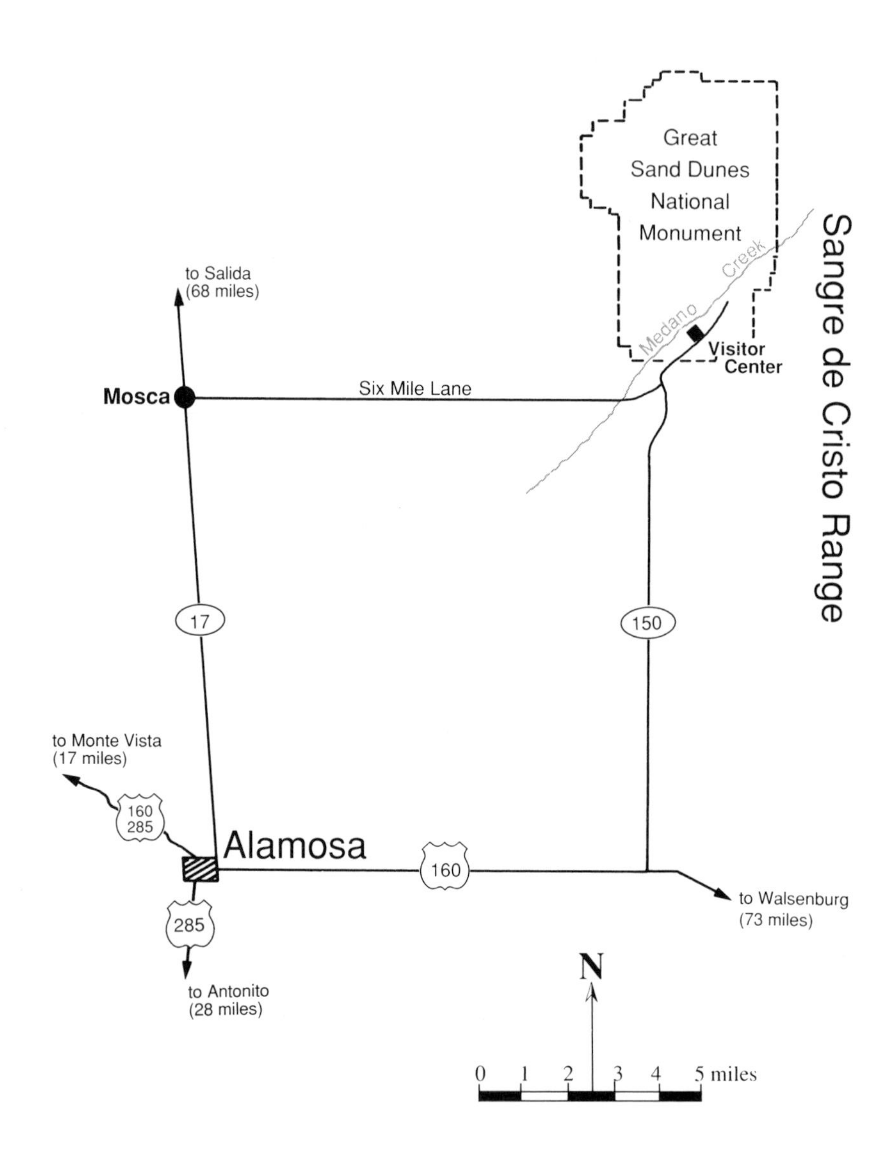
Great
Sand Dunes
National
Monument
Sangre de Cristo Range
Medano Creek
Visitor
Center
to Salida
(68 miles)
Mosca
Six Mile Lane
17
150
to Monte Vista
(17 miles)
160
285
Alamosa
160
to Walsenburg
(73 miles)
285
to Antonito
(28 miles)
N
0 1 2 3 4 5 miles

Trace amounts of gold in the Great Sand Dunes triggered one of Colorado's strangest gold rushes.

and orthoclase feldspars. The small remaining portion is composed of a variety of minerals, including traces of two that played an interesting role in the history of the dunes—magnetite and gold.

Because of its high density, magnetite, the familiar black magnetic oxide of iron present in "black sand" gold-pan concentrates, accumulates in dark streaks beneath the lee crest of the dunes. In the late 1920s, prospectors dry-panned the dune magnetite and assays revealed the presence of extremely fine particles of gold.

Newspapers announced the dune sand contained gold worth "three dollars to the ton." When promoters calculated the gold in the Great Sand Dunes was worth a staggering $8 billion, a gold rush was inevitable.

Miners hammered claim stakes, only to watch them disappear in the shifting sands; others set up sluice boxes in Medano Creek, a "disappearing" river that changed course daily. One company erected a small amalgamation mill at the edge of the dunes.

But reality soon turned boom to bust. Although gold was indeed present in the dune sand, it was microscopic and occurred in trace amounts that were almost impossible to recover.

Concerned that large-scale gold mining might compromise the tourism potential of the dunes, San Luis Valley residents petitioned Washington for help. In 1932, President Herbert Hoover proclaimed the Great Sand Dunes a national monument.

Although some mineral surveys still list the Great Sand Dunes as a gold occurrence, prospecting, mineral collecting, or gold panning is prohibited.

REFERENCES: 54

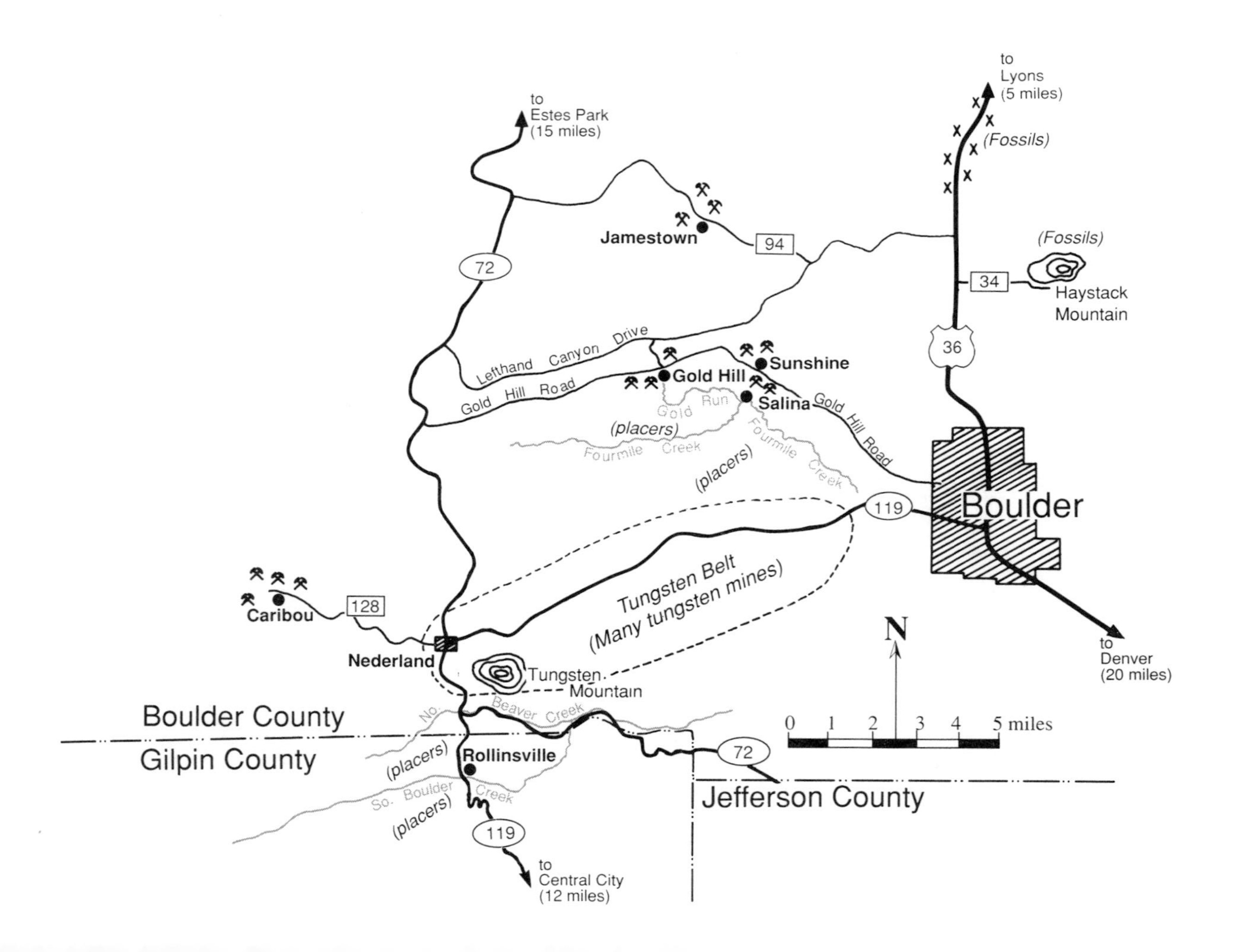
to Estes Park (15 miles)
to Lyons (5 miles)
(Fossils)
(Fossils)
Haystack Mountain
34
36
94
72
Jamestown
Lefthand Canyon Drive
Gold Hill Road
Gold Hill
Sunshine
Salina
Gold Hill Road
Gold Run
(placers)
Fourmile Creek
(placers)
Fourmile Creek
Boulder
119
Tungsten Belt
(Many tungsten mines)
Caribou
128
Nederland
Tungsten Mountain
to Denver (20 miles)
N
0 1 2 3 4 5 miles
Boulder County
Gilpin County
No.
Beaver Creek
Rollinsville
(placers)
So. Boulder Creek
(placers)
72
119
Jefferson County
to Central City (12 miles)

BOULDER COUNTY

PLACER GOLD

Boulder County has produced about 2,500 troy ounces of placer gold, most from Gold Run, Fourmile Creek, and North Beaver Creek, all tributaries of Middle Boulder Creek. In January 1859, prospectors discovered gold at the confluence of North Beaver Creek and South Boulder Creek, three miles southeast of Nederland near the Boulder-Gilpin county line. The strike, named the Deadwood Diggings after a nearby profusion of fallen timber, was among Colorado's first profitable placer strikes. Along with strikes in Clear Creek and Gilpin counties, the Deadwood Diggings helped "save" the Pikes Peak rush. Although Boulder County placer mining thrived but briefly, it led to important discoveries of lode minerals.

REFERENCES: 53

BOULDER COUNTY LODE-MINING AREAS

Hundreds of Boulder County mines have produced about $75 million in gold, silver, lead, zinc, tungsten, barite, and fluorite. Gold accounts for half of the total production.

JAMESTOWN AREA MINES

The Jamestown mines, eighteen miles northwest of Boulder, originally produced gold and by-product base metals. Mineralization was associated with fluorite in veins and breccia fillings. Fluorspar mining began in 1900 and boomed during World War II when steel-making demand for fluorspar flux soared. When the Jamestown mines shut down in the 1960s, cumulative fluorspar production topped $2 million.

Some Jamestown fluorspar veins, containing banded massive white and purple fluorite carrying small crystals of pyrite and galena, were over a quarter-mile long. Fluorite octahedrons up to two inches in size occurred in vein seams and pockets and ranged in color from clear to deep violet. Collectors still dig fluorspar and fluorescent fluorite specimens from the mine dumps.

Several prominent open-cut mines with purple fluorspar in the dumps are located immediately north of Jamestown above Boulder County Road 94.

REFERENCES: 18, 80

Abandoned tungsten mine and dump in Boulder County tungsten belt. Dumps are a source of fluorescent ferberite and scheelite.

BOULDER COUNTY TUNGSTEN BELT

Boulder County's tungsten mines are grouped within a narrow, ten-mile-long belt extending from west of Boulder west-southwest to Tungsten Mountain near Nederland. Mineralogists identified ferberite, an iron tungstate, in 1900. Mining boomed during World War I, when tungsten-steels were in great demand. Until World War II and development of large low-grade tungsten deposits in California and Nevada, Boulder County was the primary U.S. source of tungsten. The Boulder County tungsten belt has over one hundred mines, including thirty major producers. The last mines shut down in the late 1950s.

Tungsten ores occur as fissure vein fillings of quartz and ferberite. The quartz is usually gray, but may be almost black when rich in disseminated ferberite. Accessory minerals include galena, sphalerite, pyrite, argentite, scheelite and huebnerite, barite, and fluorite. Scheelite and huebnerite, also ores of tungsten, were not recovered in milling until World War II.

Ferberite forms streaks and lenses in banded quartz veins. Massive ferberite is common in the eastern part of the Boulder County tungsten belt, and coarsely crystalline ferberite occurs near Nederland. Ferberite crystals are black and generally chisel-shaped. Massive ferberite turns a distinctive chocolate brown when scratched by a knife blade. Some tungstates are strongly fluorescent. Collectors have been particularly successful at night, using portable ultraviolet light sources to identify specimens of ferberite and scheelite on the mine dumps.

Colorado Route 119 between Boulder and Nederland passes through the heart of the Boulder County tungsten belt.

REFERENCES: 18

GOLD HILL AREA MINES

Gold Run placer miners discovered lode mineralization at nearby Gold Hill in the late 1860s. Mine development began in 1872 after mineralogists identified gold tellurides. The Gold Hill mines produced over $14 million in gold and lesser amounts of silver, lead, zinc, and tungsten.

Free gold and high-grade telluride ores, primarily sylvanite and petzite, occurred in breccia fillings and thin white quartz veins bearing either pyrite-chalcopyrite or pyrite-galena-sphalerite.

About eighty small mines and the old camps of Salina and Sunshine are located in a ten-square-mile area. Mine dumps are a good source of a wide variety of minerals. The Gold Hill area is located eight miles northwest of Boulder on Boulder County Road 52.

REFERENCES: 18, 41, 66, 67

CARIBOU AREA MINES

In 1869 prospectors discovered rich silver lodes near timberline at Caribou, six miles northwest of Nederland. Caribou was among Colorado's first important silver producers, with ores averaging twenty to fifty troy ounces of silver per ton. Quartz veins carried galena, argentite, chlorargyrite, pyrite, and chalcopyrite, along with considerable free gold and wire silver. Metal sulfide specimens are abundant on mine dumps. Some Caribou mine properties are active and posted.

From Nederland, follow County Road 128 six miles northwest to Caribou. In Nederland, the Nature's Own rock and science shop has a good collection of minerals, including some local material.

REFERENCES: 23, 66, 67, 81

BOULDER COUNTY FOSSILS

North of Boulder, U.S. 36 (North Foothills Highway) passes along and through a series of low, discontinuous, Cretaceous sandstone and limestone hogback ridges. Exposures of dark Pierre Shale appear often to the east of the highway. Pelecypod fossils are locally common in both limestone and shale exposures.

Caribou, one of Colorado's oldest silver camps, has many old mines and dumps.

Collectors find Cretaceous pelecypod and cephalopod fossils on the slopes of Haystack Mountain, a prominent butte northeast of Boulder. Haystack Mountain is three miles east of U.S. 36 along County Road 34.

In Boulder, the Henderson Museum of the University of Colorado has excellent displays of Colorado minerals and fossils.

REFERENCES: 13, 68

CHAFFEE COUNTY

PLACER GOLD

Chaffee County placers have yielded about 90,000 ounces of gold, fifth among Colorado counties. Most has come from Cache Creek Park near Granite on the Chaffee-Lake county line (see Lake County). Cache Creek Park and the Arkansas River channel near Granite are popular panning and recreational mining sites. Placer gold is present in the Arkansas River for a distance of twenty miles below Granite, but production has been limited.

REFERENCES: 54, 97, 108

RUBY MOUNTAIN

Ruby Mountain, one of Colorado's most popular collecting locales, is known for its small, well-developed, brilliant crystals of deep red spessartine garnet and wine-colored topaz. Ruby Mountain is six miles south of Buena Vista and just east of Nathrop and U.S. 285.

Ruby Mountain, along with adjacent Sugarloaf Mountain and Dorothy Hill, are Tertiary rhyolite dikes intruded through Precambrian granite and gneiss country rock. The rhyolite, gray-white with faint parallel pink and purple banding, contains numerous cavities ranging from one to ten inches long lined with sugary white crystals and crusts of sanidine (orthoclase feldspar). Many also contain crystals of spessartine garnet and topaz. Ruby Mountain is named for the garnets, which resemble faceted red "rubies."

Most garnet crystals are less than one millimeter in diameter, but specimens two to four millimeters are not uncommon. The largest specimens measure ten millimeters, or nearly one-half-inch in diameter. Crystal colors range from brown-red to deep blood red; smaller crystals are bright hyacinth red.

The garnets occur as brittle dodecahedrons, with as many as six smooth, glassy faces exposed. Lapidaries have fashioned attractive jewelry by simply mounting the rough crystals in silver.

Less common topaz occurs in slender prisms about three millimeters long. Exceptional specimens may measure twelve millimeters, or about one-half inch, and are faceting quality. Topaz colors range from light yellow to sherry, but fade slightly upon prolonged exposure to light.

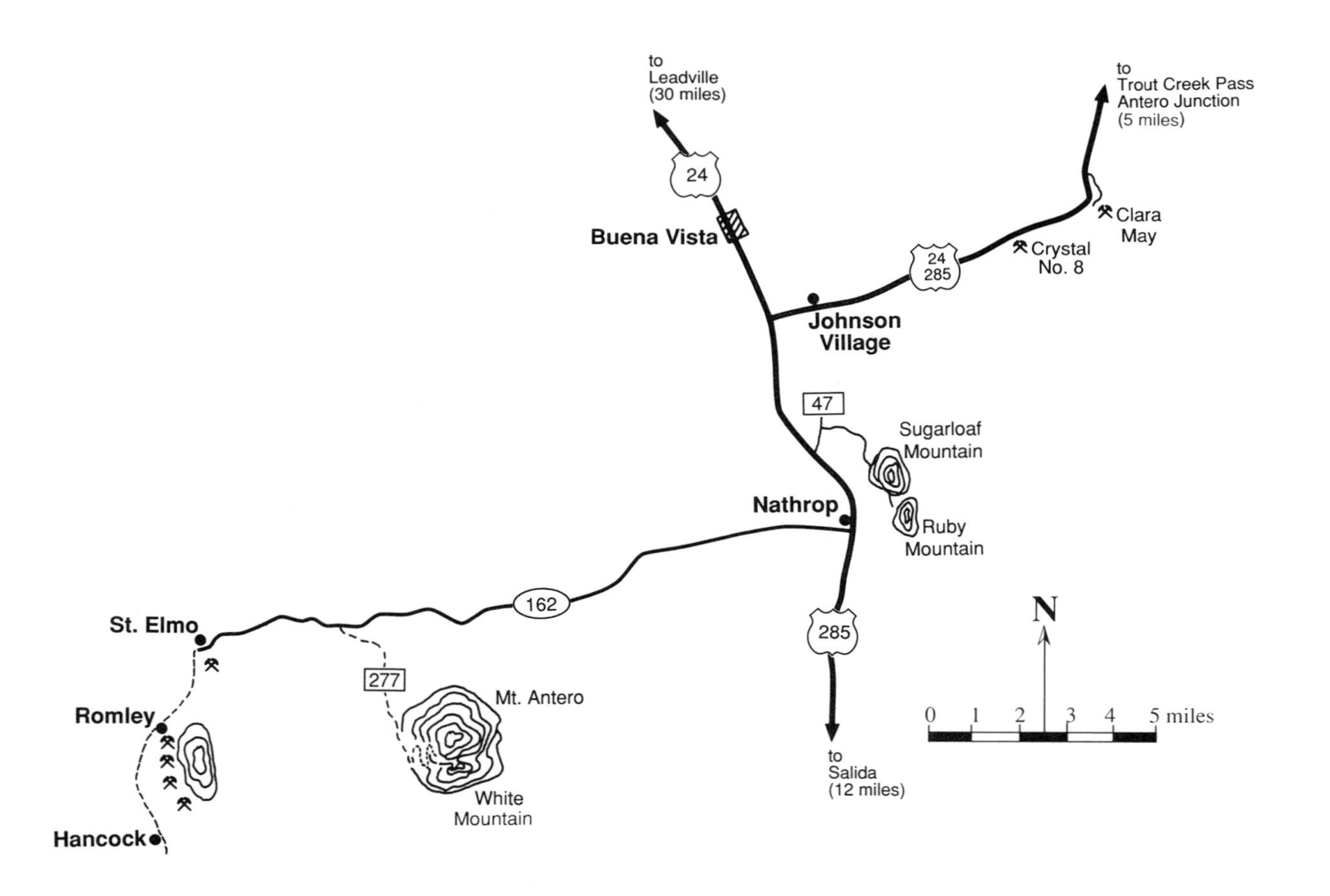
to
Leadville
(30 miles)
24
Buena Vista
to
Trout Creek Pass
Antero Junction
(5 miles)
Clara
May
Crystal
No. 8
24
285
Johnson
Village
47
Sugarloaf
Mountain
Nathrop
Ruby
Mountain
162
St. Elmo
285
N
0
1
2
3
4
5 miles
277
Mt. Antero
Romley
White
Mountain
to
Salida
(12 miles)
Hancock

Collectors find garnet and topaz by sorting through the talus rock at the base of the cliffs. Most, however, prefer to break the rhyolite to expose new vugs. The rhyolite is quite durable, and sledges and eye protection are necessary. Trimming the rhyolite matrix to desired specimen size and shape without damaging or dislodging the garnet and topaz crystals requires some patience and experience.

Obsidian nodules occur in perlite pockets at the north and south ends of Ruby Mountain. The small nodules, called "Apache tears," weather free from the perlite and collect in the bottoms of pockets.

From the Nathrop General Store, take U.S. 285 north for 1.7 miles, then turn east on Chaffee County Road 47. Cross the Arkansas River and proceed one-half mile to a small sign indicating the direction to Ruby Mountain. Turn south onto the unmarked gravel road and continue 2.5 miles to a parking area on the Arkansas River. Ruby Mountain is the tall hill immediately to the south. Light-colored freshly broken rhyolite covers the slopes. On summer weekends, Ruby Mountain echoes with the sound of collectors' hammers on rhyolite.

Both garnets and topaz are sufficiently dense to form loose alluvial concentrations in nearby sandy gullies. Some collectors have dug shallow holes and panned bedrock gravels to recover garnets that have weathered free from the rhyolite.

Ruby Mountain, which includes both private and BLM land, represents an excellent relationship between a private landowner and mineral collectors. Collectors have traditionally kept the popular, privately owned west slope of Ruby Mountain very clean, one reason why the owner, who resides locally in Nathrop, permits collecting, recently for a modest fee.

REFERENCES: 4, 5, 41, 55, 64, 66, 67, 73, 74

THE MOUNT ANTERO AQUAMARINE LOCALITY

Mt. Antero, ten miles south-southwest of Buena Vista, is a classic Colorado gemstone-collecting locale. The collecting area, just below the 14,245-foot summit, is North America's highest gemstone locality.

Nelson Wanemaker (also Wanamaker), of nearby Salida, discovered gem aquamarine, phenakite, beryl, fluorite, and smoky quartz in pockets of granite pegmatite high on Mt. Antero in 1881. Wanemaker collected crystals commercially, and George Frederick Kunz reported the occurrence in 1885. By 1890, lapidaries had cut $5,000 worth of finished gems, including some of twelve carats.

Wanemaker, the first serious collector of Mt. Antero crystals, countered the extreme elevation, inaccessibility, and short collecting season by building a crude stone shelter near the summit.

During the 1930s, collectors Arthur Montgomery and Edwin Over regularly visited the summits of Antero and adjoining White Mountain. In a 1938 *Rocks & Minerals* article, Montgomery presented this unforgettable account of collecting at 14,000 feet:

> Seventy years of mineral collecting at one limited locality—think of it! Of course the collecting has been spasmodic at best, and nine-tenths of the collectors entirely inexperienced and absurdly ineffective. However, a few of them were bound to be good, to be experienced, persistent, and lucky; and the earliest of these did not go away empty-handed.
>
> It is certain that some of the finest aquamarines ever found on Antero or White were picked up as surface "float." So that luck could, and did, play a part in the story. It is certain that some of the early searchers accomplished much thorough and effective rock-moving to shallow depth and secured many fine gem crystals from cavities or "pockets" hidden underneath the surface. All our stories of the early days on Antero are bound to be uncertain and vague, but one we do know beyond doubt.
>
> Old Wanamaker was one of the first and most successful of all the gem hunters on the mountain, and he left behind him a monument which still endures. This is a small stone cabin, partly now in ruins, but built right up at about 13,500 feet in the middle of the great south cirque, directly underneath the main summit. How many summers Wanamaker lived in this cabin no one knows, but he must have stayed there many years. On three sides he was ringed around by tremendous cliffs; only toward the east the cirque floor dropped off abruptly to the canyon far below and left an unobstructed and almost limitless view into the far distances where Pike's Peak would rise up on clear days like a mirage, fully 70 miles away. . . . For wood he had nothing nearer than timber-line, 1,500 feet below. Otherwise, he lived entirely in a medium of rock, with nothing but millions of boulders for his front yard and millions more for the back. Yet the old man had solved the problem of this mineral locality's inaccessibility, and he must have enjoyed the easiest and finest early pickings on Antero. I envy in that, despite the loneliness which must have weighed upon him up there at that altitude, alone with the rocks and the clouds and the stars. . . .

> Over showed me the very spot on the ridge of White where Wanamaker had opened his huge pocket of aquamarines. I was impressed not only by the height, 14,000 feet, but also by the thought that he had done a steady six-week's job here under almost impossible circumstances. He had lived in a cabin at the foot of the mountain and had to climb up 5,000 feet every morning and down 5,000 feet every evening just as a preliminary and afterthought to his work! That is real mineral collecting.

Montgomery then told of their own finds during the summer of 1937:

> Over opened three fine aquamarine pockets on White. . . . Pocket No. 1 opened out of a small seam of pegmatite in a vertical cliff face. A small pocket, it ran almost entirely to aquamarine, yielding more than 200 crystals altogether. Over fairly raked out the crystals when he got into the richest part. Most of them were small and not of too deep a color, but they were beautiful long prisms at their best. One specimen was wholly unique. Have you ever seen a cluster of six or seven aquamarines, with individual terminations merging together near the base into a lovely moss of the richest blue color? It was an exquisite fan-shaped delicacy from another world.
>
> The second pocket even surpassed the first. Over got it in the side of a crevice away back under a huge, overhanging mass of rock. Before he was through, he had dynamited off the whole affair, but to begin with he could not risk the slightest movement. Many crystals were loose in the crack, and they would have been crushed to dust beneath the rocks. The best crystals had perhaps the finest depth of blue color ever seen in any Antero aquamarine, and they showed beautiful modified terminations besides. . . . The crystals lacked size alone, not longer than three or four inches in the biggest, but their superb quality well made up for this.

Collecting on Antero did—and still does—involve a few risks. Montgomery recalled this lightning storm on a July morning:

> A tremendous array of black clouds was coming out of the west . . . right in my direction. I knew there was very little time, and I seized a pick and set to work at a hole. . . .
>
> Suddenly I began to hear a faint humming in the air, something like a giant dynamo in the distance. I threw down the

> tools, grabbed my pack, and made for the edge of the ridge; I knew the signs. Without warning my hair began to sizzle and buzz inside my hat like a swarm of bees. The small prospector's pick in my hand was now whining like a live thing and I chucked it from me as I went over the edge at top speed. . . .
>
> The sky was now like pitch overhead, almost a blue-black. Streamers of mist, dead-white against the dark background swarmed around and past the point of the east ridge and cut it off from view. A livid flame of lightning etched itself against the clouds and the terrific explosion of thunder was practically simultaneous. . . . The lightning began to play along the tops of the ridges with a furious intensity, and the roars of the big guns sounded so close I shivered in my boots. I looked out just in time to see a titanic arc of yellow flame leap out between the top of my ridge and the next one. . . .
>
> The noise of the hailstones, as large as mothballs, sounded with a swishing and rapid fire drumming on the rocks. In a minute or two everything was white. . . .

Any modern-day mineral collector who has witnessed a summer thunderstorm on Antero's summit knows that Arthur Montgomery was not exaggerating, and will agree with his closing thoughts:

> All this can never be forgotten. Life is worth living for such moments alone, for it is in such a high mountain world as Antero's, . . . timeless and separate from outside things, that one can put in each proper place bad weather, bad luck, hard work, and even mineral collecting. If we had not collected a single mineral, still it all would have been very much worthwhile.

Today, even after a century of collecting, Mt. Antero still yields gem aquamarine, phenakite, smoky quartz, and fluorite. Smoky quartz is the most abundant of Antero's pegmatite minerals. Some smokies have weighed fifty pounds; others were a base for clusters of aquamarine and phenakite crystals. Clear quartz also occurs, usually lining hollow seams in the large white quartz veins that lace the granite country rock.

Mt. Antero is the top U.S. phenakite locality, and some specimens are gem quality. Fluorite, in whites, pale greens, and deep purples, occurs in large octahedrons and dodecahedrons. Some topaz is also present, along with accessory muscovite and biotite mica, pyrite and goethite pseudomorphs after pyrite, and blue and blue-green common beryl.

Mt. Antero (right) is the highest gemstone-collecting site in North America. Aquamarine occurs in pegmatites at an elevation of 14,000 feet.

Although fragmented and abraded, many loose crystals are found in talus. Although uncovering intact pegmatite pockets requires removal of talus rock and weathered in-situ granite, collectors find new pockets almost every year. Some contain gem-quality and specimen crystals worth thousands of dollars.

From Nathrop and U.S. 285, take Colorado Route 162 east for eleven miles, then turn south on Forest Service Road 277, which is rough, steep, and four-wheel-drive-only. Proceed three and a half miles, take the left fork, ford Baldwin Creek, and continue the remaining three miles. There is parking room below the summit and short trails lead to the collecting areas.

The collecting season on Antero lasts only about four months. In June, when snowdrifts still block the road, collectors park at a wide switchback turnout at 12,500 feet, then hike the remaining way. In July, August, and September, the road is passable to the high saddle between Mt. Antero and Mt. White. Pegmatite pockets occur in a one-square-mile area below the Antero and White summits, and float crystals can be found in talus slopes 1,000 feet below.

Climbing to the summit leaves little time for collecting. From Colorado Route 162, the seven-mile-long climb (really a steep hike) has a vertical ascent of 5,000 feet. A person in excellent condition,

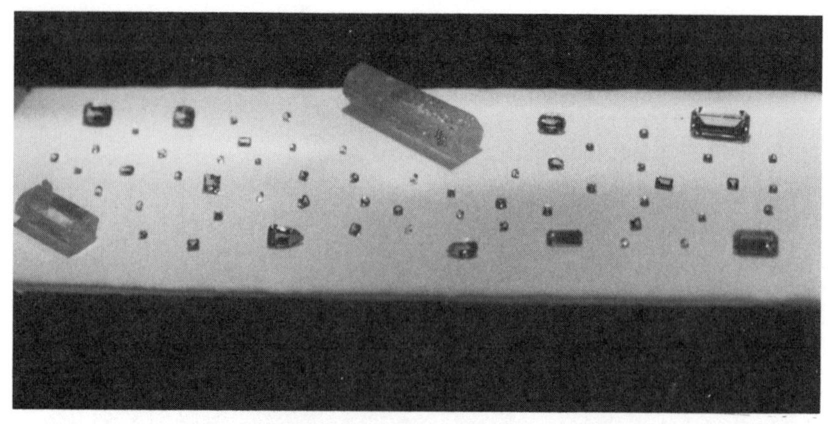

Aquamarine crystals and cut gems from Mt. Antero are displayed at the Denver Museum of Natural History. Aquamarine is the Colorado State Gem.

Cut gems of Mt. Antero aquamarine.

accustomed to mountain foot travel, and carrying only a light pack may require four hours to reach the summit and still face the prospect of a long descent.

Four-wheel-drive vehicles should be well maintained and adjusted for altitude. Carry all food, water, and other supplies necessary for a day-long trip, including proper clothing for sudden and dramatic weather changes. Get off the peak if electrical storms approach. Most severe summer thunderstorms develop during afternoons or evenings, bringing sharply colder temperatures and hail or snow. For that reason, experienced Antero collectors begin work as early as possible.

Aquamarine crystal from Mt. Antero.

Antero is not a site for anyone prone to altitude sickness, with a heart condition, or simply in poor health. Drink plenty of liquids, for dehydration may occur rapidly at high elevations. Wear sturdy boots suitable for steep and tricky talus slopes. Just as Arthur Montgomery implied over a half-century ago, whether you find aquamarine and phenakite or not, this is one Colorado collecting trip you won't forget.

REFERENCES: 1, 5, 19, 32, 36, 37, 41, 47, 52, 64, 66, 67, 73, 74

BROWNS CANYON FLUORSPAR MINES

The Browns Canyon fluorspar-mining district, eight miles northwest of Salida, covers six square miles between the Arkansas River and U.S. Highway 285. Between 1929 and 1950, Browns Canyon ranked among the top U.S. fluorspar districts, with twenty mines producing over $5 million in fluorspar.

Fluorite mineralization occurs along a four-mile-long fault section as veins of fluorite associated with chalcedony; accessory minerals are barite, calcite, pyrite, iron and manganese oxides, and small amounts of opal. Browns Canyon fluorspar ores are white or gray, with light shades of red, brown, and green.

Occasional fluorite octahedrons and dodecahedrons a few millimeters in size are found in small vein seams. The fluorite, mostly botryoidal and nonfluorescent, ranges in color from pink and yellow to green, purple, and white. Nodular growths, with fluorite forming concentric bands around breccia fragments, also occur. In the 1950s, lapidaries cut and polished nodular specimens into attractive display pieces.

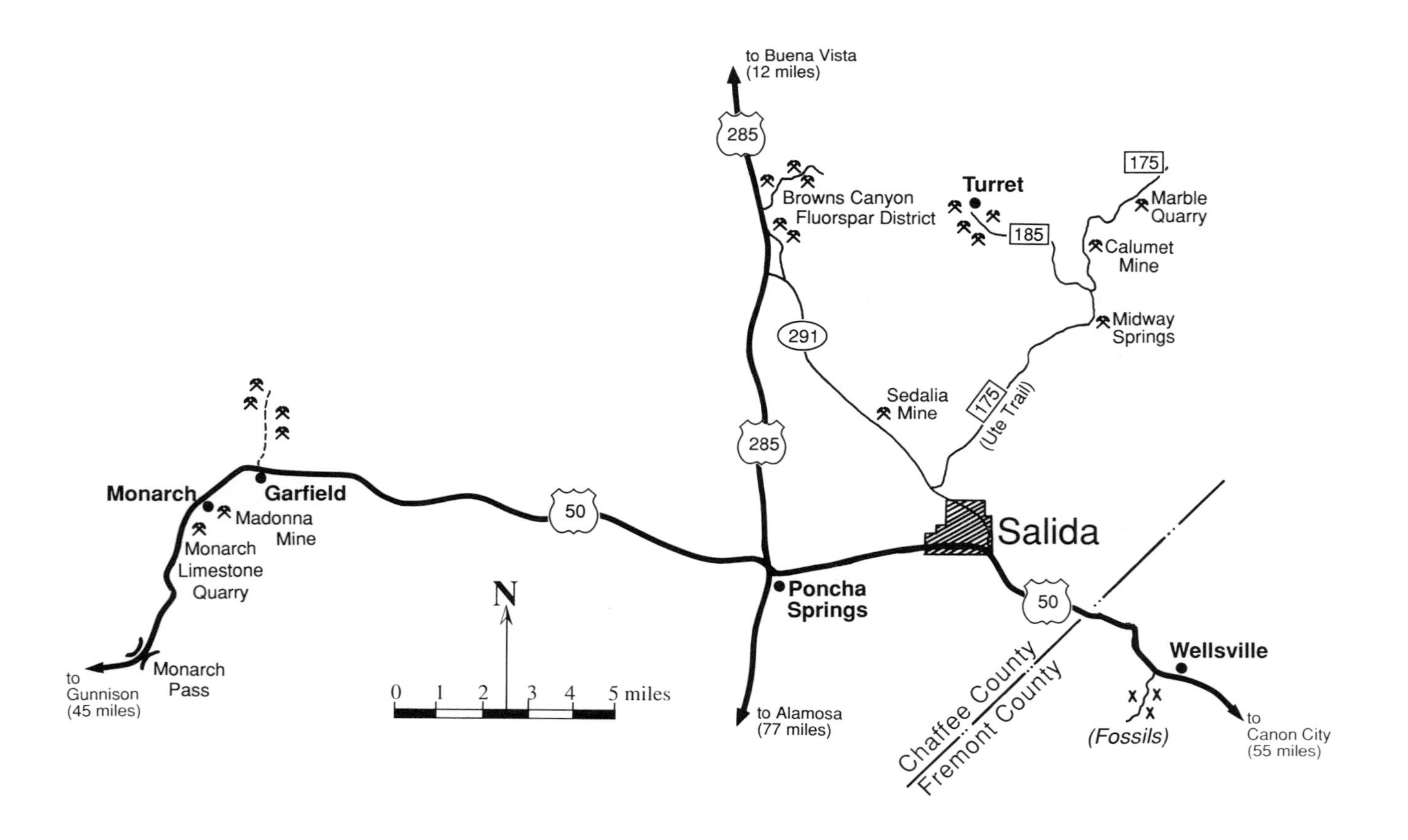

to Buena Vista
(12 miles)
285
Browns Canyon
Fluorspar District
Turret
185
175
Marble
Quarry
Calumet
Mine
Midway
Springs
291
Sedalia
Mine
175
(Ute Trail)
285
Monarch
Garfield
Madonna
Mine
Monarch
Limestone
Quarry
50
Salida
Poncha
Springs
50
Wellsville
N
0 1 2 3 4 5 miles
to
Gunnison
(45 miles)
Monarch
Pass
to Alamosa
(77 miles)
Chaffee County
Fremont County
(Fossils)
to
Canon City
(55 miles)

From U.S. 285, follow Chaffee County Roads 194 and 193 east through the district, where most unmarked roads lead to mines. Road 194 (Browns Canyon Road) passes four old fluorspar mine sites within a mile of U.S. 285. Fluorspar and small specimens of botryoidal fluorite can be dug from dumps or open cuts or found near ore bins.

REFERENCES: 4, 18, 19, 25, 80, 81

ST. ELMO AREA MINES

Prospectors discovered gold and metal sulfides in the 5,000-foot-long Mary Murphy Vein in 1870. The Mary Murphy and six other mines, two smelters, and one of Colorado's first aerial tramways were operating by the early 1880s. The Mary Murphy produced continuously until 1926, finally closing in 1952. By then, district production of gold, silver, lead, zinc, and copper exceeded $25 million, mostly from the Mary Murphy group on the west side of Chrysolite Mountain.

Oxidation-zone minerals include smithsonite and cerussite in a limonite and quartz gangue. Primary sulfide ore minerals are pyrite, chalcopyrite, galena, and sphalerite. Barren veins of fine-grained mixtures of rhodochrosite, rhodonite, and quartz were common, and early reports note fine rhodochrosite specimens on the dumps of the Pat Murphy and Mary Murphy mines. Because of rugged, remote country and historical obscurity, nice ore and gangue mineral specimens are still found on the dumps.

From Nathrop, take Colorado Route 162 sixteen miles west to St. Elmo, a picturesque ghost town with a few summer residents. Then follow Forest Service Road 295 three miles south to numerous mines near the sites of Romley and Pomeroy Gulch.

REFERENCES: 4, 18, 19, 23, 41, 81

TROUT CREEK PASS PEGMATITES

Pegmatite and rare earth minerals are found at several old pegmatite quarries near Trout Creek Pass, west of Buena Vista along U.S. Highway 24-285. Trout Creek pegmatites typically have cores of quartz, pink microcline feldspar, and biotite and muscovite mica, with intermediate zones containing quartz, potash feldspar, and albite, as well as niobium and thorium minerals.

The Crystal No. 8 Mine is located on U.S. 24-285 5.9 miles east of the junction of U.S. 24 and 285 near Johnson Village. The bright white quartz of the mine dump appears one-quarter mile south of the highway on a low hill across Trout Creek. The mine produced rose quartz, pink microcline feldspar, and monazite, a cerium-lanthanum-thorium phosphate occurring as tiny clusters of mildly radioactive, dark red-brown cubic crystals. Small crystals of green fluorite and red-brown garnet are also present.

The Clara May Mine is located a mile east of the Crystal No. 8 Mine. At the Trout Creek bridge on U.S. 24-285, follow Forest Service Road 215 for almost one mile. Park in the clearing at the base of the wooded hill to the south. The Clara May Mine is 250 feet to the south and up the hill along an old haulage road that is blocked to motorized traffic.

The Clara May, a 100-foot-long open cut, produced pink microcline feldspar and by-product bismuthinite, as striated prismatic crystals up to an inch long. Attractively banded rose quartz and foot-long books of muscovite mica are abundant in the dumps.

REFERENCES: 1, 18, 25, 64, 82

THE SEDALIA COPPER-ZINC MINE

The Sedalia Mine, northwest of Salida and just east of Colorado Route 291, has provided many unusual and interesting mineral specimens.

Prospectors discovered the deposit in 1881; production of copper, silver, and small amounts of gold began in 1884. The first oxidized ores contained 50 percent copper and ten troy ounces of silver per ton. By World War I cumulative production had topped $500,000, mostly in zinc and copper, and the Sedalia, with 8,000 feet of workings, was Colorado's largest copper mine. The copper-zinc-lead-silver ore was a mineralized schist. Garnetiferous chlorite

Almandine garnet dodecahedron from the Sedalia Mine.

schists, pegmatite minerals, and copper minerals provided an array of specimens. The Sedalia dumps were famed for spectacular crystals of almandine garnet in beautifully formed dodecahedrons weighing up to fifteen pounds.

Copper minerals included chalcopyrite, chrysocolla, malachite, azurite, cuprite, and chalcocite. Other ore minerals were pyrite, sphalerite, and cerussite. Apatite, epidote, calcite, tourmaline, corundum, and actinolite were accessory minerals.

The mineralized schist is cut by a granite pegmatite dike yielding specimens of green beryl, white quartz, muscovite mica, microcline feldspar, and several rare earth minerals.

From the Arkansas River bridge at Salida, follow Colorado Route 291 northwest for 2.5 miles. A dirt road leads to the mine, one-half mile north on the hillside. Gates are locked and permission must be secured before collecting.

REFERENCES: 4, 5, 23, 25, 41, 64, 74

TURRET AREA MINES

Turret, eight miles north of Salida, has been a productive mineral-collecting area for decades. After prospectors discovered a large vein of magnetite in 1882, the Calumet Mine was an important source of iron ore for the Colorado Fuel & Iron Company works in Pueblo. Calumet declined by 1900, but prospectors discovered small deposits of gold, silver, and base metals and founded the town of Turret. In 1904 Turret's population peaked at 400, but the mines soon closed and Turret joined Colorado's growing list of mining ghost towns. Independent miners returned in the 1950s to work numerous small surface pegmatites for beryl and columbite-tantalite.

From the Colorado Route 291 bridge over the Arkansas River just north of Salida, take Chaffee County Road 175 (Ute Trail) north for 7.5 miles to Midway Springs. At Midway Springs, a meadow in the piñon-juniper woodlands, a small open cut exposes thin beds of compact, white-and-brown-banded aragonite that was worked by local lapidaries and sometimes sold as "onyx." Bladed actinolite-tremolite and goethite pseudomorphs after pyrite are also present.

From Midway Springs, bear right onto County Road 185 toward Turret. Proceed 1.5 miles to another fork near several mine dumps. The left track leads to the Rock King Mine, a 1960s pegmatite open-cut mine. The dumps have nice specimens of muscovite and biotite mica, white albite, pink microcline feldspar, and very coarse graphic granite. The Rock King produced beryl crystals over six feet long, but only occasional fragments remain on the dumps. Most are pale green-blue and show one or two faces.

The right fork leads to the Homestake Mine, a large 1950s feldspar producer with extensive old workings and dumps. The western edge of the quarry, a 200-foot-high cliff, drops precipitously into a deep, flooded open pit. Dump specimens include biotite and muscovite mica, pink microcline feldspar, some clear quartz, and abundant dull green, coarsely bladed actinolite.

Continue one-half mile north on County Road 175 to the Calumet Iron Mine, which is not visible from the road. Park near the small grove of aspen trees and the old loading-dock timbers and concrete foundations that appear in the creek bed. A steep trail leads east toward the mine, ascending 600 vertical feet in a quarter mile.

The Calumet is a contact metamorphic deposit, altered by intensive volcanic activity in the nearby Thirty-Nine Mile Volcanic Field (see Park County). The Calumet iron ore comes from a massive, near-vertical vein of magnetite that cuts through beds of limestone, marble, and schist. Fine-grained, massive magnetite, and occasional gray metallic crystals up to one-quarter inch long are found in the dumps. Also present are many nicely radiating sheaves of light and dark green bladed actinolite-tremolite.

The Calumet Mine is Colorado's most noted source of epidote, which occurs in massive, granular, and crystalline form within partially altered limestone. Colors range from light green to near black, and beautiful two-inch crystals associated with white quartz have been collected from the dumps. The same limestone-marble beds also yield grossular garnet, usually massive, but occasionally in brown dodecahedrons up to one inch in diameter.

The Calumet is Colorado's only sapphire occurrence. Small crystals of blue corundum are found in a layer of dark green schist. Most crystals lack clarity, but collectors have reported some gem-quality material.

An open-cut marble quarry is located a half mile farther on County Road 185. The dumps below the three bench cuts contain many specimens of snow white, coarsely crystallized marble.

Retracing the route to the Homestake quarry, County Road 174 leads two miles west to the site of Turret. A dozen cabins, privately owned and posted, are still standing. Excavations on surrounding hillsides expose small pegmatite bodies with fragmented common beryl specimens. Collecting at Turret, elevation 10,000 feet, is often restricted in winter.

REFERENCES: 4, 23, 25, 41, 43, 64

MONARCH AREA MINES

Monarch is located thirteen miles west of Poncha Springs on U.S. Highway 50. Prospectors found silver ores in 1878 and founded the boom towns of Maysville, Monarch, and Garfield.

The Great Monarch and the Madonna mines were the top producers. The Great Monarch, where thin, erratic veins graded as high as 200 troy ounces of silver per ton, had a steeply inclined 3,000-foot-long aerial tramway to convey ore to the railhead. The Madonna closed after the 1893 silver market collapse, but soon reopened as a zinc mine that operated sporadically until 1953.

Other important Monarch mines included the Lilly, Garfield, and New York in Taylor Gulch. The Monarch district produced over $16 million in lead-silver, gold, zinc, and copper ores.

Local mineralization includes both replacement and contact metamorphic ore bodies. Ore minerals are argentiferous cerussite, pyrite, native copper, chalcopyrite, bornite, covellite, magnetite, galena, and smithsonite. Gangue minerals include calcite, dolomite, and some epidote. The old Colorado Fuel & Iron Company's Monarch limestone quarry, located on U.S. 50 one mile above Monarch, was Colorado's largest limestone quarry.

From Salida and Poncha Springs, follow U.S. Route 50 thirteen miles west. A jeep road at Garfield leads north to the old Taylor Gulch mines.

REFERENCES: 23, 25, 81

SWISSVALE BRACHIOPOD FOSSILS

Finely detailed brachiopod fossils are abundant in several narrow canyons along the Arkansas River Canyon and U.S. 50 about six miles east of Salida. One particularly fossil-rich canyon is located three miles east of the Chaffee-Fremont county line, and three-quarters of mile west of the little town of Swissvale (Fremont County). Park near the culvert on the south side of the highway and walk several hundred feet up the canyon. Most of the rock in the creek bed is fossiliferous. Brachiopod fossils, about one inch wide and retaining bits of their original shell material, are present in a fine-grained, gray siltstone of Pennsylvanian age. Collecting is a matter of selecting and trimming desirable specimens.

REFERENCES: 13

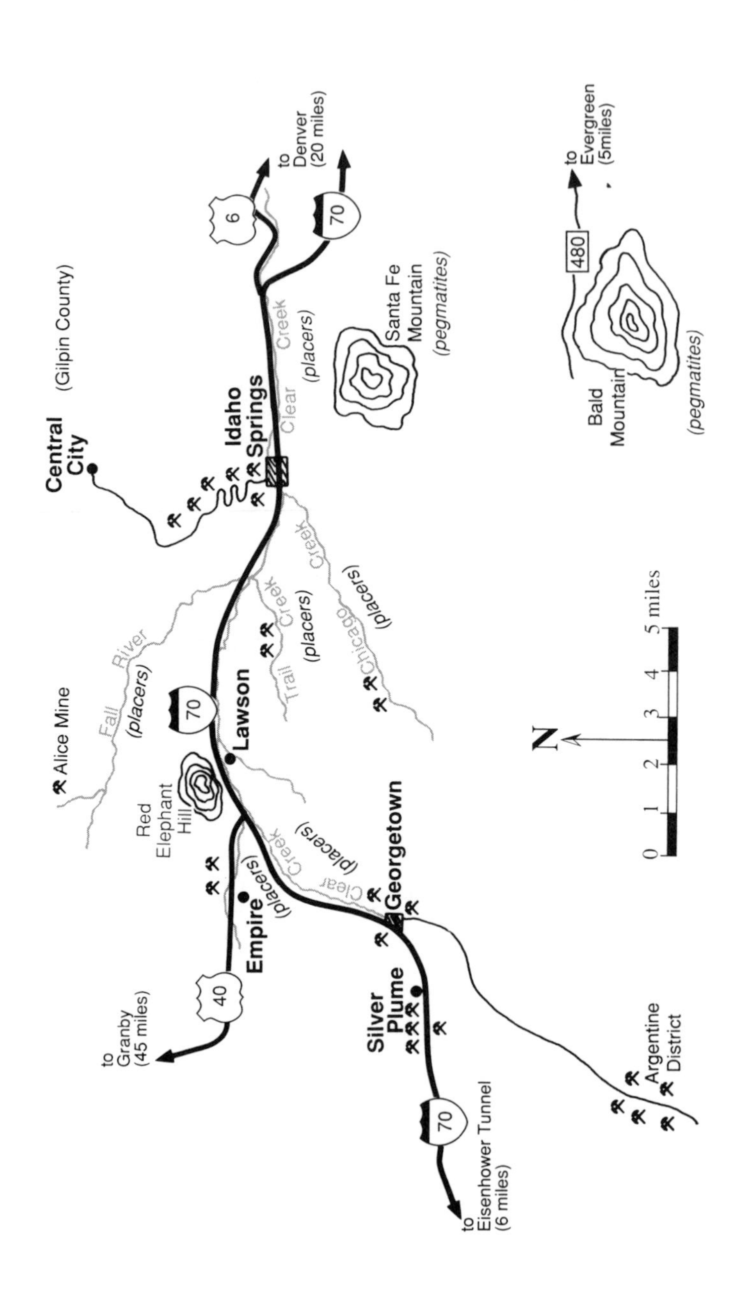
Central City
(Gilpin County)
Idaho Springs
Clear Creek
(placers)
Santa Fe Mountain
(pegmatites)
to Denver (20 miles)
6
70
to Evergreen (5miles)
480
Bald Mountain
(pegmatites)
Alice Mine
Fall River
(placers)
70
Lawson
Trail Creek
(placers)
Chicago Creek
(placers)
Red Elephant Hill
Empire
(placers)
Clear Creek
(placers)
Georgetown
N
0 1 2 3 4 5 miles
to Granby (45 miles)
40
Silver Plume
70
to Eisenhower Tunnel (6 miles)
Argentine District

CLEAR CREEK COUNTY

PLACER GOLD

Clear Creek County has produced 140,000 troy ounces of placer gold worth $3 million, ranking fourth among Colorado counties. The placers are in Clear Creek and its tributaries.

George A. Jackson discovered placer gold at the site of Idaho Springs near the confluence of Chicago Creek and Clear Creek on January 7, 1859, a strike that helped redeem the Pikes Peak gold rush. Jackson dug gravel from a frozen bar with his hunting knife and panned it in a tin cup. That night, alone by his campfire with only his dogs for company, Jackson wrote in his diary:

> I jumped up and down and told myself the story I would tell. . . . After a good supper of meat—bread and coffee all gone—I went to bed and dreamed of riches galore in that bar. If only I had a pick and a pan instead of a hunting knife and the cup, I could dig out a sack full of the yellow stuff and carry it down to the boys. My mind ran upon it all night long. I dreamed all sorts of things—about a fine house and good clothes, a carriage and horses, travel, what I would take down to the folks in old Missouri and everything you could think of—I had struck it rich!

Jackson kept his strike secret, found backers, and returned in early May to wash out 100 troy ounces of placer gold in one week—more than had been found in the first nine months of the Pikes Peak rush.

Other prospectors found gold in tributaries like upper Chicago Creek, Fall River, and Mill Creek, and all the way up Clear Creek to Georgetown and Empire. Miners recovered gold by bedrock drifting, ground sluicing, and, later, hydraulicking and draglining.

Panning is popular along many tributaries. The main channel of Clear Creek, however, is a poor location, due to extensive rechanneling for highway and rail construction projects. A commemorative plaque overlooking the site of Jackson's 1859 discovery is located near the Clear Creek Secondary School along Colorado Route 103 one-half mile south of I-70.

REFERENCES: 53, 98,108

IDAHO SPRINGS–GEORGETOWN–SILVER PLUME AREA MINES

Clear Creek County lode mining began with simple crushing and washing of oxidized outcrops of gold-bearing quartz at Idaho Springs and Empire. Prospectors next discovered silver in the Argentine district eight miles above Georgetown. Georgetown and Silver Plume became milling, smelting, and transportation centers for many silver mines in western Clear Creek County. In 1876, Georgetown became Colorado's first "Silver Queen" when annual production value topped $1 million.

After the 1893 silver-market crash, Georgetown and Silver Plume mines survived on lead and zinc. Idaho Springs mines continued with silver-lead-zinc-gold ores. Construction of Idaho Springs' Argo Tunnel and Mill provided haulage for many mines in both Clear Creek and Gilpin counties and contributed greatly to operational longevity. When most mines closed in the 1950s, total county production of silver-gold-lead-zinc-copper exceeded $175 million.

Clear Creek County has about 600 mines, 150 of which were major producers. Mine dumps still yield an array of sulfide ore minerals, including sphalerite, galena, pyrite, chalcopyrite, and argentite. Many Clear Creek County metal sulfide ores are associated with small, clear, terminated quartz crystals.

Gangue minerals include some rhodochrosite and amethyst. Collectors have found pale rhodochrosite in mine dumps on Red Elephant Hill just north of Lawson, and some amethyst in mine dumps along Trail Creek west of Idaho Springs.

The Alice Mine, a large open pit eight miles northwest of Idaho Springs on Fall River Road, was a noted source of superb pyrite and chalcopyrite crystals associated with clear quartz and siderite. In the late 1980s, however, the pit was filled as part of the Mined Land Reclamation Program of the Colorado Division of Natural Resources.

Along with hundreds of mine dumps, Clear Creek County has interesting mining-related attractions and museums. The Underhill Museum on Miner Street in Idaho Springs has displays of early mining equipment and local ores. The Edgar Mine, on the north side of Idaho Springs, is the experimental mine of the Colorado School of Mines and offers tours during the summer months.

The prominent Argo Gold Mill overlooks Idaho Springs. The Argo Mill and Tunnel, the center of Idaho Springs mining and milling activity from 1913 to 1943, processed over $100 million in ores, mostly of gold. Exhibits represent a century of milling technology and include stamp mills, mill feed decks, grinding equipment, classifiers, amalgamation tables, flotation cells, concentrating tables, and cyanidation systems. Visitors may also tour the nearby underground Double Eagle Mine and pan for gold.

The Phoenix Mine, a small working gold-silver mine just west of Idaho Springs, offers public underground tours and displays of both frontier-era and modern underground-mining equipment. The Phoenix, which dates to 1871, has produced over $1 million in metals. The Lebanon Silver Mine offers an underground tour at the halfway point of the Georgetown Loop steam railroad between Silver Plume and Georgetown. The mine is accessible by rail only.

REFERENCES: 18, 23, 32, 41, 66, 67, 81, 98, 108

CLEAR CREEK COUNTY PEGMATITES

Granite pegmatites occur southeast of Idaho Springs near Santa Fe Mountain and Bald Mountain. Although not mined commercially, they have provided interesting and attractive specimens of black tourmaline, lepidolite, garnet, and feldspar.

Several small pegmatites on the northeast slope of Santa Fe Mountain, four miles southeast of Idaho Springs, are reached by a rough four-wheel-drive road up Sawmill Gulch. The Bald Mountain pegmatite area is reached from Evergreen (Jefferson County) on Road 480.

Santa Fe and Bald mountains are a general area of pegmatite occurrences, some of which are on private property and closed to collecting. Adjacent National Forest lands, however, provide prospecting opportunities.

REFERENCES: 29, 66, 67

Section of two-inch pyrite cube from Idaho Springs.

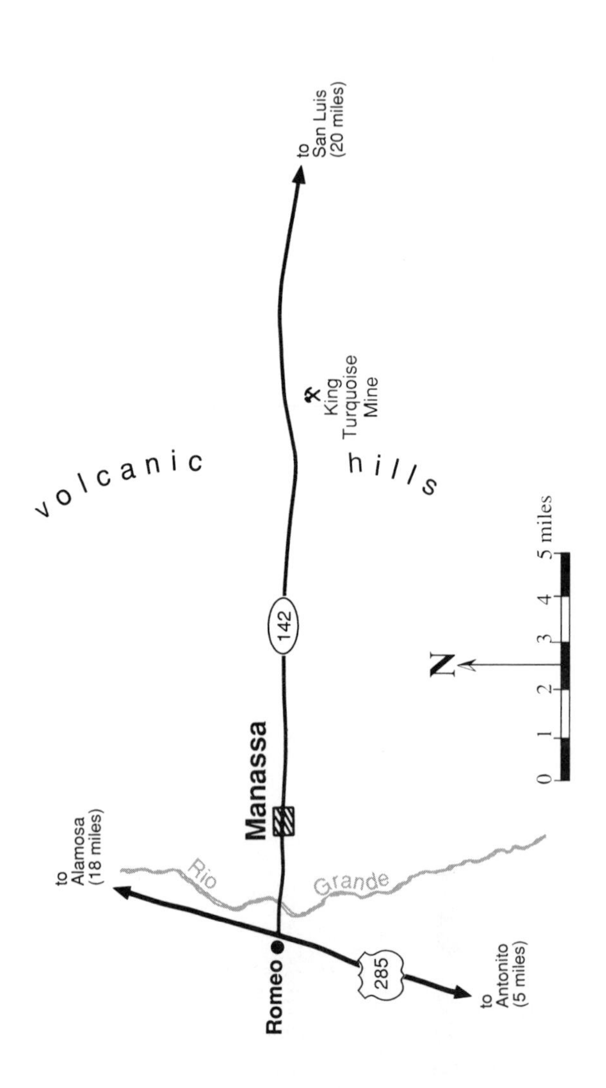
to
San Luis
(20 miles)
King
Turquoise
Mine
volcanic
hills
142
Manassa
N
0
1
2
3
4
5 miles
to
Alamosa
(18 miles)
Rio
Grande
Romeo
285
to
Antonito
(5 miles)

CONEJOS COUNTY

THE KING TURQUOISE MINE

The historic King Turquoise Mine, eight miles east of Manassa, is located on a low volcanic hill just west of the Rio Grande. Many colorful Indian and Spanish legends are tied to the King Turquoise Mine. The Spanish probably learned of the mine by the late 1700s, for it was a well-known source of Indian turquoise. "Of all the hills that dot the grass-covered vistas in the southern San Luis Valley," legend tells us, "there is only one upon which a man can stand on a summer's evening and have his shadow cast all the way to the Rio Grande. This is the hill that contains the turquoise."

In 1890 settler Israel Pervoise King discovered an old narrow tunnel containing Indian tools of stone and deer antler. King, hoping he had found an ancient gold mine, was initially disappointed. He recognized copper mineralization, but grades were too low for mining. King collected a few colorful specimens for his mantle, apparently unaware that they were turquoise.

This volcanic hill is the site of the King Turquoise Mine. According to Spanish legend, it was the only hill that cast an evening shadow on the Rio Grande.

Traditional silver and turquoise squash blossom necklace with turquoise from the King Turquoise Mine.

A friend finally identified the specimens as turquoise in 1900. King staked twelve claims, formed the Colorado Turquoise Mining Company, named the deposit the "Lick Skillet" in reference to hard economic times, and began mining. In 1909, new operators leased the mine and opened a lapidary shop in Colorado Springs to market the turquoise. King's son, Charles, eventually took over the mine and, by the 1930s, developed it into a major American turquoise source.

Charles King's twelve employees produced about 5,000 carats (roughly three and one-half pounds) of gem turquoise each month, ranging from single-carat "chips" the size of split peas to 100-carat pieces measuring two by three inches. Top-quality turquoise then sold for fifty cents per carat. Indians from New Mexico and Arizona, the biggest buyers, favored solid stones with lighter colors. Tourists sought darker colors with matrix-veining, the latter as assurance of authenticity.

Extensive mining through pits, tunnels, and shafts revealed the ancient Indian workings as "catacombs reaching hundreds of yards into the mountainside." In 1936, the overhead rock in a tunnel collapsed and antler tools, stone hammers, and human bones, including parts of skulls, came tumbling down. King presented the relics to the Colorado Historical Society for display at the old State Museum in Denver.

W. P. "Pete" King, Israel Pervoise King's grandson, took over the mine in the late 1930s and opened a spectacular turquoise vein. By 1946, annual production soared to 2,000 pounds of rough turquoise worth more than $30,000, including one of the largest turquoise "nuggets" ever found. In June 1947, Richard Pearl, writing in *The Mineralogist*, described the $8,000 piece of turquoise:

> Weighing eight and three-quarter pounds, the King nugget has an attractive blue-green color and the popular spider-web pattern. The network that constitutes this natural design is formed partly by brown veinlets of iron oxide, and partly by darker veinlets of turquoise.

Pete King's son, Bill, took over in 1960. The mine was idle until operations resumed in 1980. The mine remains active and is an important commercial source of southwestern American turquoise.

The surrounding volcanic hills covering forty square miles have numerous surface occurrences of copper mineralization. Prospect holes and small shafts and tunnels dot the hillsides, testimony to the continuing search for other possible turquoise veins.

The King Turquoise Mine is posted and gates are locked. Collecting is prohibited. For information contact Bill King, King Turquoise Mine, P.O. Box 237, Manassa, Colorado 81141.

REFERENCES: 19, 56, 59, 60, 61, 66, 67, 74, 81

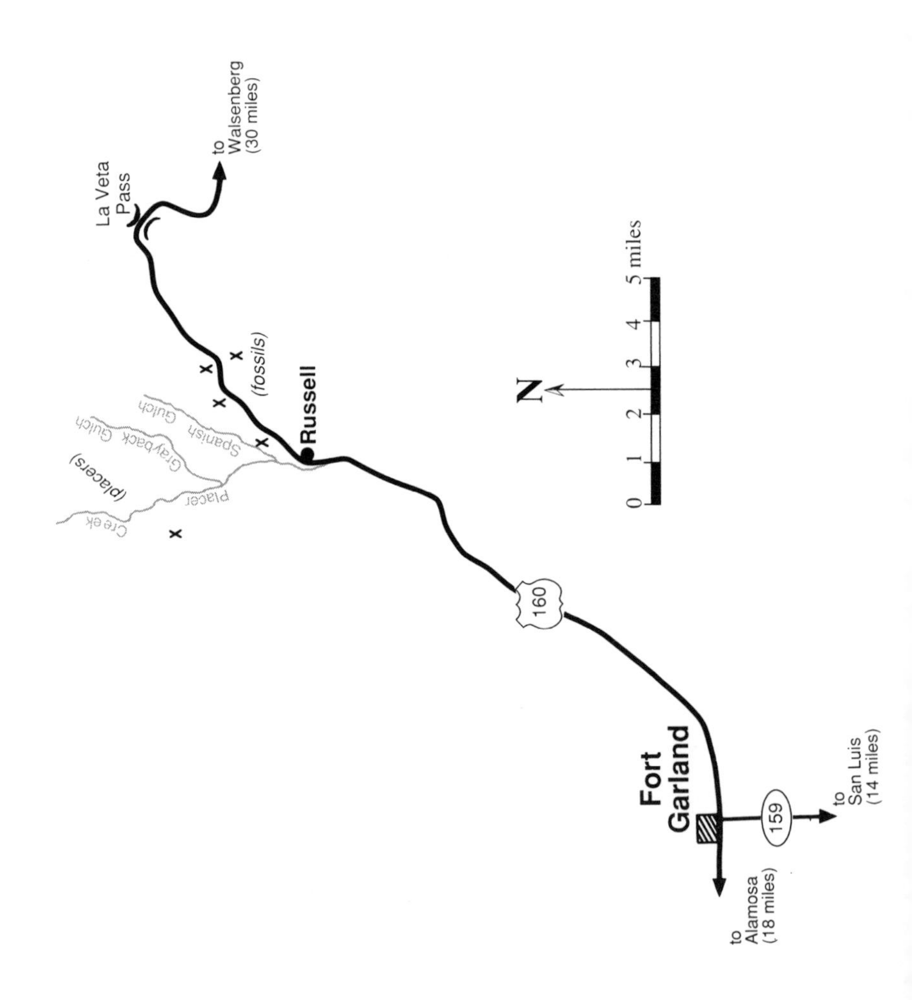
to
Walsenberg
(30 miles)
La Veta
Pass
(fossils)
Russell
Spanish Gulch
Grayback Gulch
(placers)
Placer
Cre ek
160
Fort
Garland
159
to
San Luis
(14 miles)
to
Alamosa
(18 miles)
N
0
1
2
3
4
5 miles

COSTILLA COUNTY

PLACER GOLD

Costilla County has produced about 1,500 troy ounces of placer gold. Although only a minor source of gold, the placers are historically interesting, for they were first worked six years before the Pike's Peak rush. Grayback Gulch, Placer Creek, and Spanish Gulch are located in northern Costilla County, twelve miles east of Fort Garland and just north of the old U.S. 160 highway settlement of Russell.

U.S. Army soldiers worked the placers soon after establishing nearby Fort Massachusetts in 1852. Mining apparently continued until 1858, when Fort Massachusetts was replaced by Fort Garland. The original mining site at the mouth of Grayback Gulch is still known as Officers Bar.

Miners sluiced the placers sporadically until 1898, then turned to steam shovels and, in 1910, a small floating-bucketline dredge, which accounted for most of the district's production. Independent small-scale miners returned during the 1930s.

The placers are on posted private property and not open to recreational mining and panning.

REFERENCES: 54, 81

COSTILLA COUNTY FOSSILS

Brachiopod, gastropod, coral, and tiny radiolaria fossils are found in a series of steep road cuts from one to four miles northeast of Russell along U.S. 160. Russell is an abandoned highway settlement twelve miles east of Fort Garland and seven miles west of La Veta Pass. Most exposures of the steeply tilted sediments are on the north side of the highway.

Fossils occur within thin, crumbly beds of gray Madera Shale of Pennsylvanian age. The most productive site is the road cut just east of the culvert and old cabin where U.S. 160 crosses Sangre de Cristo Creek one mile north of Russell. Seams in the shale occasionally contain well-developed white calcite crystals.

REFERENCES: 12, 13

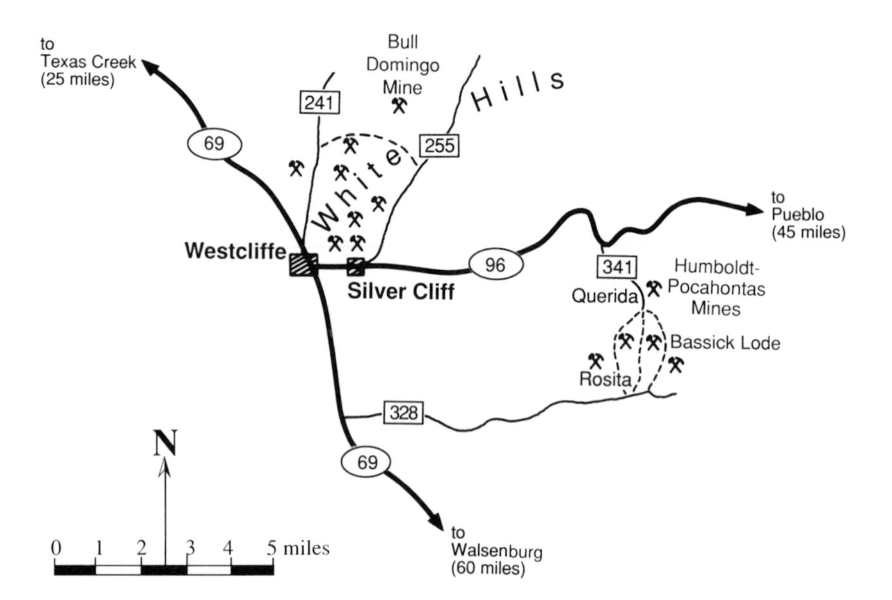
to
Texas Creek
(25 miles)
Bull
Domingo
Mine
Hills
241
255
69
White
Westcliffe
Silver Cliff
96
to
Pueblo
(45 miles)
341
Humboldt-
Pocahontas
Mines
Querida
Bassick Lode
Rosita
328
69
N
0 1 2 3 4 5 miles
to
Walsenburg
(60 miles)

CUSTER COUNTY

SILVER CLIFF AREA MINES

Prospectors discovered small, erratic veins of silver-lead sulfides in the northern Wet Mountain Valley in Custer County in 1872. Development began in 1874 near Rosita, after prospectors found outcrops of the million-dollar Humboldt-Pocahontas vein that carried azurite, malachite, and native silver. Miners found the Bassick Lode, another million-dollar deposit, two miles north of Rosita, in 1877.

Silver Cliff sprang up in 1878 when prospectors struck chlorargyrite (horn silver) in the nearby White Hills. The next bonanza was the Bull Domingo Lode two miles to the north. Mineralization occurred as small, near-vertical veins emplaced in the rhyolite and tuff country rock and in the underlying granite.

By 1885, production topped $3 million in silver and copper. But the rich veins were quickly exhausted and mining became sporadic. Nevertheless, cumulative production of silver, gold, lead, zinc, and copper from the Silver Cliff-Rosita-Querida mines exceeded $8.5 million by 1923. The district is Colorado's largest multimetal deposit outside the Mineral Belt.

Galena, sphalerite, azurite, malachite, cerussite, argentite, and chlorargyrite are present in the White Hills mine dumps north of

The mining town of Silver Cliff was named for a massive outcrop of chlorargyrite, or "horn silver."

Mine dumps in the White Hills north of Silver Cliff contain sulfide minerals, clear quartz, and some native silver.

Silver Cliff, as are one-inch-long clear quartz crystals. Pyrite occurs in small "plates" of unusually bright cubic crystals. Some collectors use metal detectors to help locate small specimens of gray-black native silver in the dumps. Hundreds of prospect holes pockmark the White Hills, but the most productive dumps are those of larger mines. The dumps are heavily oxidized and digging is necessary.

Although few metal minerals remain on dumps near the extensively mined cliffs just north of Silver Cliff, collectors can find interesting metamorphic specimens, including epidote and garnetiferous schist.

Silver Cliff and adjacent Westcliffe are located fifty miles west of Pueblo on Colorado Route 96. Good gravel roads lead north to the White Hills mines.

REFERENCES: 18, 23, 66, 67, 81

DELTA COUNTY

DELTA COUNTY AREA FOSSILS

DRY MESA *SUPERSAURUS* SITE

In 1971, Delta residents Vivian and Eddie Jones, prospecting for uranium in the rugged Dry Mesa-Escalante Canyon area west of Delta, discovered in-situ, fossilized bones in mudstone and siltstone beds of the Brushy Basin Member of the Morrison Formation. Realizing the potential scientific importance of vertebrate remains, they reported their find to paleontologists at Utah's Brigham Young University, who identified the fossils as dinosaur bones.

Excavation has since uncovered thousands of bones. In 1988, paleontologists announced that the bones confirmed the suspected existence of *Supersaurus*, possibly the longest dinosaur that ever lived. The highly evolved *Supersaurus* was over ninety feet long, fifty feet high, and weighed eighty tons. It had small forelegs, a very long tail with at least seventy-three vertebrae, an extended drooping neck, and a small head with tiny, pencil-like teeth. *Supersaurus* lived 135 million years ago, thus straddling two major geologic periods, and is a "benchmark" indicator of life during the Jurassic-Cretaceous transition.

The 10,000 bones recovered from the Dry Mesa site fill the entire basement of the Brigham Young University sports stadium in Provo. Among the fourteen dinosaurs identified to date, four are new to science. Complete excavation and study will take decades, and paleontologists believe the bones may even include those of a behemoth tentatively named "Ultrasaurus." The name given to the discovery species of *Supersaurus* (*Supersaurus vivianae*) honors its discoverer, Delta rockhound and amateur paleontologist Vivian Jones.

The fossil site is on public land near the East Fork of Escalante Creek on the western slope of Dry Mesa. It is twenty miles west of Delta near the Montrose-Mesa county line. The site is open to the public when excavation is in progress. For information, contact the Bureau of Land Management District Office in Grand Junction.

THE YOUNG EGG LOCALITY

The Young Egg Locality, one of Colorado's more unusual paleontological sites, is located near U.S. 50 and the confluence of Wells

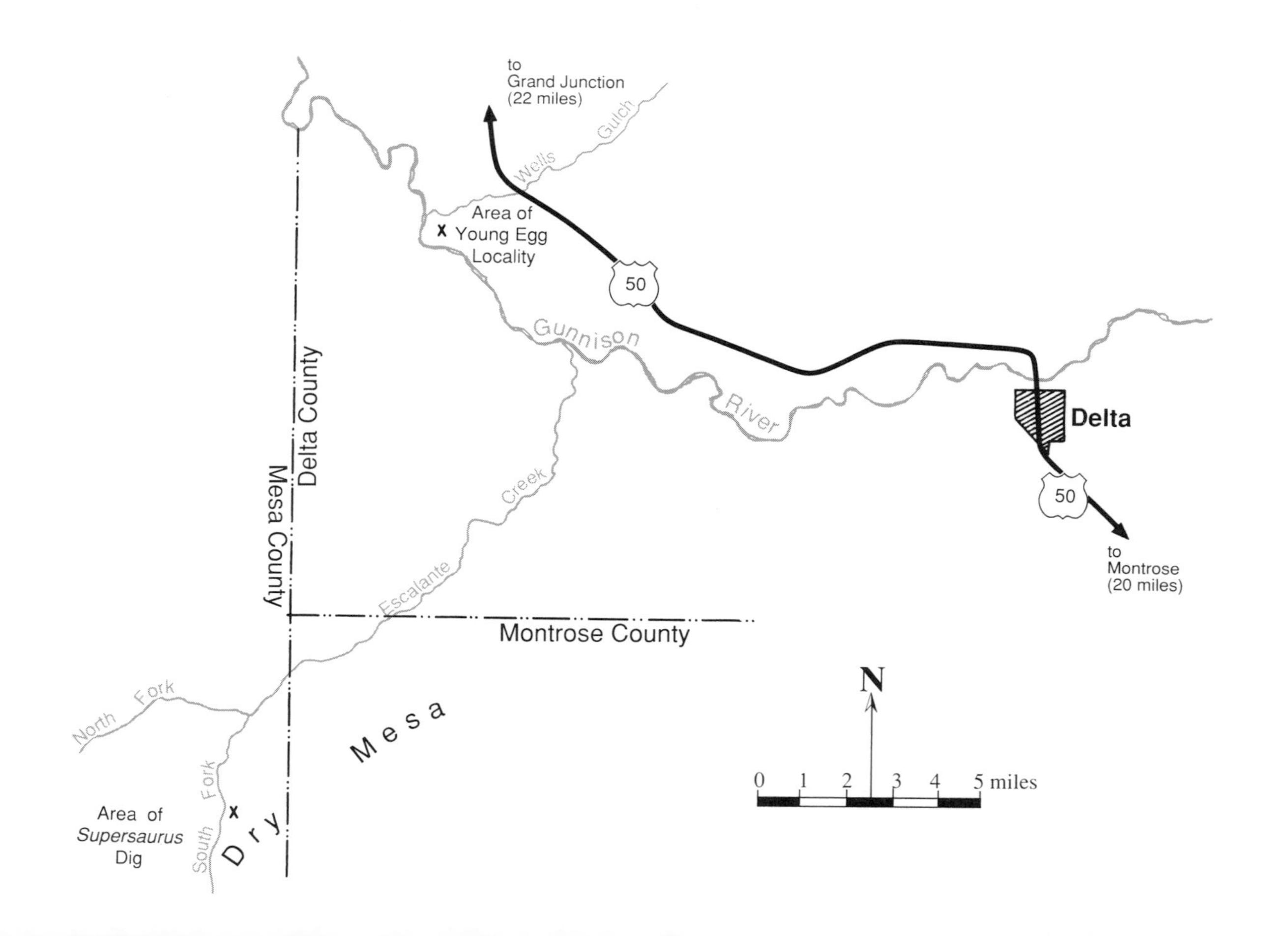
to
Grand Junction
(22 miles)
Wells Gulch
Area of
Young Egg
Locality
50
Gunnison River
Delta
50
to
Montrose
(20 miles)
Delta County
Mesa County
Escalante Creek
Montrose County
North Fork
South Fork
Mesa
Dry
Area of
Supersaurus
Dig
N
0 1 2 3 4 5 miles

Gulch and the Gunnison River, about twelve miles northwest of Delta. The area contains thousands of black fossilized eggshell fragments from a well-used nesting site from the upper Jurassic period 150 million years ago. The fossil eggshells, along with bones and teeth, occur in Morrison sandstones. The eggshell fragments are the oldest found in the Northern Hemisphere.

Since Dr. Robert G. Young discovered the site in 1987, it has been surface collected, quarried three times, and studied by geologists, palynologists, and experts in soils, paleodating, and eggshells. Further discoveries are expected to provide even greater understanding of dinosaur nesting patterns and the paleoecology of Jurassic nesting sites.

The Bureau of Land Management plans to protect the site as an area of critical environmental concern and a research natural area. For information on the Young Egg Locality, contact the Bureau of Land Management District Office in Grand Junction.

REFERENCES: 3

Visitors at the Denver Museum of Natural History view the Campion Gold Collection. The crystallized gold specimens were mined in Summit County in the late 1880s and 1890s.

Diorama at the Denver Museum of Natural History recreates a typical pegmatite exposure in Pikes Peak granite.

DENVER CITY AND COUNTY

PLACER GOLD

Because of its intensive urban and industrial development, Denver may seem an unlikely area for gold. Nevertheless, the city and county of Denver, Colorado's smallest county, has produced about 300 troy ounces of placer gold. The South Platte River contains fine placer gold from the confluence of Big Dry Creek in Littleton (Arapahoe County) north to the present 6th Avenue bridge in Denver.

Miners recovered most of the gold not during the Pikes Peak rush, but the Depression years. In a 1931 public works project, experienced placer miners and old prospectors taught hundreds of Denver's unemployed to pan gold. Denver's "gold-mining schools" received national publicity and were quite successful. Many jobless individuals panned or sluiced a dollar's worth of placer gold per day from the South Platte River.

There is no recreational panning in Denver, however, because flood control and landfill projects have rerouted much of the original channel of the South Platte River.

REFERENCES: 53

THE DENVER MUSEUM OF NATURAL HISTORY

The Denver Museum of Natural History houses the world's premier collection of Colorado minerals, gemstones, and fossils. The museum was founded in 1900 as the Colorado Museum of Natural History. The founding gifts included mining magnate John F. Campion's spectacular 600-piece collection of crystallized gold from Farncomb Hill (Summit County).

Today, the Denver Museum of Natural History is the seventh-largest natural history museum in the United States, with office and exhibit space of nearly a half-million square feet. The museum now hosts about a quarter-million visitors annually.

About 2,500 of over 16,000 catalogued mineral specimens are displayed in the Coors Mineral Hall, which was constructed in 1982 through a $300,000 grant from the Coors Foundation. Colorado minerals and gemstones are the featured display of the Coors Mineral Hall. They include superb specimens of all the minerals

Reconstructed Stegosaurus *skeleton at the Denver Museum of Natural History. The fossilized bones were excavated from Garden Park in 1937.* Stegosaurus *is the Colorado State Fossil.*

and gemstones for which Colorado is especially noted: amazonite, topaz, turquoise, aquamarine, barite, quartz, rhodochrosite, native silver and gold, lapis lazuli, molybdenite, galena, sphalerite, and other ore minerals from many of the state's greatest mines.

Gold exhibits in the dramatically lighted, triple-security Gold and Silver Room include "Tom's Baby," the largest specimen of crystallized gold ever found in Colorado (Summit County); the Campion collection; the Summitville "gold boulder" (Rio Grande County); the Penn Hill and "turtle" nuggets, the two largest known Colorado nuggets (Park County); ram's horn wire gold from the Ground Hog Mine (Eagle County); gold tellurides from Cripple Creek (Teller County); and gold nuggets from Cache Creek (Chaffee County).

Field gemstone collectors may be especially interested in a life-sized diorama depicting a granite pegmatite cavity filled with smoky quartz and amazonite, typical of those found in Teller, Park, El Paso, and Douglas counties.

The many excellent paleontological exhibits at the Denver Museum of Natural History include a reconstructed *Stegosaurus*, recovered from Garden Park (Fremont County) in 1937. Other exhibits feature many remarkably detailed insect and leaf fossils from Florissant (Teller County), fish fossils from the Green River Formation in Wyoming, Pliocene turtle fossils from Yuma County, Oligocene mammal fossils and reconstructed skeletons from Weld County, Eocene mammal fossils from Moffat County, and dinosaur tracks from Prowers County. The Denver Museum of Natural History is located at 2001 Colorado Boulevard in City Park.

THE DENVER GEM AND MINERAL SHOW

The annual Denver Gem and Mineral Show celebrated its twenty-fifth anniversary in 1992 and now ranks second in size and importance in the United States.

In 1967, the Federated Council of Gem and Mineral Societies, composed of five Denver-area clubs, staged Colorado's first joint-sponsored mineral show. Growth was slow but steady until 1976, when the modest ten-dealer show ambitiously hosted the annual conventions of both the regional and national mineral-society federations at the newly built Merchandise Mart Exposition Hall. The rapid expansion resulted in a financial loss that seriously jeopardized the show's future.

But the show survived, went into the black in 1982, and has grown rapidly since. The 1992 Denver Gem and Mineral Show had 100 dealers and exhibitors and a record paid attendance of over 11,000. There are also four concurrent dealer shows, bringing the total number of exhibitors to over 400.

The Denver Gem and Mineral Show is now sponsored by the Greater Denver Area Gem and Mineral Council. The nine member organizations include the Colorado Mineral Society, Denver Gem and Mineral Guild, Flatiron Mineral Club, Colorado Chapter of the Friends of Mineralogy, Gates Rock and Mineral Club, Littleton Rock and Mineral Club, Mile-Hi Rock and Mineral Society, North Jeffco Gem and Mineral Club, and the Western Interior Paleontological Society.

Each annual show features a mineral theme with dozens of outstanding specimen exhibits provided by major universities and museums from the United States, Canada, and Europe. The Denver Gem and Mineral Show is held annually during the third week of September at the Merchandise Mart Exposition Hall on 58th Street in North Denver, one mile west of I-25.

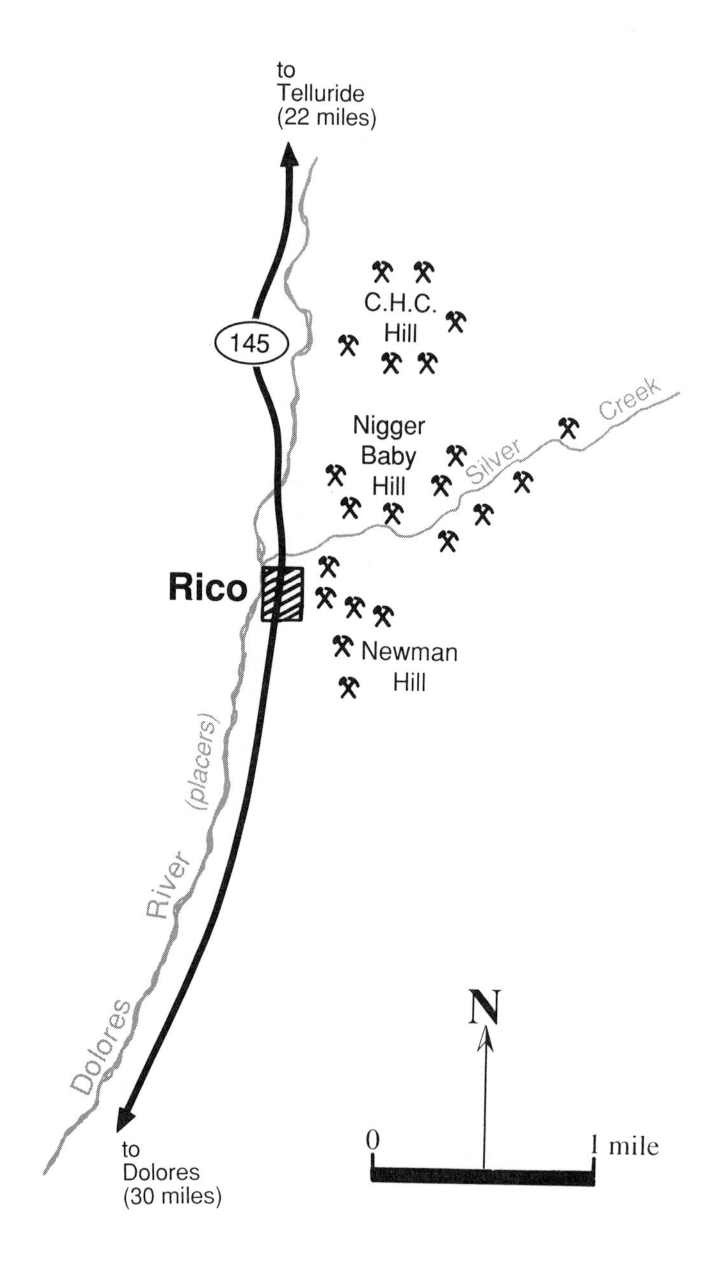
to
Telluride
(22 miles)
145
C.H.C.
Hill
Nigger
Baby
Hill
Silver
Creek
Rico
Newman
Hill
(placers)
Dolores
River
to
Dolores
(30 miles)
N
0
1 mile

DOLORES COUNTY

PLACER GOLD

Dolores County is a minor source of placer gold, with production estimated at only twenty troy ounces. Prospectors discovered small concentrations of placer gold in the upper Dolores River in 1882. The gold, derived from lode deposits at Rico, occurs in the Dolores River channel from Rico five miles downstream to the Montezuma County line.

REFERENCES: 54

RICO AREA MINES

In 1833, American mountain men found early Spanish diggings and the ruins of a crude smelting furnace near the present site of Rico. Prospectors arrived in 1861, but were discouraged by the Utes, isolation, and severe winters. Claims staked in 1869 were illegal, as was a small furnace built by daring miners in 1872. Prospectors arrived in numbers only after the Brunot Treaty legalized mineral entry in 1873.

Prospectors found rich oxidized silver ore on Nigger Baby Hill, C.H.C. Hill, Newman Hill, and along Silver Creek in 1879. Within a year, the population soared to 10,000. Promoters proclaimed Rico "the new Leadville," but isolation and high shipping costs delayed development.

Rico boomed in 1892, when Newman Hill miners discovered the rich Enterprise ore body with select ores grading as high as 500 troy ounces of silver per ton. That year, fifty-nine mines operated and the Enterprise alone produced $1.5 million in silver. Prosperity was brief, for few mines survived the 1893 silver-market crash.

During World War I, Rico's mines produced copper, lead, and zinc. By the 1920s, cumulative production included 100,000 troy ounces of gold, 11 million troy ounces of silver, 6 million pounds of copper, 5,000 tons of zinc, and 20,000 tons of lead worth more than $15 million.

In 1938, the Rico Argentine Company constructed a new mill on Silver Creek as mines reopened to contribute to wartime metal demand. Miners reworked many old dumps in the 1950s for pyrite, valuable as a raw material for sulfuric acid manufacture. The Rico Argentine consolidated mines finally closed in 1971.

Rico is one old Colorado mining town that has not been altered by tourism.

In Rico, old mines and mine-dump mineral collecting begin right in town.

Mineralization at Rico occurs as fissure veins, massive sulfide replacement deposits in limestone, and contact metamorphic deposits. Ore minerals include sphalerite in well-developed crystals; argentite, stephenite, and other silver sulfides; galena, usually argentiferous; wire and crystallized silver; tetrahedrite; chalcopyrite; proustite ("ruby silver"); and some native gold. Well-developed calcite rhombohedrons are present in the limestone replacement deposits. Massive pink rhodochrosite and clear and white crystalline quartz are common gangue minerals.

Rico is an uncommercialized Colorado mining town complete with original Victorian architecture in a beautiful mountain setting. Collectors can find small specimens of metal sulfide ores and gangue minerals literally on the main street, where the headframe of the Atlantic Cable Mine is marked by a commemorative plaque. Other mine dumps begin only three blocks away at the eastern edge of town and extend up Nigger Baby Hill, C.H.C Hill, Newman Hill, and along Silver Creek. Most of the old mining-district roads require four-wheel drive.

REFERENCES: 7, 18, 23, 66, 81

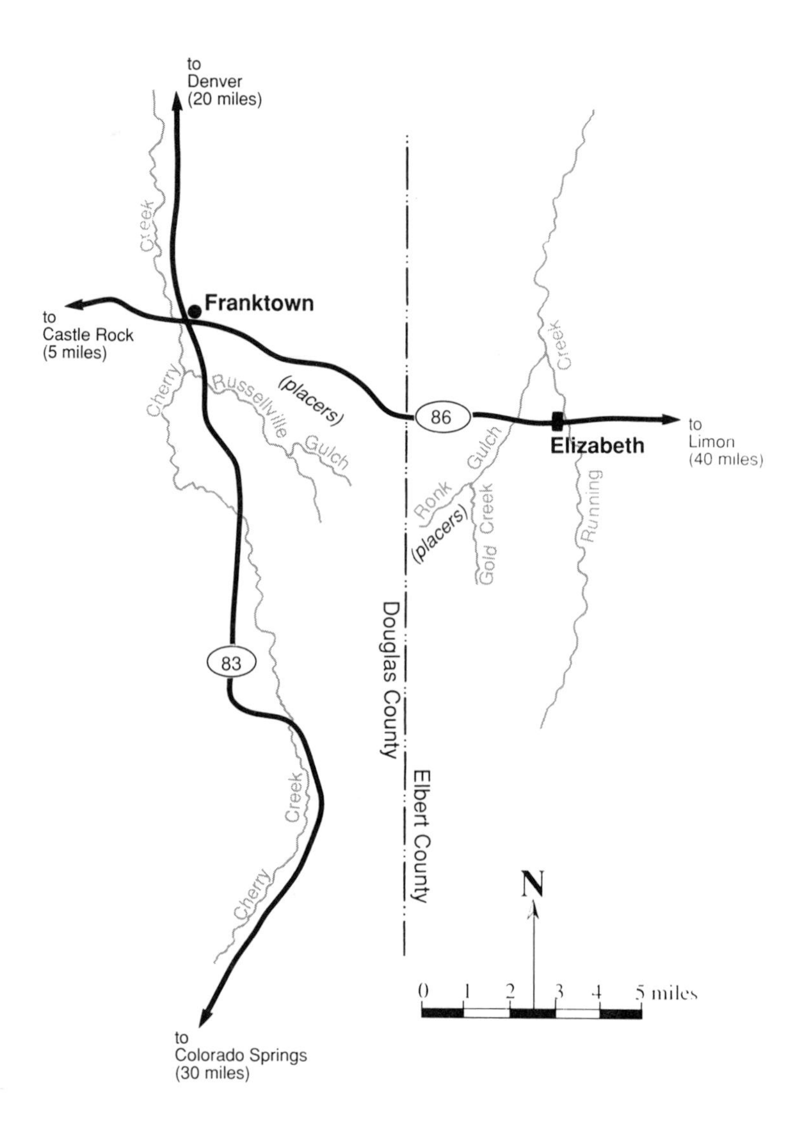

to
Denver
(20 miles)
Creek
Franktown
to
Castle Rock
(5 miles)
Cherry
Russellville
Gulch
(placers)
86
Elizabeth
to
Limon
(40 miles)
Creek
Ronk
Gulch
(placers)
Gold Creek
Running
83
Douglas County
Elbert County
Creek
Cherry
N
0 1 2 3 4 5 miles
to
Colorado Springs
(30 miles)

DOUGLAS COUNTY

PLACER GOLD

Douglas County has produced about 750 troy ounces of placer gold. While only a minor producer, its deposits are historically significant and also are Colorado's easternmost commercial placers. The Russellville Gulch, Ronk Gulch, and Gold Creek placers are located near Franktown in southeastern Douglas County on the Elbert County line.

William Green Russell and his party of Georgian and Cherokee miners discovered the placers in May 1858, a month before their Dry Creek strike near Denver triggered the Pikes Peak rush. Miners occasionally worked the placers until the 1950s. The Douglas County placers, because of poor concentrations and deep overburden, have little potential for recreational miners or gold panners.

REFERENCES: 53, 108

DOUGLAS COUNTY PEGMATITES

The Pikes Peak Batholith granite in western Douglas County contains hundreds of pegmatites with fine specimens of smoky quartz, topaz, and amazonite. The two primary collecting areas, Devils Head and Pine Creek, are in southwestern Douglas County. Both are about twenty miles west of Sedalia and north of Woodland Park. Devils Head may be reached on Rampart Range Road; Pine Creek is located near Colorado Route 67.

DEVILS HEAD

Devils Head is a prominent, isolated, 9,348-foot-high peak and ridge. Prospectors discovered topaz and smoky quartz there in the early 1880s, and Devils Head quickly gained recognition as a classic locality for fine crystal specimens. Topaz color ranges from reddish brown and sherry to colorless and slightly bluish. Some topaz crystals have weighed hundreds of carats, although most are much smaller. Devils Head smoky quartz crystals are some of Colorado's largest, although most are semiopaque and fractured. Gem-quality smoky quartz crystals also occur, but are rarely larger than four inches.

Devils Head amazonite is typically pale green to greenish white and about an inch or two in length. Exceptional amazonite crys-

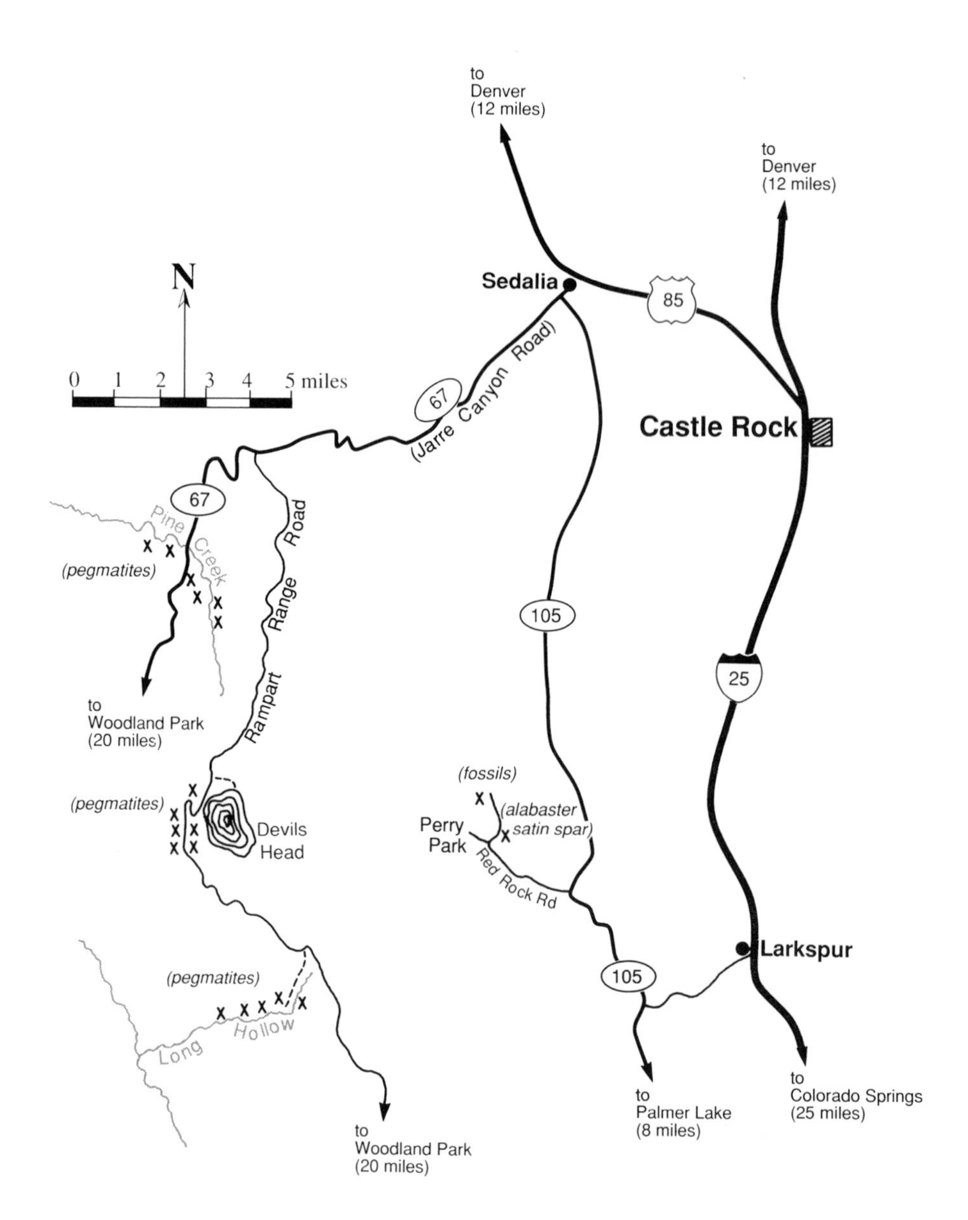

to
Denver
(12 miles)
to
Denver
(12 miles)
N
Sedalia
85
0 1 2 3 4 5 miles
67
(Jarre Canyon Road)
Castle Rock
67
Pine
Creek
(pegmatites)
Rampart Range Road
105
25
to
Woodland Park
(20 miles)
(fossils)
(alabaster
satin spar)
Perry
Park
Red Rock Rd
(pegmatites)
Devils
Head
Larkspur
105
(pegmatites)
Long Hollow
to
Woodland Park
(20 miles)
to
Palmer Lake
(8 miles)
to
Colorado Springs
(25 miles)

Orthoclase crystals from pegmatite near Devils Head.

tals are an attractive, intense blue-green and can be as long as six inches. Goethite, hematite, and fluorite also occur locally.

The collecting area extends from Devils Head Campground Road five miles south along Rampart Range Road to Long Hollow, where large smoky quartz and topaz crystals also occur. Scattered white quartz and pink feldspar make many diggings easily visible through the pine forests. Individuals and several mineral clubs hold pegmatite claims in the area. Crystals may be found in clay-filled pockets in the in-situ granite, in talus slopes, or by digging and screening decomposed granite gravel. Prospecting opportunities are excellent.

PINE CREEK

Pine Creek, four miles northwest of Devils Head, is reached from Colorado Route 67. A pegmatite-collecting area, covering four square miles and marked by numerous diggings, has produced clear and smoky quartz crystals to twelve inches in length, along with crystals of greenish white amazonite.

REFERENCES: 5, 19, 32, 41, 51, 64, 66, 67, 70, 74

PERRY PARK AREA FOSSILS AND SATIN SPAR

Perry Park, six miles northwest of Larkspur and I-25, is located at a gap in the Dakota hogback ridge. Exposures include strata of the

Smoky quartz crystals from pegmatites near Devils Head.

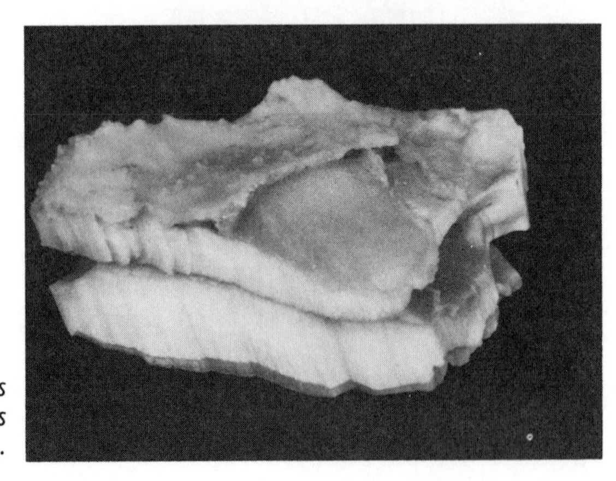

Satin spar from Lykins Formation exposures at Perry Park.

Permian Lykins Formation and Cretaceous Pierre Shale and Dakota Sandstone. The Lykins Formation, exposed for several miles north and south of Perry Park, contains gypsum beds with white-and-pink-mottled alabaster and seams of white and pinkish orange satin spar up to six inches thick.

From Sedalia, take Colorado Route 105 thirteen miles south, then follow Red Rock Road west for just over two miles. Turn right on Perry Park Road. One-third of a mile farther, opposite the Perry Park Golf Club, a road cut on the right exposes Lykins Formation sediments and gypsum beds with alabaster and satin spar. Exposures of gypsum rock, alabaster, and satin spar extend south for at least one-quarter mile.

Northwest of the golf club lake, exposures of Cretaceous sandstone and shales contain marine fossils. Fossil leaves occur in the Dakota Sandstone, the uppermost strata of the hogback.

REFERENCES: 21